INFORMATION THEORY

INFORMATION THEORY

By

STANFORD GOLDMAN

Professor of Electrical Engineering
Syracuse University

New York

DOVER PUBLICATIONS, INC.

Published in Canada by General Publishing Company, Ltd., 30 Lesmill Road, Don Mills, Toronto, Ontario.
Published in the United Kingdom by Constable and Company, Ltd., 10 Orange Street, London WC 2.

This Dover edition, first published in 1968, is an unabridged and unaltered republication of the work originally published by Prentice-Hall, Inc., in 1953.

Library of Congress Catalog Card Number: 68-8882

Manufactured in the United States of America
Dover Publications, Inc.
180 Varick Street
New York, N. Y. 10014

To My Mother and Father

PREFACE

The primary objective of this book is to present a thorough discussion of the principles of information theory. The presentation is based on the pioneering and classical work of C. E. Shannon and, to a lesser extent, on the earlier and also classical work of Norbert Wiener. If, at times, the author takes exception to the opinions or points of view of these pioneers, it should not be interpreted as disrespect for their fine work.

It has been the purpose of the author to carry out the foregoing, and yet retain a mathematical level that is understandable to first-year graduate students in electrical engineering. The author has also tried to present the basic and general aspects of the subject at a sufficiently elementary mathematical level so as to be intelligible to those readers who work in the non-mathematical sciences and are interested in information theory. For the latter, Chapters I and IX and parts of VII have been written at a more elementary level than the remainder of the book. The mathematical developments related to Chapter I are presented in Appendices I, II, III, IV, and VIII. For the reader with a mathematical background these Appendices should be read concurrently with Chapter I.

Much of the material and many of the results in the book are new, or at least are new to the author. The work of Shannon, for example, has been modified so that the information aspects of individual messages can be treated more readily and so that the semantic aspects of information as well as its language aspects may be considered. In addition, a detailed treatment is given of the transformation of both information and constraints from the time to the frequency domain, and a general development is given of various sampling theorems. There is also a thorough treatment of the information theory aspects of random noise.

It should also be mentioned that the definitions of fundamental terms sometimes do not agree with those of Shannon or other authorities. Among these terms are "random noise," "white noise," "information," and "ergodic ensemble." Certain new words and expressions have been introduced, such as "liniva," "quadiva," and "ergodic range." The author has been reluctant to make changes in definitions or to introduce new words and has done so only when he felt that they were necessary for consistency or clarity.

The book is written at about the same mathematical level as the author's two previous books, *Frequency Analysis, Modulation and Noise* and *Transformation Calculus and Electrical Transients*. It so happens that much of the material in the former will serve as background material for the present book. Generally, basic material which may be found in *Frequency Analysis, Modulation and Noise* is not repeated here, but is assumed to be available to the reader.

The book includes many problems which should be useful to the teacher and the general reader who is using the book for self-study. To help the latter reader, complete solutions to some of the problems are given in Appendix XII, and answers to some others accompany the problems. A number of the problems contain useful additional theorems and matters of information, so that they may well be read by the general reader even if he does not intend to solve them.

It was originally intended that the book should contain a bibliography of the important publications in the field of information theory. In the meantime F. L. Stumpers has written *A Bibliography of Information Theory*, published by the Research Laboratory of Electronics, Massachusetts Institute of Technology, Feb. 2, 1953. This bibliography is very extensive and well organized, and since it has been published the author has decided not to include a formal bibliography in the present text. There are, however, detailed references to all publications which have been used by the author in developing the material presented here.

Certain parts of the book were previously distributed in mimeograph form, and the author is indebted to various persons who found errors or suggested improvements during this period, particularly Dr. E. W. Samson and D. W. Spence. The intelligent assistance of Mrs. Dorothy D. Lett and Mrs. Mary Rizika is also appreciated. The illustrations were prepared by Mr. B. B. Muvdi.

Finally, the author wishes to thank his student and colleague, Charles R. Cahn, for his loyal help. Mr. Cahn read all of the material at several stages of its development, and he has been responsible for many additions and improvements.

STANFORD GOLDMAN

Syracuse University
Syracuse, New York

CONTENTS

INFORMATION THEORY

CHAPTER 1

INFORMATION THEORY OF DISCRETE SYSTEMS

1.1 Introduction

The transmission of information by means of the human voice, by means of letters, newspapers, books, and by countless other means is one of the most important aspects of our everyday life. Practically all of us have used terms such as "a lot of information," but few people have suspected that it is really possible to measure information quantitatively. Nevertheless, as we will show before long, an amount of information has a useful numerical value just like an amount of sugar or the amount of a bank balance.

Before introducing numbers or formulas, let us get our bearings by considering a few examples. Suppose you are listening to the local weather forecast on the radio in Boston on the 3rd of July and you hear the weatherman say, "It will not snow tomorrow." Since it has probably not snowed in Boston on the 4th of July for a century or more, you would consider that such a statement by the weatherman gave you very little information. If, however, you knew the weatherman to be very reliable, and you heard him say, "It will snow tomorrow," then you would consider that he gave you a great deal of information. Thus according to our usual way of looking at information, if something is very likely to occur, the statement that it will occur does not give much information. On the other hand, if something is unlikely to occur, the statement that it will occur gives a good deal of information.

Let us consider another example. A man goes to a new community to rent an apartment and asks an unreliable renting agent, "Is this apartment cool in the summertime?" If the agent answers, "Yes," the man has received very little information because more than likely the agent would have answered, "Yes," regardless of the facts. If, on the other hand, the man has a friend who lives in the apartment building, he can get more information by asking his friend the same question because the answer will be more reliable.

If we consider the foregoing examples, we see that the effect of the information in a message is to change the probability concerning a situation, as far as the receiver of the message is concerned, from its value before the message is received to what is usually a larger value after the message is received. In a general way, it would appear that the amount of information in the message should be measured by the extent of the change in probability produced by the message. We will now show that it is possible to replace the rather vague meaning of the word information, as we have so far used it, by a more precise definition which will allow us to assign a numerical value to an amount of information. We will find, furthermore, that this precise definition is a most useful tool in analyzing the content of messages and in evaluating a wide variety of situations.

1.2 Probability

If an event is certain to occur, the mathematical probability of its occurrence is, by definition, one. If the event is certain not to occur, the mathematical probability of its occurrence is zero. The mathematical probability of the occurrence of any event whose occurrence or non-occurrence cannot be predicted with certainty lies somewhere between zero and one. In order to see how numerical probabilities are assigned in this range, let us consider some examples. If a coin is tossed at random and if it is as likely to fall "heads" as "tails," the probability of the occurrence of "heads" is said to be $\frac{1}{2}$. If a standard six-sided die is tossed, the probability of its falling with a "one" up is $\frac{1}{6}$. The probability of its falling with a "two" up is likewise $\frac{1}{6}$ and so on for "three," "four," "five," and "six." In general, we may say that the probability of the occurrence of an event is the fractional number of times that it will occur in a large number of repetitions (independent trials) of the original situation.

It may be shown that the probability of the simultaneous occurrence of two independent events is the product of their individual probabilities. Thus the probability of obtaining a pair of heads when two coins are tossed is

$$\tfrac{1}{2} \text{ times } \tfrac{1}{2} = \tfrac{1}{4}$$

Furthermore, if two events are mutually exclusive (*i.e.*, if one occurs the other cannot), the probability that one or the other will occur is equal to the sum of their individual probabilities. Thus the probability that either two heads or two tails will result when two coins are tossed is

$$\tfrac{1}{4} + \tfrac{1}{4} = \tfrac{1}{2}$$

As a final example let us calculate the probability of occurrence of "seven" when two dice are tossed. Now seven can be obtained in the following mutually exclusive ways:

First die yields one and second die yields six.
First die yields two and second die yields five.
First die yields three and second die yields four.
First die yields four and second die yields three.
First die yields five and second die yields two.
First die yields six and second die yields one.

Each of these individual possibilities has a probability of occurrence of

$$\tfrac{1}{6} \times \tfrac{1}{6} = \tfrac{1}{36}$$

For example the probability that the first die will yield *one* is $\frac{1}{6}$, and the probability that the second die will yield *six* is $\frac{1}{6}$. The probability of the simultaneous occurrence of these two events is therefore $\frac{1}{36}$.

Since six mutually exclusive possibilities, each having a probability of $\frac{1}{36}$, will give seven, the probability of occurrence of seven is

$$\tfrac{1}{36} + \tfrac{1}{36} + \tfrac{1}{36} + \tfrac{1}{36} + \tfrac{1}{36} + \tfrac{1}{36} = \tfrac{1}{6}$$

1.3 A mathematical definition of information

We are now ready to give a mathematical definition of information. The definition which we shall give involving the logarithms of probabilities is not the only one possible.[1] It has certain virtues, however, which recommend it. Thus it is simple, it has useful

[1] An alternative definition in terms of entropy or choice is used by Shannon. "A Mathematical Theory of Communication," *Bell System Tech. Jour.* vol. 27, p. 379–423, July, 1948; p. 623–656, October, 1948. The relation between the two definitions is discussed in our Appendix IV.

properties in analysis,[1] it does not lead to contradictions, and it corresponds very well with our usual methods of thinking. The truth of the foregoing statements will gradually become evident.

Let us suppose that an event occurs and a message is transmitted telling about the occurrence. We will define the amount of information received in the message as

$$\text{information received} = \log \left[\frac{\text{probability at the receiver of the event after the message is received}}{\text{probability at the receiver of the event before the message is received}} \right] \tag{1}$$

We will begin our analysis by considering what is called the "noiseless" case. In the noiseless case the receiver is certain that the message received is correct. Therefore, the probability of the event after the message is received is unity and Eq. (1) reduces to

$$\text{information received (in the noiseless case)} = -\log \left[\text{probability at the receiver of the event before the message is received} \right] \tag{2}$$

As an example, let us suppose that a baby has just been born at a neighbor's house and the question is asked, "What is the baby?" The answer, "It's a boy," then gives a specific amount of information, in accordance with Eq. (2), namely,

$$\text{quantity of information} = -\log\left(\tfrac{1}{2}\right) = \log 2 \tag{3}$$

In the above equation, it has been assumed that the baby was equally likely to have been a boy or a girl, so that the apriori probability of its being a boy is $\frac{1}{2}$. The numerical value of the amount of information in the above equations depends upon what logarithmic base is used. If 10 is used as the base of logarithms, a unit of information may be called a decimal digit. If 2 is used as the base of logarithms, a unit of information may be called a binary digit. We shall use the contractions *decit* for decimal digit and *binit* for binary digit.[2]

[1] The use of the logarithm makes the quantity of information in independent messages additive.

[2] Shannon uses the contraction *bit* for binary digit (which he attributes to

The use of 2 as the logarithmic base has certain advantages in information theory, which will become apparent later on. We note in passing that since

$$\log_2 2 = 1 \tag{4}$$

exactly one *binit* of information tells which of two equally likely possibilities has been chosen.

We will next illustrate the meaning and use of the foregoing definitions with the aid of another example. In Fig. 1-1 is shown a large field divided into 64 squares. A cow has wandered into this field in the dark and is located by a member of a searching party who sends back a message giving the location of the cow as the 43rd square. The amount of information obtained in the reception of this message will now be calculated. Before the message was received, the cow was just as likely to be in any one of the 64 squares as any other. Therefore the probability that the cow was in the 43rd square was

$$p = \tfrac{1}{64} \tag{5}$$

We will assume that the message is completely reliable, so that the location of the cow is certain after the message is received. The amount of information received with the message is thus

$$\text{information} = \log \frac{1}{p} = \log 64 = 6 \text{ binits} \qquad (6)^1$$

The content of the foregoing message could equally well have been sent in several different forms. For example, the message may have stated that the cow was in the square of the 6th column and 3rd row of the field. Let us see how much information is received with this message according to our definitions. Before the message was received, the cow was equally likely to have been in any of the different columns. Accordingly, the probability that it was in the sixth column was

$$p_c = \tfrac{1}{8} \tag{7}$$

J. W. Tukey, and his terminology has been widely used. Since, however, the phrase "a bit of information" has a very inexact meaning in ordinary English, we prefer to use the less ambiguous term "binit." The author does not know who originated the term *binit*, but he first heard it used by Dr. E. W. Samson.

[1] In this chapter and throughout the book when dealing with discreet information, unless otherwise stated, the base 2 is used for logarithms.

Similarly, the probability that it was in the third row was

$$p_r = \tfrac{1}{8} \tag{8}$$

If it is agreed, when the searching party is sent out, that a two-symbol message will be sent in which the first symbol stands for

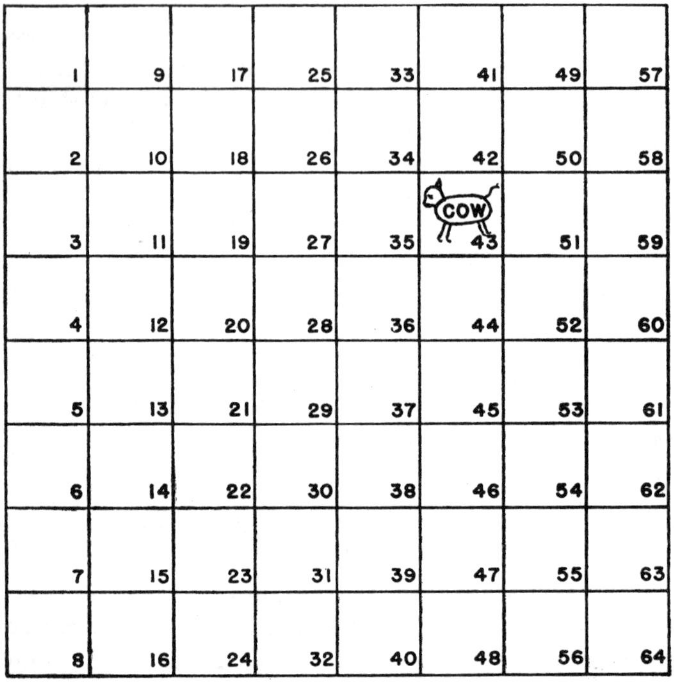

NUMBERING OF SQUARES OF FIELD INTO WHICH COW HAS STRAYED

Fig. 1-1.

the column and the second symbol stands for the row of the square in which the cow is located, then the message "6, 3" locates the cow. In this message, the first symbol gives

$$\log \frac{1}{p_c} = \log 8 = 3 \text{ binits} \tag{9}$$

of information, and the second symbol gives

$$\log \frac{1}{p_r} = \log 8 = 3 \text{ binits} \tag{10}$$

of information. The total amount of information necessary to locate the cow is thus 6 binits, in agreement with Eq. (6).

In the foregoing paragraph, we have added the information in the two separate symbols in order to get the total information of the message. This is possible because the information given by each symbol is independent of the information given by the other. In case there is some redundancy of the information in the two symbols, it is not possible simply to add the information given by separate symbols. The methods which must be used in the latter case will shortly be developed.

PROBLEMS[1]

1. A man is informed that, when a pair of dice were rolled, the result was a seven. How much information is there in this message?

Ans. 2.58 binits.

2. In a certain community 25% of all girls are blondes and 75% of all blondes have blue eyes. Also 50% of all girls in the community have blue eyes. If you know that a girl has blue eyes, how much additional information do you get by being informed that she is blonde? *Ans.* 1.42 binits.

1.4 Exponential growth of the number of different messages

Suppose we have a message consisting of a sequence of symbols drawn from an alphabet of L symbols. Then it is a simple matter to show that the number of different messages which are possible in a sequence n symbols long is

$$\text{number of different possible messages} = L^n \tag{11}$$

As an example, if the alphabet consists of the three symbols A, B, and C and if $n = 2$, then

$$L^n = 3^2 = 9 \tag{12}$$

messages are possible. In particular, the nine different possible messages are

AA	BA	CA
AB	BB	CB
AC	BC	CC

[1] A complete solution of Prob. 2 is given in Appendix XII.

As another example, the arabic system of numbers has an alphabet of 10 symbols, namely, 0, 1, 2, 3, 4, 5, 6, 7, 8, and 9. According to Eq. (11) there should then be precisely

$$L^n = 10^6$$

different six-digit numbers from 000,000 to 999,999. This is, of course, clearly true.

Equation (11) is capable of an important generalization. Instead of measuring the duration of a message by the number of symbols, let us measure it as a time duration. We will furthermore not assume that the durations of the L different symbols are necessarily equal, but we will assume that these durations are $t_1, t_2, \ldots, t_L$, respectively. It may then be shown[1] that, if a message has a duration T which is very large in comparison with L times the duration of the longest symbol, the number of different messages which are possible is

$$\text{number of different possible messages} = AX_0^T \tag{13}$$

where A and X_0 are constants depending upon L and the symbol lengths. In particular X_0 is the largest real solution of

$$\frac{1}{X^{t_1}} + \frac{1}{X^{t_2}} + \cdots + \frac{1}{X^{t_L}} = 1 \tag{14}$$

In case the symbol lengths are all equal and of length t_0, then Eq. (14) reduces to

$$\frac{L}{X_0^{t_0}} = 1 \quad \text{or} \quad X_0 = L^{1/t_0} \tag{15}$$

In this case it may be shown that $A = 1$, so that Eq. (13) becomes

$$\text{number of different possible messages} = AX_0^T = L^{T/t_0} = L^n \tag{16}$$

in agreement with Eq. (11).

In addition to inequality of symbol lengths, there is an additional type of complication which is often imposed upon a symbol sequence by the organization of the coding system. Thus in one representation of telegraphy we have the four symbols: dot, dash, letter space, word space. However, the organization of the code forbids a letter space or a word space being followed by a letter space or a word

[1] See Appendix I.

space. Instead it must be followed by a dot or a dash. Such restrictions (which we will call fixed constraints) reduce the number of different possible messages which may be represented by a sequence of given duration. It may, however, be shown that in any system having only fixed constraints, for a sequence of long duration T,

$$\text{number of different possible messages} = aW^T \tag{17}$$

where a and W are constants, depending on the number of symbols, the symbol lengths, and the nature of the fixed constraints.[1]

The general result that runs through Eqs. (11), (13), and (17), namely, that *the number of different possible messages in a sequence of long duration T increases exponentially with T*, holds for all coding systems having fixed constraints provided only that the constraints are of finite[2] duration. Systems for transmitting information are, however, not limited to those having fixed constraints. Thus, as will be discussed in more detail in a later section, the constraints in languages are probability constraints. We will, nevertheless, be able to show that, even in systems controlled by probability constraints, the number of different possible messages in a sequence of long duration T increases exponentially with T. The fact that this same type of exponential law holds for the number of possible messages, regardless of the type of the constraints, is the basic mathematical law which makes it possible to develop a relatively simple mathematical theory of the transmission of information. We will now begin our study of probability constraints by considering the *law of large numbers*.

1.5 The law of large numbers

In an earlier section we mentioned the fact that, when a coin is tossed, the probability of its landing heads up is $\frac{1}{2}$, and when a six-sided die is tossed, the probability of its landing with, say, the 4-side up is $\frac{1}{6}$. Furthermore, we stated that the probability of the simultaneous occurrence of a number of independent events is the product of their individual probabilities. In a way, it may appear as

[1] Appendix I shows how to calculate W.

[2] In saying that the constraints are limited to finite duration, we mean that the influence of a symbol or combination of symbols in a sequence in disallowing or otherwise constraining some future symbol in the sequence is not effective for more than a finite range from its point of origin.

though this last statement is inconsistent with the meaning of probability. Thus the meaning of a probability of $\frac{1}{2}$ in the coin case is that a coin when tossed is just as likely to land heads as tails, which is considered to be equivalent to the statement that if a very large number (N) of coins are tossed essentially half of them will be heads and half tails. On the other hand, since the probability of getting heads is $\frac{1}{2}$ for each coin, our rule above for the probability of the simultaneous occurrence of independent events says that there is a probability of

$$\frac{1}{2^N}$$

that all N coins will land heads. Even though $(1/2^N)$ is a very small number, the fact that there is a finite probability that all the N coins will land heads leads us to inquire further into what we really mean by a numerical probability.

In order to clarify our ideas of probability, we will consider some examples. Let us first consider the tossing of four coins. The possible results are listed in Table I, where H stands for heads and T for tails. In each group, the first symbol stands for the landing side of the first coin, the second symbol for the second coin, and so on. Thus the group HHTH stands for the possibility that the first coin landed heads, the second heads, the third tails, and the fourth heads.

TABLE I

POSSIBLE RESULTS WHEN FOUR COINS ARE TOSSED

HHHH	HTHH	THHH	TTHH
HHHT	HTHT	THHT	TTHT
HHTH	HTTH	THTH	TTTH
HHTT	HTTT	THTT	TTTT

We note in passing that there are 16 possibilities, in agreement with Eq. (11).

Each of these 16 possibilities has a probability of

$$\tfrac{1}{2} \times \tfrac{1}{2} \times \tfrac{1}{2} \times \tfrac{1}{2} = \tfrac{1}{16}$$

However, there is only one possibility of getting 4 heads while there are four possibilities of getting 3 heads, six possibilities of getting 2 heads, four possibilities of getting 1 head, and one possibility of getting no heads. The probabilities of getting 4, 3, 2, 1, and 0 heads are thus $\frac{1}{16}$, $\frac{4}{16}$, $\frac{6}{16}$, $\frac{4}{16}$, and $\frac{1}{16}$, respectively. We see by inspection that two heads is more likely than any of the other numbers.

Following this introduction, let us next consider the tossing of 20 coins. The number of possible results in this case, according to Eq. (11), is

$$2^{20} = 1{,}048{,}576$$

We will not list all these individual possibilities, but in Table II we have tabulated the number of possibilities having 0,1,2, . . . , up to 20 heads. These are calculated by a well-known formula in the theory of combinations which gives in this case

$$n = \begin{bmatrix} \text{number of possibilities hav-} \\ \text{ing } N_H \text{ heads and } N_T \text{ tails} \\ \text{in a sequence of 20 symbols} \end{bmatrix} = \frac{20!}{N_H!N_T!} \qquad (18)^1$$

Examination of Table II shows that about 74% of the possible results in the tossing of 20 coins have between 8 and 12 heads. Stated in other words, about 74% of the possibilities lie within 20% of 10 heads. We thus see that as the number of coins tossed is increased, from 4 to 20, the probability that a result will be obtained in which *about* $\frac{1}{2}$ of the coins are heads has increased markedly. If the number of coins to be tossed is greatly increased, over 99.999% (say) of the possibilities will lie within 0.0001% (say) of half heads and half tails. If the number of coins tossed is made sufficiently large, as close to 100% as we please of these equally likely possibilities will lie within any narrow range, no matter how small, of half heads and half tails. This then is what we mean when we say that the probability of getting heads in the tossing of a coin is $\frac{1}{2}$.

As a final example in clarifying our ideas about probability, let us consider a sequence of 20 symbols in which each symbol is either an A or a B. Let us suppose that the probability of getting an A is $\frac{3}{4}$ and the probability of getting a B is $\frac{1}{4}$, and there is no interdependence in the occurrence of the symbols in a sequence.

The total number of different possible sequences 20 symbols long is again, of course,

$$2^{20} = 1{,}048{,}576$$

and the number of different possible sequences having 1,2, . . . ,20 A's is again given by column 3 of Table II. However, in Table II,

[1] The sign (!) stands for the factorial, that is, $n! = 1 \times 2 \times 3 \times \ldots \times (n-1)n$. Furthermore, by special definition, $0! = 1$.

the probability of getting any particular sequence was

$$p = (\tfrac{1}{2})^{N_H}(\tfrac{1}{2})^{N_T} = (\tfrac{1}{2})^{20} \tag{19}$$

These probabilities were all equal, since the probability of getting an H was the same as the probability of getting a T. In the present

TABLE II

NUMBER n OF DIFFERENT POSSIBLE RESULTS HAVING 0,1,2, . . . ,20 HEADS WHEN 20 COINS ARE TOSSED

N_H (number of heads)	N_T (number of tails)	$n = \dfrac{20!}{N_H!N_T!}$ [see Eq. (18)]	Probability of getting N_H heads
0	20	1	0.000001
1	19	20	0.000019
2	18	190	0.00018
3	17	1,140	0.0011
4	16	4,845	0.005
5	15	15,504	0.015
6	14	38,760	0.037
7	13	77,520	0.074
8	12	125,970	0.120
9	11	167,960	0.160
10	10	184,756	0.176
11	9	167,960	0.160
12	8	125,970	0.120
13	7	77,520	0.074
14	6	38,760	0.037
15	5	15,504	0.015
16	4	4,845	0.005
17	3	1,140	0.0011
18	2	190	0.00018
19	1	20	0.000019
20	0	1	0.000001
	Totals	1,048,576	1.000

case the symbol probabilities are not equal. Accordingly, the probability of getting any particular 20-symbol sequence depends on the number of A's and the number of B's in it. Thus we may write

$$\text{probability of sequence} = (\tfrac{3}{4})^{N_A}(\tfrac{1}{4})^{N_B} \tag{20}$$

If the number of different sequences having N_A of symbol A (see column 3, Table III) is multiplied by the probability of such a sequence (given in column 4), the product will give the probability that a 20-symbol sequence will have N_A of symbol A without specifying the order in which the A's and B's occur. These products are tabulated in column 5 of Table III. We note in passing that

TABLE III

Number of A's N_A	Number of B's N_B	Number of different sequences of given N_A $n = \frac{20!}{N_A!N_B!}$	Probability of each particular sequence $p = \left(\frac{3}{4}\right)^{N_A}\left(\frac{1}{4}\right)^{N_B}$	Probability of obtaining a sequence with N_A of symbol A and N_B of symbol B without specifying the order of symbols $p \times n$	
0	20	1.	9.095×10^{-13}	9.095×10^{-13}	
1	19	20.	2.729×10^{-12}	5.457×10^{-11}	
2	18	190.	8.186×10^{-12}	1.555×10^{-9}	
3	17	1,140.	2.456×10^{-11}	2.800×10^{-8}	
4	16	4,845.	7.337×10^{-11}	3.555×10^{-7}	
5	15	15,504.	2.210×10^{-10}	3.426×10^{-6}	
6	14	38,760.	6.630×10^{-10}	2.570×10^{-5}	
7	13	77,520.	1.989×10^{-9}	1.542×10^{-4}	
8	12	125,970.	5.967×10^{-9}	7.517×10^{-4}	
9	11	167,960.	1.790×10^{-8}	3.007×10^{-3}	
10	10	184,756.	5.371×10^{-8}	9.923×10^{-3}	
11	9	167,960.	1.612×10^{-7}	2.706×10^{-2}	
12	8	125,970.	4.833×10^{-7}	6.088×10^{-2}	
13	7	77,520.	1.450×10^{-6}	1.124×10^{-1}	
14	6	38,760.	4.350×10^{-6}	1.686×10^{-1}	
15	5	15,504.	1.305×10^{-5}	2.023×10^{-1}	*max
16	4	4,845.	3.915×10^{-5}	1.897×10^{-1}	
17	3	1,140.	1.174×10^{-4}	1.338×10^{-1}	
18	2	190.	3.523×10^{-4}	6.694×10^{-2}	
19	1	20.	1.057×10^{-3}	2.114×10^{-2}	
20	0	1.	3.171×10^{-3}	3.171×10^{-3}	
	Totals	1,048,576.		1.000	

the sum total of column 5 is unity. This is as it should be since it means that a 20-symbol sequence is certain to have either 0,1,2, . . . ,19, or 20 A's. We see in Table III that column 5 has its maximum value for the group of sequences having 15 A's, that is, when $\frac{3}{4}$ of the symbols are A's and $\frac{1}{4}$ are B's. If the number of symbols in the sequence were increased from 20 to some very large number, the probability of obtaining a sequence having almost exactly $\frac{3}{4}$ A's and $\frac{1}{4}$ B's would rapidly approach one. As a consequence, we would also obtain the result that the probability approaches zero of obtaining a long sequence in which the number of A's differs appreciably (on a percentage basis) from $\frac{3}{4}$ of the number of symbols as the length of sequence is increased.

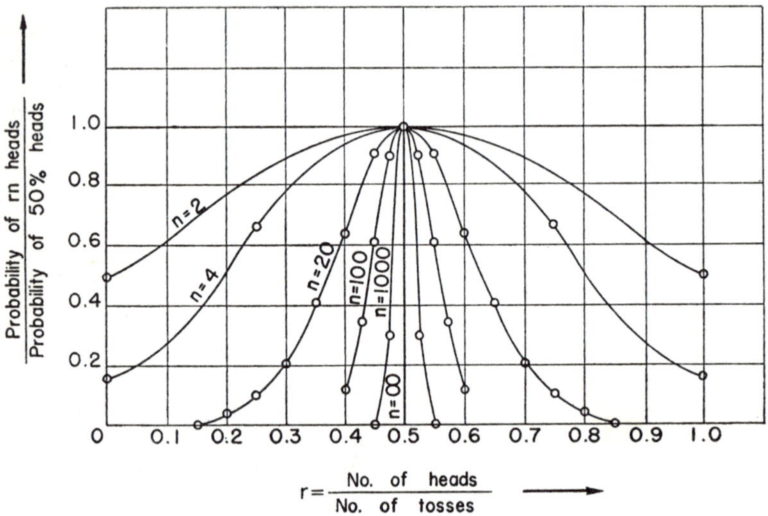

Fig. 1-2. Normalized plot illustrating the law of large numbers. When n, the number of tosses, is small, only the marked points have significance. (*Courtesy the author's former student, W. L. Hurford.*)

The foregoing examples are special cases of a general mathematical phenomenon which may be called the *law of large numbers.*[1] This is the phenomenon that, if a sequence of independent trials is made of a process controlled by probability, the fraction of the trials which will yield any result approaches a probability of one (*i.e.*, certainty) of being equal to the probability of obtaining the result in question, as the sequence becomes longer and longer. Any particular sequence whose percentage composition corresponds with the probability laws of the process may be called a *typical*

[1] Certain related mathematical phenomena are also sometimes called the law of large numbers.

sequence of the process. The law of large numbers thus tells us that it is certain that a sufficiently long sequence will be typical.

In sequences of finite length, there will usually be deviations from exact correspondence with the probabilities of the process. These deviations are called *fluctuations*.[1] The law of large numbers tells us that the percentage fluctuations approach zero as the sequence of independent trials becomes sufficiently long. An illustration of the way in which the law of large numbers works is shown in Fig. 1-2. This shows how the probability distribution of heads varies with the number of tosses of a coin.

PROBLEM

1. Let a message consist of only A's and B's. Let

$$p_A = 90\% \qquad p_B = 10\%$$

For the possible messages 10 symbols long, tabulate

Number of A's in message	Number of messages	Probability of each message	Total probability of all messages of class
0			
1			
2			
3			
4			
5			
6			
7			
8			
9			
10			

1.6 Coding systems with probability constraints—Languages

Most human means of transmitting information include the use of a language. An important characteristic of languages, and one in which they differ from artificially devised codes, is that languages have constraints controlled by probabilities. For example, it is known that in English the different letters do not all occur with the

[1] A discussion of fluctuations related to random-noise problems is given in Stanford Goldman, *Frequency Analysis, Modulation, and Noise*, McGraw-Hill, New York, Chap. VII.

same frequency, but rather they occur with the approximate frequencies indicated in Table IV. Furthermore, pairs of letters (digrams) such as AA, AB, TH, etc., have known approximate frequencies of occurrence in English, which have also been tabulated. In general, there is a large number of letter and letter-combination frequencies and of word and word-combination frequencies and similar characteristics which taken together describe the structure of the English language.[1] All these frequencies (*i.e.*, probabilities of occurrence) reduce the number of different messages which are

TABLE IV[2]

FREQUENCY OF OCCURRENCE OF LETTERS IN ENGLISH, IN 1,000 LETTERS

Letter	Frequency	Letter	Frequency	Letter	Frequency
E	131	D	38	W	15
T	105	L	34	B	14
A	86	F	29	V	9.2
O	80	C	28	K	4.2
N	71	M	25	X	1.7
R	68	U	25	J	1.3
I	63	G	20	Q	1.2
S	61	Y	20	Z	.77
H	53	P	20		

possible in an English message of given length, as we shall soon show. They are, nevertheless, quite worth while since they greatly increase the reliability of transmission of English, that is, they greatly decrease its susceptibility to error. Thus, a man who is well acquainted with English will know that the message

AVZ FOMW TZQRA

is probably not English, and that

CORING HOME FY TRAIM TLIGHT. MEET ME AL THE STATIAN.

probably was meant to be

COMING HOME BY TRAIN TONIGHT. MEET ME AT THE STATION.

[1] In Fletcher Pratt, *Secret and Urgent*, Blue Ribbon Books, Garden City, N.Y., (1942), a valuable group of frequency tables is given concerning the structure of English. Some are also given for French, Spanish, and German.

[2] Data taken from Pratt, *Secret and Urgent*.

We will deal with reliability later on in the book, but we now wish to treat the way in which the probability constraints reduce the number of different possible messages in a sequence of given length.

A language is a good approximation to what we shall call an *ergodic system* of symbols, a term which will now be defined.

> DEF.: *An ergodic system of symbols is one in which the occurrence of symbols is controlled by probability but no appreciable inter-symbol influence extends over more than a number of symbols whose top value is finite.*[1]

In an ergodic sequence the relation between symbols separated by more than this top value may be considered purely random. In Sec. 1.5, we dealt with sequences of symbols whose occurrence was controlled by probability, but the only constraint existing was the set of symbol probabilities. There was no intersymbol influence. In that case we pointed out that, as a sequence became longer and longer, the probability approached unity that the frequency of actual occurrence of symbols in a particular sequence would equal the probability of occurrence of the symbols. It may likewise be shown, with the aid of probability theory, that a similar result holds for sufficiently long sequences in an ergodic system. In the latter case, there is unit probability (*i.e.*, certainty) that a sufficiently long sequence will have a percentage composition of symbols, bigrams, trigrams (and all the other combinations which comprise the intersymbol influences) which corresponds exactly to the system probabilities. In other words, there is unit probability that a sufficiently long sequence in an ergodic system will illustrate exactly the symbol probabilities and intersymbol influences of the system. A sequence which does so illustrate the system constraints will be called a *typical sequence of the ergodic system* or simply an *ergodic sequence.*[2] Non-ergodic sequences are possible (similar to the case of non-typical sequences discussed in Sec. 1.5), but the sum total of all their probabilities approaches zero for long sequences, while the

[1] In order for a system of symbols to be ergodic, it may be shown that the intersymbol influence must ultimately fall off at at least an exponential rate.

[2] An ergodic sequence in which there is no intersymbol influence may be called a pure random sequence, and the process which generates it may be called a random process. In this book we shall also use the term *ergodic range* for the range of the intersymbol influence in an ergodic system. A further discussion of ergodic systems is given in Sec. 3.3.

total of the probabilities of all the ergodic sequences approaches unity.[1]

Actual languages, such as English, are good approximations to ergodic systems, and as a practical convenience the mathematics of ergodic systems may be used in dealing with languages. If English were truly ergodic, a very long sequence, such as a book, would have exactly the percentage composition of symbols, digrams, etc., which are characteristic of English and these percentage compositions would be the same for all books. Actually, if we took a book such as the volume *Plants* to *Raym* of the Encyclopaedia Britannica, it would, for example, almost certainly have more *q*'s and more *qu*'s than is usual in English. The reason is not hard to find. The symbol structure of the entire book is interrelated because of the fact that all parts of the book are concerned with words alphabetically ordered from *Plants* to *Raym*. The range of the intersymbol influence is thus not relatively small as required for ergodic systems, but extends over the entire book. Ordinary books will usually be coherent in meaning from cover to cover, but this coherence ordinarily does not carry over into language structure such as symbol frequencies. Ordinary books are therefore good approximations to ergodic systems as far as language structure is concerned. A library of books on unrelated topics is an even better approximation to an ergodic system.

In Appendix II a result of great importance in information theory is derived. It is there shown that, in an ergodic system, even though the constraints are controlled by probability, the number of different messages which are possible in a sequence of given duration grows exponentially with the length of the sequence. In particular it is shown that, in an ergodic sequence m symbols long, the number of different possible messages $M(m)$ is

$$M(m) = A2^{hm} \tag{21}$$

where A is a constant of the sequence and is independent of its length m while

$$h = -\frac{1}{q} \sum_i p_i \log p_i \tag{22}$$

[1] Needless to say, we are here assuming that the probability laws of the system do not change as we move along the sequence. The sequences with which we are dealing correspond to what are usually called stationary time series.

In the derivation of Eq. (21), it is assumed that, since the sequence is ergodic, and thus has a finite range of intersymbol influence, it can be separated into independent symbol groups or elementary messages. The quantity q in Eq. (22) is approximately the average number of symbols per elementary message of this type, q being greater than the range of the intersymbol influence.[1] The probability of occurrence of the ith elementary message is p_i. The notation $\sum_i p_i \log p_i$ stands for the summation over all values of i of $(p_i \log p_i)$. Thus

$$\sum_i p_i \log p_i = p_1 \log p_1 + p_2 \log p_2 + p_3 \log p_3 + \cdots + p_n \log p_n$$

where n is the total number of elementary messages.

Expressed on a time basis, the number of different possible ergodic[2] messages of any long duration T is

$$M(T) = A2^{(h/b)T} \tag{23}$$

where b is the average elementary message duration.

As a very simple example, suppose we have only two possible messages in a system X or Y. Suppose that the probability of transmitting X is $\frac{3}{4}$ and the probability of transmitting Y is $\frac{1}{4}$. Suppose that each symbol is of one-unit duration (say 1 sec). Then for this system

$$q = 1 \tag{24}$$

$$b = 1 \tag{25}$$

$$p_X = \tfrac{3}{4} \tag{26}$$

$$\log p_X = \log \tfrac{3}{4} = \log 3 - \log 4 = (\log 3 - 2) \tag{27}$$

$$p_Y = \tfrac{1}{4} \tag{28}$$

$$\log p_Y = \log \tfrac{1}{4} = -\log 4 = -2 \tag{29}$$

$$h = -\frac{1}{q}\sum_i p_i \log p_i = -\frac{1}{1}\left[\frac{3}{4}\log\frac{3}{4} + \frac{1}{4}\log\frac{1}{4}\right] = 0.811 \tag{30}$$

$$M(T) = A2^{(0.811)T} \tag{31}$$

[1] If $q' > q$, then $h' = \frac{1}{q'} \sum_{i'} p_{i'} \log p_{i'} = h$, where the primed sequences just are longer sequences. This shows that the value of h does not change, once the independent sequences are chosen sufficiently long.

[2] Since the sum total of the probabilities of the non-ergodic messages is zero, these need not be considered.

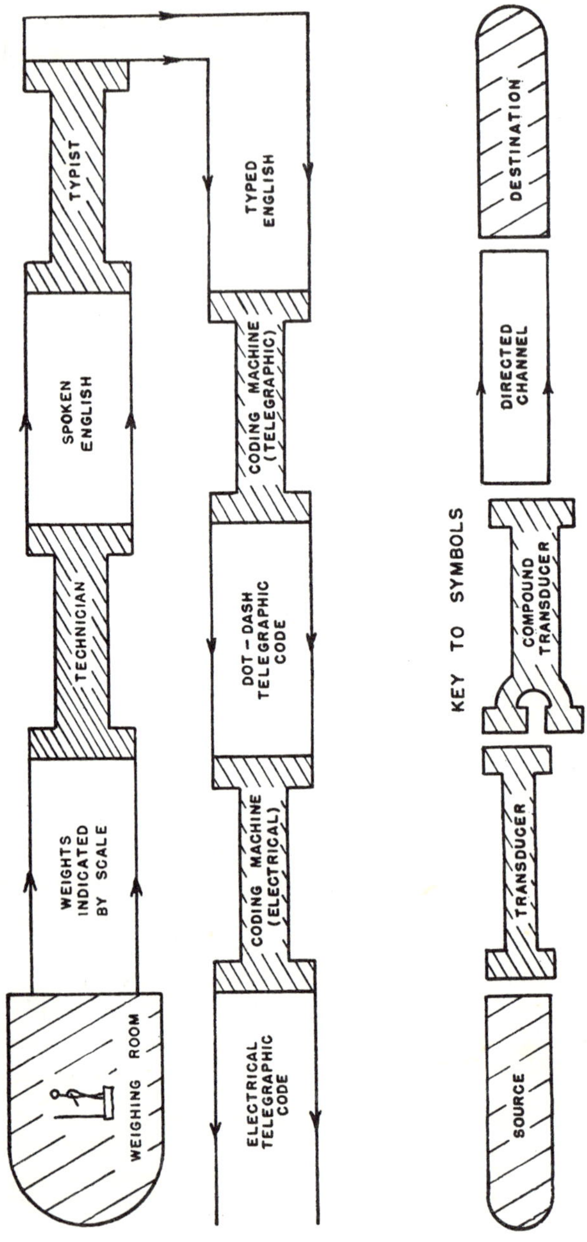

Fig. 1-3. Block diagram of part of an information transmission channel chain.

Although the value of A is actually found in Appendix II, we will find that it is not necessary to know its value in practice. We therefore never bother to calculate it.

It should be remembered that Eq. (23), and therefore Eq. (31), hold only for large values of the duration T. In the above example this might require that T be greater than about 1000.

Equation (23) shows that in an ergodic system with probability constraints the number of different messages which are possible grows exponentially with the duration T of the sequence. This is the same type of law of growth as held for the systems having fixed constraints discussed in Sec. 1.4. With the aid of these results we are now prepared to discuss the use of symbol sequences for the transmission of information.

1.7 Code capacity, language transmission capacity, and the transmission of information

We next wish to consider the coding of information into symbol sequences. In general it may be said that a code is a means of expressing information in a form suitable for use in some transmission system.[1] The codes in which information is expressed in the form of spoken or written languages, or as telegraphic messages, or audio, or radio signals all express information as symbol sequences. That this is true for written languages and telegraphic messages is fairly obvious. That it is also true for spoken languages and audio and radio signals will become clearer when we consider continuous information. An example of a type of information which is not directly expressed as a symbol sequence in time is a picture.[2] Even this, however, can be translated into such a symbol sequence, as is actually done in television by application of the principle of scanning.

In order for a code to be completely suitable for the expression of the information arising in an information source, it is necessary that every possible message which can arise in the source can be expressible in terms of the code. Consider the block diagram in Fig. 1-3 of an idealized information transmission system. Here the

[1] A code consists of an alphabet plus a system of fixed constraints. It may be pointed out here that the verb "to encode" means to identify the symbols or symbol groups of one code with those of another code in a particular way.

[2] This is a two dimensional sequence in space rather than a one dimensional sequence in time.

primary information source is a number of men who are being weighed by a technician at a medical-examination center. The results are reported by the technician in spoken English to a typist. The typist translates the spoken description into a typed message which is next translated into a telegraphic code by a machine, and is then further translated into an electric signal which is sent over a telegraph transmission line to a data bureau. For our present purposes we need not follow the message further. Let us now consider what codes are used in these operations and whether they are suitable for their purpose.

The first code used which is indicated in the diagram is spoken English. The technician translates the information appearing on the scale into a sequence of sounds of the kind used in English. These sounds are then translated into the code of symbols (*i.e.*, the alphabet) of typed English. The next code used is the sequence of dots and dashes of telegraphic code into which the typed English is translated. The fourth code used is the series of on and off electrical signal blocks into which the telegraphic code is translated.

We now wish to consider the suitability of these codes for their intended purpose. Let us first consider the English description. As far as the reporting of weights to any desired accuracy, say one pound, is concerned, there is no question but that this can be expressed exactly in English. If it were necessary to report the spirit and local color of the weighing room, there might be some question about the suitability of the language to give a complete description; but since we are limiting our interest to the reporting of weights, the language is quite suitable.

The translation of spoken English into typed English can be accomplished completely, in theory at least. By this we mean that, since every word has a correct spelling, it can be expressed exactly in type. The accent or dialect of the technician is lost in the process, but for our present purposes, those characteristics of spoken language carry no information.

When it comes to the next two codes, dot-dash telegraphic code and electrical telegraph code, there is no question, whatever, but that they can completely and exactly express all the English messages which it may be desired to send.

Now every code, into which messages can be translated has an alphabet of symbols, and all messages in any code are made up of sequences of symbols of the alphabet of that code. The alphabets of the codes used in Fig. 1-3 are tabulated in Table V. The reader may wonder whether the number of symbols in the alphabet of spoken English messages is finite, that is, whether there is only a finite number of possible spoken English messages. The answer is "yes," for practical and for our theoretical purpose. The reason is that all messages are of finite length, and since the number of words in the language is finite, the number of possible messages is also finite. In our particular example, in which we are using the English language solely for the purpose of reporting weights of men to the nearest whole number of pounds, the alphabet of possible messages is limited to the whole numbers in the range from, say, fifty to five hundred. The constraints of this specialized English are, of course, different from those of general English prose, and it is thus a different language for our purposes.

TABLE V

ALPHABETS OF CODES USED IN FIG. 1-3

Spoken specialized English	Typed specialized English	Dot-dash telegraphic code	Electrical telegraphic code
50	1	(dot)	mark (+)
51	2	— (dash)	
52	3	letter space	space (−)
...	4	word space	
...	5		
...	6		
...	7		
...	8		
...	9		
...	0		
...	word space		
499			
500			

When we are dealing with two codes, one of which is translated into the other in the transmission of messages, we will call the one which is closer to the source the primary (or input) code, and the other will be called the secondary (or output) code. Thus in Fig. 1-3, spoken English is a primary code for which typed English is the secondary, with the typist plus typewriter being the transducer.

At the next transducer, the telegraphic coding machine, typed English is the primary code and dot-dash telegraphic code is the secondary code. In most cases, the codes with which we deal are compound, such as electrical telegraphic code carrying dot-dash code. In this case, for example, all the constraints of dot-dash code as well as the inherent constraints of electrical telegraphic code limit the number of allowable sequences in a message of given length. In the future, we will use the term *code* for either a simple or a compound code if the statement applies equally well to either type. A system in which a sequence of symbols in a given code is transmitted at a specified rate is called a *code channel.*

We next wish to determine a numerical measure for the maximum average rate at which a given code channel can transmit information. Consider once more the example in Fig. 1-3 in which the primary information source is used solely for the reporting of weights. In this case, if m weights are reported, the information source generates a sequence of m symbols, each standing for the weight of a man. This source may, for practical purposes, be considered an ergodic source of information. By this we mean that the weights of enough men are reported by the information source so that the percentages of those having any particular weight is proportional to the probability of occurrence of this weight in a man of the type of the group being examined.

When the information source reports the weight j, the information given is, according to Eq. (2) (for the noiseless case),

$$\begin{bmatrix}\text{information reported in} \\ \text{report of weight } j\end{bmatrix} = -\log p_j \tag{32}$$

Now all the reported weights are independent of each other, *i.e.*, John Jones' weight does not depend upon George Smith's. Therefore, the total information of the sequence can be obtained by adding the information of the separate symbols. In the report of m different weights, the number of times that the weight j is reported is

$$N_j = mp_j \tag{33}$$

provided that m is a very large number. The total information carried by these N_j symbols of the sequence is thus

$$N_j(-\log p_j) = -mp_j \log p_j \tag{34}$$

The total information carried by all the m symbols is then

$$-m \sum_j p_j \log p_j \tag{35}$$

where the summation is taken over all possible weights which can occur. From Eq. (35) it follows that

$$\text{average information/symbol} = -\sum_j p_j \log p_j \tag{36}$$

Now for sufficiently long T, according to Eq. (18) of Appendix II, the number of different possible message sequences coming from the ergodic information source is

$$M(T) = A\left[2^{-\frac{1}{b}\sum_j p_j \log p_j}\right]^T = A_v V^T \tag{37}$$ [1]

Furthermore, it was pointed out in Sec. 1.4 that in practically all cases of interest the number of different sequences allowed by all the fixed constraints of a code channel in time T is

$$N(T) = A_w W^T \tag{38}$$

If $W < V$, then $M(T)$ must be greater than $N(T)$, at least for large values of T. In this case the code channel is inadequate to carry the desired information since it cannot have a different method of expression for every different possible message sequence.

In case $W \geq V$, then it is at least possible that the code channel may be adequate to transmit the information which is generated from the source or comes from the primary (input) side of the transducer feeding it. In fact, if

$$W = V + \epsilon \tag{39}$$

where ϵ is a small quantity, no matter how small, $N(T)$ will be greater than $M(T)$, if T is sufficiently large. Therefore, every possible different message of length T can have a different coded expression. Consequently *if we do not mind waiting till the end of the time T to find out the meaning of what is expressed in the code channel,* there is certainly a method of encoding all the information arising from the primary of the transducer.

[1] V is used here for the quantity $\left[2^{-\frac{1}{b}\sum_j p_j \log p_j}\right]$. The notation A_v is used for A to distinguish the latter from the multiplier of W^T in Eq. (38).

Let us define the *code capacity* of the code channel as

$$C = \log W \tag{40}$$[1]

Then since

$$\log V = -\frac{1}{b}\sum_j p_j \log p_j = \begin{bmatrix}\text{average information per unit}\\ \text{of time arising from primary}\end{bmatrix} \tag{41}$$

we may restate the conclusion of the foregoing paragraphs as

THEOREM A: *If the code capacity of a code channel exceeds the average information per unit of time arising in the information generator or transducer feeding it (i.e., if $W > V$), there is a method of encoding all the information into the channel. There is no method of encoding all the information if it comes at a rate exceeding the code capacity (i.e., if $V > W$).*[2,3]

Our derivation above was limited to the case in which the individual messages are independent. However, so long as the source is ergodic, the proof can easily be generalized to cover the case where messages are interrelated, by using methods of grouping of messages and averaging similar to the procedure used in Appendix II.

The foregoing theorem shows that $\log W$ is the maximum possible rate at which information can be sent through a code channel. The derivation of the theorem, however, was able to show only that the rate $\log W$ could be approached if relatively long time delays [up to time T in Eqs. (37) and (38)] were allowed before learning the meaning of the coded symbol groups. In Appendix III

[1] The code capacity may be defined more generally as

$$C = \lim_{T\to\infty} \frac{1}{T} \log N(T)$$

Examples and problems in the calculation of code capacities are given at the end of this section and in Appendix I.

[2] When noise is present, this same relation between the code capacity and $\log V$ still holds. However, $\log V$ will no longer be the average information per unit of time in the noisy case, as discussed in Sec. 1.9.

[3] The derivation of Theorem A assumes that the probability of an event is the same as the probability of the message reporting it. This will be true if the language reports only one matter, such as the weights in Fig. 1-3. When this is not true, the term "average information" in the theorem should be replaced by "average language information" or "entropy." The latter terms will be introduced in the next few pages.

it is shown that the rate log W can be achieved without the handicap of long time delays provided that a particular relation[1,2] holds between symbol lengths and the probabilities of the symbols occurring in the particular ways in which they are used. This relation represents a condition of "match" between the probability constraints of the language in which the information is expressed and the fixed constraints of the code channel which transmits the information. If there are no intersymbol constraints, it represents the choice of symbol frequencies which gives the maximum rate of transmission of information for a given alphabet of symbols having specified symbol lengths. It has of course, been well known for a long time that in order to get a high rate of transmission of information the shortest code groups should be used for those symbols which occur most frequently. Equation (10) of Appendix III, however, makes the relationship precise.[3]

The foregoing matters are of considerable importance. Their principal value for practical purposes is that they set limits or objectives for what can be done. However, they leave the details of many important questions unanswered. For example, if we wish to transmit English through a dot-dash code channel to be followed by an electrical telegraphic code channel, we do not have any design formulas which will tell us which method of encoding will give us the maximum rate of transmission of information. More work needs to be done along this line, but none will be attempted in the present book.

It is well to point out here how fixed constraints of the primary channel can eliminate possible available symbol sequences of the secondary channel and thus make the compound-code capacity less than that of the simple-code capacity. For example, the dot,

[1] Equation (10) of Appendix III.

[2] If each of the long symbol sequences considered in Eqs. (37) and (38) is considered to be a single composite symbol, then these composite symbols are all of the same duration T and have the same probability, namely $1/M(T)$. These composite symbols thus will also satisfy Eq. (10) of Appendix III.

[3] When there are no intersymbol constraints, Eq. (10) of Appendix III reduces to

$$p_{11}^{(s)} = W^{-b_{11}^{(s)}}$$

where $p_{11}^{(s)}$ is the symbol probability and $b_{11}^{(s)}$ is the symbol length of the symbol s; and as above, log W is the code capacity of the channel. A numerical example is solved in Prob. 15 at the end of this section.

dash, letter space, and word space of ordinary telegraphic code have the equivalents in electrical telegraphic code shown in Table VI.

TABLE VI

MORSE CODE SYMBOLS AND THEIR ELECTRICAL EQUIVALENTS[1]

dot	+ −
dash	+ + + −
letter space	− − −
word space	− − − − − − −

Accordingly, such electrical code groups as

$$+ - - + \quad \text{or} \quad + - - - - +$$

have no significance in dot-dash code and the possible signals which they represent are wasted. The effect of such unusable signals is to make the compound-code capacity less than the code capacity of the secondary alone.

One important question which we will attempt to treat is to determine at what rate information can be sent through a channel with the transducer, the language, and the coding system as actually used. As a first step in dealing with this problem, we wish to inquire further into the properties of languages.

For our purposes, we may say that the alphabet plus the probability constraints of an ergodic sequence (or assemblage[2]) of coded messages in a particular code constitute a language. A language is thus a code to which probability constraints have been added. Consider the English language as an example. This has the system of alphabets shown in Table VII. The code channel upon which the probability constraints of English have been superimposed is thus a compound-code channel. The basic code of this compound code is that of the English letter code. Superimposed upon this is the English word code. The fixed constraints imposed by this code are primarily that sequences of letters which do not form English words are not allowed. Superimposed upon the English word code is the English sentence code. The fixed constraints of this code

[1] The symbol (−) can mean either absence of electrical signal or reversed polarity.

[2] An ergodic assemblage of messages is a set of messages (such as the set of all the spoken conversations going on during an hour in a large crowded room) which are not actually in the form of a single sequence, but which if put together at random in a single sequence would form an ergodic sequence.

eliminate all sequences of words which do not form sentences in accordance with a reasonable approximation of English grammar. Finally, superimposed upon the English sentence code is the English message code. The fixed constraints of this code eliminate all sequences of sentences which do not have meaning.

TABLE VII

THE SYSTEM OF ALPHABETS[1] OF THE ENGLISH LANGUAGE

English letter alphabet		English word alphabet	English sentence alphabet	English message alphabet
A	O	All words in an unabridged English dictionary	All possible English sentences that can be made using the words in an unabridged English dictionary	All possible complete English messages.
B	P		(The numbers of symbols in these two alphabets is extremely large)	
C	Q			
D	R			
E	S			
F	T			
G	U			
H	V			
I	W			
J	X			
K	Y			
L	Z			
M	word space			
N				

All the codes of a language change slowly with time and vary somewhat among the different groups of people using the language. However, for our purposes we may consider that, for a particular group of people at a particular time, the fixed constraints of a language are constant.

Languages, as we all know, are used in transmission channels to transmit information. The first step in this process is the coding of the messages at the information source into the (English) message alphabet. Thus, for example, an event occurs at the information source, and its description for transmission is its coded equivalent in the message alphabet. We have used the same word "message" for both the event and its description. When it is desirable to make a distinction, we shall call the former, *the event*, and we shall call its coded equivalent in the language *the message*.

[1] The alphabet of a code consists of a set of symbols or symbol groups which can be formed into sequences to form all possible messages which can be expressed in the code.

In order that the message should carry information, there must be a probability at some receiver concerning the occurrence of the event which can be changed by the reception of the message. What we mean by this probability at the receiver is the following.

Definition: In accordance with the knowledge of the receiver, if N identical worlds were contemplated having all the characteristics of which he was aware, but in which those characteristics which he did not know about were distributed at random, then if $\bar{p}$ were the probability of the event, the event would occur in $\bar{p}N$ of these worlds and would not occur in $(1 - \bar{p})N$ of them.

If the above definition means anything, it defines the probability at the receiver of the event.

Consider next the message. It is one of the symbols of the English message alphabet (or in general of the language message alphabet being used). As such it also has a certain probability of occurrence which we shall call p. If the language used reports only one matter, such as the weights in Fig. 1-3, then

$$p = \bar{p} \tag{42}$$

If, however, the language reports a great variety of matters, as is customary in a general spoken language such as English, then

$$p \ll \bar{p} \tag{43}$$

The probabilities of the messages in the message alphabet determine all the probability constraints of the language, since all the symbol probabilities and intersymbol influences of the sentence, word, and letter alphabets can be determined from the probabilities of occurrence of the various possible transmitted messages. In Eq. (41), we have the relation

$$\begin{bmatrix}\text{average information/} \\ \text{unit length of sequence}\end{bmatrix} = -\frac{1}{b}\sum_j p_j \log p_j \tag{44}$$

The derivation of Eq. (41) and thus of (44) assumed that the information in the jth message was

$$(-\log p_j) \tag{45}$$

According to Eqs. (42) and (43), we can then say that Eq. (44) represents the maximum average information per unit length of message sequence in a language where the p_j represent the message probabilities of the language. This maximum is ordinarily attained only when the language is used to report upon a single matter as was the case in the derivation of Eq. (41). Since Eq. (44) represents the maximum rate of information transmission in a language, we will call it the *language transmission capacity*. We will find it convenient to use the term "average language information" for this "maximum average information."[1] Thus Eq. (44) is a formula for the average language information per unit length of a message sequence, where the p_j represent the message probabilities of the language.

According to our terminology, if p is the probability of a particular message in a language and $\bar{p}$ is the probability of the event which it describes, then we will say that $(-\log p)$ is the amount of *language information* in the message and $(-\log \bar{p})$ is the amount of *semantic information* in the message. Our fundamental definition of information, Eq. (1), defines semantic information. If the "probability of the message" were used in Eq. (1) instead of the "probability of the event," then Eq. (1) would give the amount of language information. Shannon and most other writers use the term "information" for what we call language information. Studies of technical communication systems will occupy most of our attention in this book, and in these studies we, also, will be dealing with language information rather than semantic information.

Equation (44) is a very simple theoretical relationship, and can actually be used for calculation in the case of a simple language such as that used in Fig. 1-3 but it is wholly impractical for a general language such as English where the number of possible messages is extremely large. In the latter case, it is practical to start with only the average language information per symbol of the letter alphabet as determined by the letter frequencies, and then to add the effects of the intersymbol influences until the approximate limiting value of the entropy (average language information) per symbol is reached. The values of the various intersymbol influences are not known accurately, but they are known to a reasonable degree of approximation. On the basis of these, B. McMillan has obtained the following table:

[1] An equivalent term for "average language information" is "entropy." The latter will be defined and discussed in Sec. 1.8.

TABLE VIII

EFFECT OF CONSTRAINTS ON AVERAGE LANGUAGE INFORMATION PER LETTER IN ENGLISH

Letters taken singly at random[1]	4.70	binits/letter
Letters taken singly with English frequencies	4.15	binits/letter
Pairs of letters with English frequencies	3.57	binits/letter
Groups of 8 letters with English frequencies	2.35	binits/letter

On the basis of work of this kind, it is believed that the language transmission capacity of general English is approximately somewhere between one and two binits per symbol.

We will use the symbol R for the language transmission capacity of a channel.[2] R is thus a measure of average language information transmitted per unit of time.[3] According to Eq. (44),

$$R = \frac{-\sum_j p_j \log p_j}{\sum_j p_j b_j} \tag{46}$$

where b_j is the amount of time required to transmit the jth message through the channel. The same formula will hold for symbols as well as messages, if there are no fixed constraints or intersymbol influences. When there is intersymbol influence, it is pointed out in Appendix III, (Eq. 16), that R can be expressed in terms of symbols and constraints of the channel by the formula

$$R = \frac{-\sum_i [P_i \sum_j (\sum_s p_{ij}^{(s)} \log p_{ij}^{(s)})]}{\sum_i \sum_j \sum_s P_i p_{ij}^{(s)} b_{ij}^{(s)}} \tag{47}$$

The interpretation of the notation used in Eq. (47) will be found in Appendix III. Another expression for R is

$$R = \lim_{T \to \infty} \frac{\log \bar{M}(T)}{T} \tag{48}$$

[1] $\log_2 26 = 4.70$.

[2] A *channel* for transmitting code messages or for transmitting information is a complete transmission system from an input location to an output location. The properties of the channel include the properties of the equipment used in it as well as the properties of the codes or languages used.

[3] Language transmission capacity is significant only for long messages; for it is only for long messages that it is always possible to match input and output messages without having an important residue of meaningless output messages.

where $\bar{M}(T)$ is the number of possible sequences of duration T in the channel when the probability constraints of the language are superimposed upon the fixed constraints of the code channel. The latter, of course, includes all the fixed constraints which may be present in the letter, word, and sentence codes of the language, as well as the fixed constraints which may be part of the code channel (say, dot-dash electrical telegraphy) before the language is superimposed. Equation (48), however, is not ordinarily a practical formula for use in calculating R.

As far as the design and interconnection of communication channels are concerned, it is the average language information per unit time rather than the average semantic information per unit time that is important. The matching of language in and language out of a transducer is a matter of language information and not of semantic information, since the matching does not depend upon what the language symbols mean. With this in mind, let us now examine the properties of networks of communication channels.

The maximum average rate at which language information can be transmitted through a chain of transmission channels in cascade (such as shown in Fig. 1-4) is equal to the smallest R in the chain. The total average language information sent through the chain is $R_n t_n$, and

$$R_n t_n \leq \ . \ . \ . \ \leq R_2 t_2 \leq R_1 t_1 \tag{49}$$ [1]

where t_1 is the duration of time that channel 1 is in operation, t_2 is the duration of time that channel 2 is in operation, and so on. If there is no loss[2] of language information in transmission, then the equals sign in Eq. (49) applies and the equation becomes

$$R_1 t_1 = R_2 t_2 = \ . \ . \ . \ = R_n t_n \tag{50}$$

If t_1 exceeds t_2, for example, in Eq. (50), then channel 2 will have to interspace stand-by time in its transmissions. During these stand-by times, channel 2 can be used for other purposes, especially if it has storage means for the signals coming in from channel 1, so that it can bunch its stand-by times into relatively long intervals. In the

[1] It is assumed that the times t, involved here, are relatively long. The case of short times is discussed following Formula (52).

[2] We are here referring to loss due to lack of language transmission capacity, not to loss due to noise.

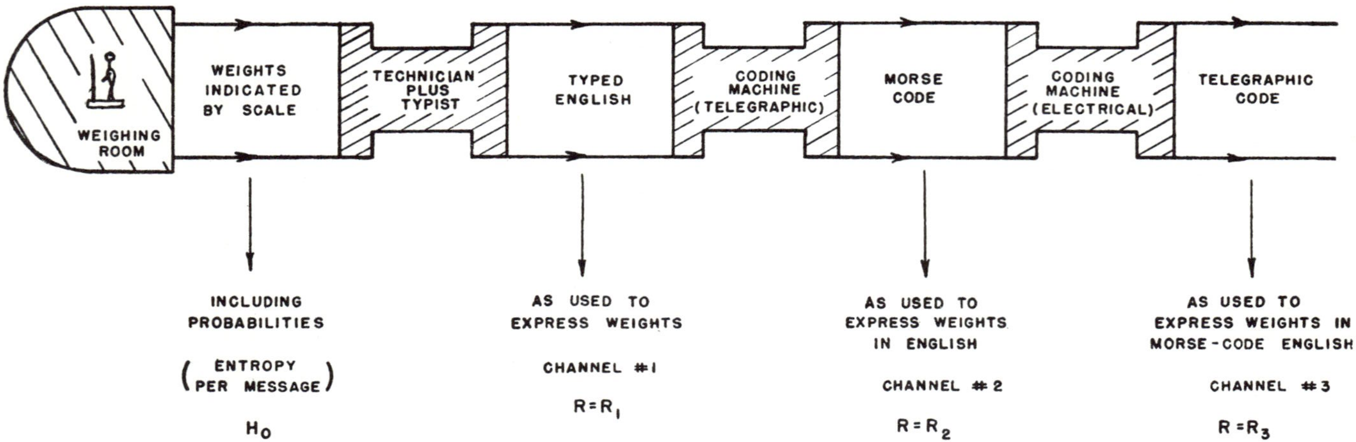

Fig. 1-4. Block diagram of a series of information transmission channels in cascade.

latter case, it might be used as a common element of several transmission channels in parallel, as shown in Fig. 1-5.

In case the inequality sign in Eq. (49) applies, for example,

$$R_1 t_1 > R_2 t_2 \tag{51}$$

then part of the language information which goes through channel 1 never gets through channel 2. For example, in the case of Fig. 1-5, if

$$R_{1a} t_{1a} + R_{1b} t_{1b} + R_{1c} t_{1c} > R_2 t_2 \tag{52}$$

then part of the typed tape has not been sent through the electric coding machine.

In any case, we can draw the following important conclusion:

THEOREM B: *If the language transmission capacity of a secondary (output) transmission channel exceeds the average language information per second at which the primary (input) is generating language information, then the secondary can transmit all the language information which is generated by the primary.*

Strictly speaking, Formulas (49) to (52) apply only if the times involved are long. For short messages, the actual rates of transmission of language information may be considerably different from the average rates; and the effective value of the language transmission capacity R is subject to the same type of fluctuation. This will necessitate channel 1 taking time out while channel 2 catches up, and vice versa. These times out due to fluctuations cannot readily be accumulated so that they are ordinarily lost times. When it is required that there shall be no delay in transmission, it is necessary that the language transmission capacity of each channel in a chain shall be somewhat greater than that of the preceding channel in order that fluctuations shall not cause appreciable delays.

It is important to note that in the derivation of Theorems A and B it was assumed that the information source was ergodic. In other words, the range of the intersymbol influence was finite and every possible message occurred a sufficient number of times so that its a priori probability of occurrence could be equated to its actual fractional frequency occurrence. A large number of sources of information met in practice actually approximate the ergodic condition. Thus, in the earlier example, the technician reporting weights is very close to an ergodic source of original information. It is espe-

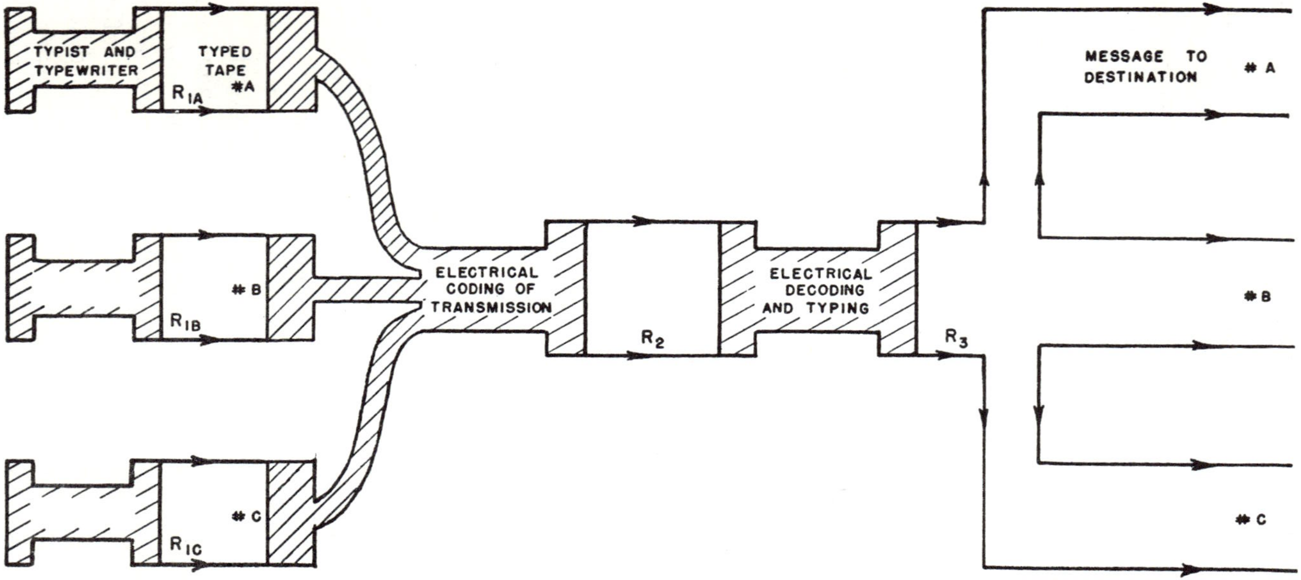

$$R_{1A}T_{1A} + R_{1B}T_{1B} + R_{1C}T_{1C} = R_2T_2 = R_3T_3 = \ldots\ldots$$

FOR NO LOSS OF LANGUAGE INFORMATION IN TRANSMISSION

Fig. 1-5. Block diagram of an information transmission network.

cially important to note that the input of equipment used for the general communication of messages in a particular language, such as public telephone or telegraph facilities, may be considered an ergodic source of, say, English language information. The language transmission capacity of this source of English symbols considered as a transmission channel is rarely if ever fully utilized in carrying the information in the ideas of the original messages. However, we know for sure from Theorem B that encoding of all messages originating in this ergodic source of English symbols will be possible in a succeeding channel, such as a telegraphic-code-carrying English channel, if the latter channel has a language transmission capacity greater than or equal to that of the source.

Once information has gotten into channels of communication via the use of language or its equivalent, the primaries of all succeeding transducers may ordinarily be considered as ergodic sources of language information. The original events, however, with which the information deals, are usually not ergodic. Thus, for example, in September of 1939, when it was reported that Hitler had invaded Poland, a considerable amount of information was conveyed by this report. However, the world in which this event occurred and which was the source of this information was certainly not ergodic. By this is meant that a sufficiently identical set of conditions had not previously occurred a large number of times so that the a priori probability of Hitler's invading Poland could not have been established by the average frequency of occurrence. Nevertheless, in the knowledge of the listener to a radio broadcast reporting this event, there was a certain a priori probability of the occurrence of this event as defined by the earlier definition of $\bar{p}$, and the broadcast certainly gave him information.

Generally, much of the semantic information which we get from the world is of a non-ergodic character.[1,2] However, the equip-

[1] If messages are chosen in accordance with their probabilities, such as by a news reporter, the average information per message may exceed that of English and the actual information per unit duration may exceed the value of R for general prose English. However, since the message probabilities are different from those of general prose English, the news reporter is really using a different language from our point of view, even though the same letter symbols are used. On the other hand his reports may also be considered a fluctuation in the use of general prose English.

[2] The ideas in an entire book are usually coherent, even though the language in which they are expressed is ergodic.

ment designed to transmit this information is designed on the basis of ergodic information, since the equipment must usually transmit language information, and the latter is of an essentially ergodic nature.

PROBLEMS[1]

(A knowledge of Appendix I and/or III is required in the working of several of these problems.)

1. Find the code capacities of the following code channels:

	Symbols in alphabet	*Symbol durations*	*Other constraints*
(a)	dot	2	None
	dash	4	
	letter space	3	
	word space	6	
			Ans. $C = 0.60$ binit/unit of time
(b)	A	1	No letter may follow itself
	B	1	
	C	1	
			Ans. $C = 1$ binit/unit of time
(c)	A	1	None
	B	2	
	C	3	
	D	4	
			Ans. $C = 0.94$ binit/unit of time
(d)	A	1	No letter may occur twice in any group of 4, *i.e.*, ACBA is not allowed
	B	1	
	C	1	
	D	1	
			Ans. $C = 0$
(e)	A	1	The A's always occur in pairs
	B	1	
	C	1	
			Ans. $C = 1.265$ binits/unit of time
(f)	X	1	There is a space after every three consecutive letters
	Y	1	
	Z	1	
	space	1	
			Ans. $C = \frac{1}{4} \log 27$ binits/unit of time

[1] Detailed solution of Prob. 1 is given in Appendix XII to illustrate the methods used.

2. If a language having only the four symbols A, B, C, and D is coded into a secondary channel, having the symbols (+) and (−) in accordance with the code

$$\mathrm{A} \to ++ \qquad \mathrm{C} \to -+$$
$$\mathrm{B} \to +- \qquad \mathrm{D} \to --$$

and the symbol probabilities are

$$p_{\mathrm{A}} = 25\% \qquad p_{\mathrm{C}} = 25\%$$
$$p_{\mathrm{B}} = 25\% \qquad p_{\mathrm{D}} = 25\%$$

what is the language transmission capacity of the secondary channel? Assume that (+) and (−) each have unit duration.

Ans. 1 binit/unit of time.

3. Assuming that weights are expressed in English type as

27 = twenty seven

443 = four forty three

what is the average information per letter in the typed English reports of Fig. 1-3? Reasonable approximations may be made.

4. Suppose that a language having only the three symbols A, B, and C is coded into a secondary channel having the symbols (+) and (−) in accordance with the encoding rules

$$\mathrm{A} \to ++ \qquad \mathrm{B} \to +- \qquad \mathrm{C} \to -+$$

Let the symbol probabilities be

$$p_{\mathrm{A}} = \tfrac{1}{3} \qquad p_{\mathrm{B}} = \tfrac{1}{3} \qquad p_{\mathrm{C}} = \tfrac{1}{3}$$

Assume that (+) and (−) each have unit duration.

(a) Find the simple code capacity of the secondary channel alone.

Ans. 1 binit/unit of time.

(b) Find the compound code capacity of the combined primary and secondary channels.

Ans. $\frac{1}{2}\log 3$ binit/unit of time = 0.79 binit/unit of time.

(c) Find the language transmission capacity of the secondary channel. *Ans.* $\frac{1}{2}\log 3$ binit/unit of time.

(d) Suppose that both (++) and (−−) are assigned the meaning A. What is the language transmission capacity of the secondary channel? *Ans.* $\frac{1}{2}\log 3$ binit/unit of time.

(e) If the symbol probabilities are changed to

$$p_{\mathrm{A}} = \tfrac{1}{2} \qquad p_{\mathrm{B}} = \tfrac{1}{4} \qquad p_{\mathrm{C}} = \tfrac{1}{4},$$

and the same meanings as in part (d) are assigned, what is the language transmission capacity of the secondary channel?

Ans. 0.75 binit/unit of time.

(f) If the symbol probabilities are changed to

$$p_A = \tfrac{1}{4} \qquad p_B = \tfrac{1}{4} \qquad p_C = \tfrac{1}{2}$$

and the same meanings as in part (d) are assigned, what is the language transmission capacity of the secondary channel?

Ans. 0.75 binit/unit of time.

5. An observation station has two meters.

The first meter has 100 scale markings (significant values), and its reading can change to any allowed value in 0.05 sec.

The second meter has 10 scale markings, and its readings can change to any allowed value in 0.01 sec.

What code capacity is required in binits per second in order to transmit all the information which may appear on the meters? *Ans.* 465 binits/sec.

If the observation station sends its signals to the base station in the form of a sequence of off-on pulses, such as shown in Fig. 1-A, what transmission bandwidth is required? Assume that a bandwidth $W = 1/2T$ is required to transmit a pulse of duration T. *Ans.* 232.5 cycles/sec.

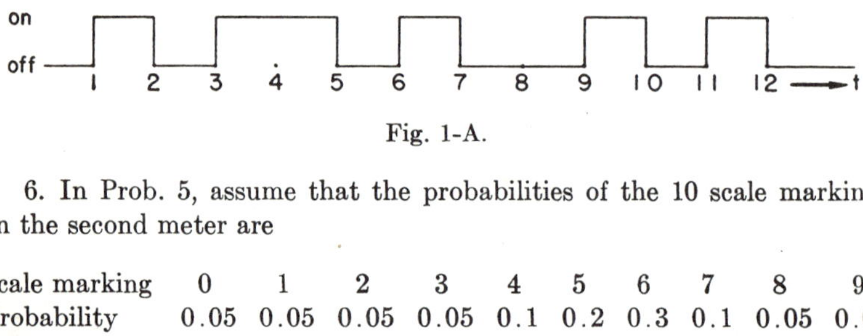

Fig. 1-A.

6. In Prob. 5, assume that the probabilities of the 10 scale markings on the second meter are

Scale marking	0	1	2	3	4	5	6	7	8	9
Probability	0.05	0.05	0.05	0.05	0.1	0.2	0.3	0.1	0.05	0.05

What language transmission capacity in binits per second is required to transmit the information on the second meter? *Ans.* 295 binits/sec.

7. In Prob. 5, assume that the quantities which the first and second meters measure are so related that the reading of the first meter is always in the range

$$10S \pm 9$$

For Prob. 7 assume that readings of the first meter can change every 0.01 sec. Suppose that the second meter has the probabilities given in Prob. 6 and that the first meter is otherwise unconstrained. What language transmission capacity in binits per second is required to transmit the information from the observation station? *Ans.* 716 binits/sec.

8. The following description is taken from a story by Margaret Cousins in *Good Housekeeping Magazine.*

"From the back, she was small and slightly toasted, with golden hair, much wind-blown. Her neck was proud, and her back was straight, and her nondriving arm was weighted with a bunch of silly gold bangles. She had on some kind of barenecked getup made out of that silky stuff—what did they call it, pongee? But a lot of her tanned skin was showing—arm, neck, and so forth. He was certain she smelled of that perfume that always reminded him of ripe fruit and champagne."

(a) Estimate the total amount of semantic information in this description, showing how you arrive at your estimate.

(b) Estimate the redundancy (See Sec. 1.8) in this description.

(c) Give a reasonable facsimile of an equivalent description with a marked reduction in redundancy.

(d) Discuss the relative merits of the original and the facsimile.

(e) Estimate the total amount of language information in the original description, showing how you arrive at your estimate. Discuss briefly the relative amounts of semantic and language information.

9. Show that, if a transmission channel is used to transmit independent semantic information concerning more than one matter, the channel will transmit semantic (as distinguished from language) information at less than its maximum language transmission capacity, because incomplete advantage is taken of all possible permutations of symbols.

10. Show that if, symbols are of equal duration, maximum language transmission capacity is obtained if the symbol probabilities are equal. Show that, in this case, the language transmission capacity is equal to the code capacity.

11. Show that the code capacity of a channel is unchanged if an alphabet of symbol groups is used instead of the elementary symbols themselves, provided that all possible groups are used.

12. (a) Show how to find the code capacity of the telegraphic code shown in Table A. Assume that the length of a dot is $\frac{1}{24}$ sec and the length of a dash is $\frac{3}{24}$ sec. The space between components of a letter equals $\frac{1}{24}$ sec, between letters is $\frac{3}{24}$ sec, and between words is $\frac{6}{24}$ sec.

Discuss the question of which items are important in determining the code capacity and find the approximate value of the latter.

(b) Find the language transmission capacity of this code carrying English. Use only the English letters and assume that the letter frequencies are the only probability constraints. Ignore the intersymbol constraints.

Table A

A	· —	J	· — — —	S	· · ·
B	— · · ·	K	— · —	T	—
C	— · — ·	L	· — · ·	U	· · —
D	— · ·	M	— —	V	· · · —
E	·	N	— ·	W	· — —
F	· · — ·	O	— — —	X	— · · —
G	— — ·	P	· — — ·	Y	— · — —
H	· · · ·	Q	— — · —	Z	— — · ·
I	· ·	R	· — ·		

(c) What is the code capacity of a system using dots and spaces only, each of 1/24 sec duration?

13. Suppose that a language has 90% A's and 10% B's. Suppose (+) and (−) each have unit duration.

(a) If we use the encoding system

$$\mathrm{A} \to + \qquad\qquad \mathrm{B} \to -$$

find the language transmission capacity.

(b) If we use the encoding system

$$\mathrm{A} \to + \qquad\qquad \mathrm{B} \to - - - +$$

find the language transmission capacity.

(c) How do you reconcile the relative values of the language transmission capacities in parts (a) and (b) with Eq. (10) of Appendix III?

14. Find the form to which Eq. (16) of Appendix III for the language transmission capacity reduces, when there are no intersymbol constraints but only different symbol lengths and different symbol probabilities. Compare this with Eq. (46) of Chap. I.

$$\textit{Ans.}\ R = \frac{-\sum_s p_{11}^{(s)} \log p_{11}^{(s)}}{\sum_s p_{11}^{(s)} b_{11}^{(s)}}.$$

15. Suppose we have a transmission channel having only the three symbols A, B, and C. Suppose also that the duration of A is one unit, that of B is two units, and that of C is three units and there are no fixed constraints.

(a) Find the code capacity of this channel.

Ans. $W = 1.84$, $C = 0.880$ binit/unit of time.

(b) Find the language transmission capacity if

$$P_{\mathrm{A}} = \tfrac{1}{6} \qquad P_{\mathrm{B}} = \tfrac{2}{6} \qquad P_{\mathrm{C}} = \tfrac{3}{6}$$

and there are no intersymbol constraints.

Ans. $\frac{6}{14}$ (log 6 − $\frac{1}{3}$ log 2 − $\frac{1}{2}$ log 3) = 0.626 binit/unit of time.

(c) Find the language transmission capacity if

$$P_A = \tfrac{3}{6} \qquad P_B = \tfrac{2}{6} \qquad P_C = \tfrac{1}{6}$$

and there are no intersymbol constraints.

Ans. $\frac{6}{10}$(log 6 − $\frac{1}{3}$ log 2 − $\frac{1}{2}$ log 3) = 0.876 binit/unit of time.

(d) What must the symbol probabilities be for the language transmission capacity to equal the code capacity?

Ans. $P_A = 0.544$, $P_B = 0.296$, $P_C = 0.161$.

1.8 Entropy and redundancy

The right side of Eq. (36) has been familiar to physicists ever since the days of the classical researches of Maxwell, Boltzmann, and Gibbs in statistical mechanics. Boltzmann used the symbol H for this type of summation and showed that, in idealized systems, it is proportional to the thermodynamic quantity entropy. These pioneers also realized that there was a close relation between entropy and information; but they did not develop a quantitative theory of information. It probably did not occur to them that a quantitative theory of information could be developed, since they never dealt with the transmission of information. However, because of this history, it was natural that Shannon, in his now classical investigations of information theory, should have used the symbol H for this summation and should have called it entropy. Accordingly, the terminology

$$H = -\sum_j p_j \log p_j = \text{entropy of the set of probabilities } p_j \qquad (53)$$

has become standard in information theory. If the probabilities p_j represent symbol probabilities, or more generally message probabilities, then it follows from the discussion following Eq. (44) that H represents the average language information per symbol or per message as the case may be. The relation between entropy and choice is discussed in Appendix IV where it is shown that entropy is a measure of the available choice in ergodic sequences. The magnitude of the entropy is a measure of the uncertainty described by a set of probabilities. The entropy H will vanish if and only if all the p_j are zero excepting one which is unity. In this case, the situation is completely determined and there is no choice. In all other cases, the entropy will be positive. Various mathematical properties of entropy are derived and discussed in Appendix VIII.

A great insight into the second law of thermodynamics may be obtained by considering Eq. (53) in the light of our discussion of Table III in Appendix III. There it is shown, in a particular example, that of the possible long sequences which can exist those which make

$$- \sum_j p_j \log p_j$$

a maximum have a much greater probability than that of all the others combined. It may be shown that this result is quite general and is not limited to the particular example cited. By an analogy, which could be justified in detail, if a physical system consists of many parts (say molecules) and the system has a number of possible conditions in which it can exist, and if the actual condition in which it does exist is controlled by a set of probabilities, then the possible conditions in which the entropy is approximately a maximum have a probability which exceeds that of all the others combined by an enormous factor. Therefore, if the system is not in a condition of maximum entropy, the possible conditions into which it may be transformed in which the entropy increases greatly exceed in probability the possible conditions in which the entropy decreases. Accordingly, if the activities of the elementary parts of the system are generally independent of one another so that the next state into which the system is transformed is apparently a matter of chance, the probability that the entropy will increase vastly exceeds the probability that it will decrease. In practical cases, for systems in which the number of elementary parts runs into millions or trillions or even more, it is practically certain that the entropy will increase. This is the equivalent of the statement of the second law of thermodynamics.

Returning now to our considerations of information theory, let us consider a language such as English in which there are symbol frequencies and intersymbol influences. In this case, the symbol frequencies alone do not represent a set of probabilities from which the entropy H representing the average language information per symbol can be calculated. However, symbols can be arranged in symbol groups of sufficient length so as to include essentially all the intersymbol influences. This is actually done in effect in

Appendix II. The probabilities of these symbol groups then do represent a set from which the entropy can be calculated. If this entropy is then divided by the number of symbols in the groups, it gives a value which may be called the *entropy per symbol.* A comparison with Eq. (13) of Appendix II and the derivation of Eq. (36) shows that this entropy per symbol is equal to the average language information per symbol. Let us use the notation

$$h \equiv \text{entropy/symbol} \equiv \text{average language information/symbol} \tag{54}$$

Equation (16) of Appendix II shows that with this definition the number of different possible ergodic messages m symbols long in the language is

$$M = A2^{hm} \tag{55}$$

where A is a constant for the language. This, of course, assumes that the logarithms in Eq. (53) are calculated using the base 2. Table VIII (page 32) illustrates how the symbol frequencies and intersymbol influences decrease the information per symbol of English.

The ratio of the entropy per symbol of a language to the maximum value it could have while still restricted to the same symbols may be called the *relative entropy* ratio. One minus the relative entropy ratio is called the *redundancy.* The redundancy thus represents the factor by which the average lengths of messages are increased due to the language structure beyond the minimum necessary to transmit the desired information. If we assume that intersymbol influences beyond 8 letters are negligible, then Table VIII gives for the redundancy of English

$$\text{redundancy of English} = 1 - \frac{2.35}{4.70} = \frac{1}{2} = 50\% \tag{56}$$

The actual redundancy of English is probably considerably greater than 50% since important intersymbol influences extend beyond 8 letters.[1]

[1] Most possible combinations of letters beyond digrams and trigrams will be ruled out because they do not make English words. Most possible combinations of English words will be ruled out because they do not have meaning. The intersymbol constraints of English thus extend over wide ranges.

It may also be pointed out that different permutations of independent messages or of independent parts of the same message usually have the same mean-

While redundancy decreases the speed of transmission of information, it is not an unmitigated evil. We shall soon see that it can increase the reliability of transmission.

In certain practical applications, where channel capacity is at a premium, the possibility of *information compression* is important. This may, for example, be accomplished by the recoding of messages into shorter symbol sequences. Such an expedient is used in the sending of "greeting" telegrams. In this case the sender can choose one of a number of listed message numbers in order to have the proper message received. It is apparent from Eq. (56) that it would, in theory, be possible to recode the entire English language using the same letters but eliminating all probability constraints and thereby cut the average length of messages by at least 50%. Offhand, it is rather surprising that the possible reduction in length by this means is not practically 100%, since the number of possible combinations of English letters which do not have meaning so vastly exceeds those combinations which do have meaning. However, if we consider messages of 1000 symbols duration, then the ratio of the number of possible sequences to the number having meaning in English is, according to Eqs. (55) and (56),

$$\frac{\begin{bmatrix}\text{number of possible sequences of} \\ \text{1000 symbols duration}\end{bmatrix}}{\begin{bmatrix}\text{number of sequences of same} \\ \text{duration having English meaning}\end{bmatrix}} = \frac{2^{4700}}{A2^{2350}} = \frac{2^{2350}}{A} \tag{57}$$

This number is so large that it is no longer surprising that the redundancy of English may be only of the order of 50%.

If the maximum compression which is possible for English for general use is only of the order of 50% if all redundancy is removed,

---

ing. Thus, "the lot was 80 ft wide and 200 ft deep" has the same meaning as "the lot was 200 ft deep and 80 ft wide." For this reason the number of semantically different messages in English or any other human language does not actually increase exponentially with the message duration. It is proper, however, to ignore the effect of this equivalence of permutations in actual languages in calculating the language transmission capacity, because these permutations represent different English messages even though their meaning is the same, for language transmission capacity is not concerned with meaning. We should also note that very complicated coding and long time delays would be involved if an attempt were made to make use of this reduction in the required language transmission capacity to transmit the meaning in English.

it seems that the improved reliability obtained by the redundancy of English is purchased for a very reasonable loss in speed of transmission. For special applications, however, in which certain messages are frequently used, it may well be worth while to recode these messages into short combinations of letters.

1.9 The transmission of quantized information in the presence of noise

So far we have considered only the so-called noiseless case of information transmission, namely, the case in which all received information is certain to be correct. We will now consider the case in which it is not certain that the received information is correct, *i.e.*, the case in which noise is present in the transmission channel.[1] In order to deal with this case, we go back to Eq. (1) in which we defined the information received in a message as

$$\begin{bmatrix}\text{information}\\ \text{received}\end{bmatrix} = \log \left[\frac{\begin{array}{l}\text{probability at the receiver of the}\\ \text{event after the message is received}\end{array}}{\begin{array}{l}\text{probability at the receiver of the}\\ \text{event before the message is received}\end{array}}\right] = \log\left(\frac{p_B}{p_A}\right) \qquad (58)$$

If there were no noise, the probability p_B would in all cases be unity.

In case there is noise, the probabilities referred to in Eq. (58) are the probabilities of the events which actually occur, regardless of whether these are correctly reported or not. For example, if an e is transmitted and an f is received, the language information received is

$$\log \left[\frac{\begin{array}{l}\text{probability at the receiver that an } e \text{ was}\\ \text{transmitted on the basis of receiving an } f\end{array}}{\begin{array}{l}\text{a priori probability at the receiver that an}\\ e \text{ was transmitted}\end{array}}\right]$$

The noise with which we will be concerned is noise which is present in the transmission channel and which may make the received message different from the transmitted message. If the received messages are all the same as the transmitted messages,

[1] A transmission channel with superimposed noise may also be considered a tranducer between the noiseless input signal and the noisy output signal. In this transducer, the coding between input and output messages is controlled by probability.

no noise is present. The effects of noise are thus effects on the transmission of language information, and we will therefore study noise in terms of the transmission of language information.[1] The probability of the event referred to in Eq. (58) is consequently the probability of the actual transmitted message having been transmitted.

We wish now to determine the value of p_B when noise is present. Suppose that the transmitted message is represented by the letter i and the received message by j. If j is the same as i, then the received message is correct, if j is not the same as i, the received message is false. Let

$$P_i = \text{probability that } i \text{ will be the transmitted message} \tag{59}$$

$$P'_j = \text{probability that } j \text{ will be the received message} \tag{60}$$

$$p_{ij} = \text{probability that } j \text{ will be the received message if } i \text{ is the transmitted message} \tag{61}$$

These probabilities are determined from long-time average values of what actually happens under the conditions present. The probability P_i is just another notation for $p_A(i)$. In other words,

$$p_A(i) = P_i \tag{62}$$

The probability P'_j is the probability that j is the received message under the noise conditions present, regardless of what message is transmitted. The probabilities p_{ij} give a detailed story of the effect of the noise. The probabilities given by Eqs. (59), (60), and (61) may be assumed to be known (or determinable) in practice from average results even though we cannot tell for sure what the particular transmitted message was from a knowledge of the received message.

It follows from the definitions in Eqs. (59), (60), and (61) that

$$\sum_i P_i p_{ij} = P'_j \tag{63}$$

$$\sum_i P_i = 1 \tag{64}$$

$$\sum_j P'_j = 1 \tag{65}$$

$$\sum_j \sum_i P_i p_{ij} = 1 \tag{66}$$

[1] In case it should be desired to study semantic information as distinguished from language information, the effects of noise on the language information can then be converted into the effects of noise on the semantic information.

When there is no noise,

$$p_{ij} = \begin{matrix} 1 & \text{if } i \text{ is the same as } j \\ 0 & \text{if } i \text{ is not the same as } j \end{matrix} \tag{67}$$

We note in passing that it is not necessary that the number of possible i messages should be equal to the number of possible j messages, except in the noiseless case.

Let us now determine the value p_B to be used in Eq. (58) when i is the transmitted message and j is the received message. We will call this $p_B(i)_j$. This is the probability that i was the transmitted message as indicated by the fact that j is the received message. Before the received message is observed, we know that the probability that the transmitted message will be i is P_i and the probability both that the transmitted message will be i and the received message will be j is $P_i p_{ij}$. After the received message is observed to be j, this probability is increased by the factor $1/P'_j$, since all the cases in which j is not the received message have been eliminated. Thus (Bayes' theorem)

$$p_B(i)_j = \frac{1}{P'_j} P_i p_{ij} \tag{68}$$

The information received with the message $i \rightarrow j$ (this is an abridged notation to express the fact that the transmitted message is i and the received message is j) is, according to Eqs. (58), (62), and (68),

$$\begin{aligned} \begin{bmatrix} \text{information received} \\ \text{with message } i \rightarrow j \end{bmatrix} &= \log \left[\frac{p_B(i)_j}{p_A(i)} \right] = -\log P_i + \log \left[\frac{P_i p_{ij}}{P'_j} \right] \\ &= \log p_{ij} - \log P'_j \end{aligned} \tag{69}$$

Since the probability of the occurrence of the message $i \rightarrow j$ is $P_i p_{ij}$, we may write for the average information per message when noise is present

$$\begin{aligned} \text{average information/message} &= \\ &= \sum_j \sum_i P_i p_{ij} [\log p_{ij} - \log P'_j] \\ &= \sum_j \sum_i P_i p_{ij} \log p_{ij} - \sum_j \sum_i P_i p_{ij} \log P'_j \\ &= \sum_j \sum_i P_i p_{ij} \log p_{ij} - \sum_j P'_j \log P'_j \\ &= \sum_i P_i \sum_j p_{ij} \log p_{ij} - \sum_j P'_j \log P'_j \\ &= \mathrm{H}(j) - \mathrm{H}_i(j) \end{aligned} \tag{70}$$ [1]

[1] This corresponds to the derivation of Eq. (35) in Appendix VIII.

We emphasize once more that information here refers to language information. The last quantities in Eq. (70) are defined by the equations

$$H(j) = -\sum_j P'_j \log P'_j \tag{71}$$

and

$$H_i(j) = -\sum_i P_i \sum_j p_{ij} \log p_{ij} \tag{72}$$

$H(j)$ is called the entropy per message of the received message sequence when the input is unknown, while $H_i(j)$ is called the conditional entropy per message of the received message sequence when the input is known. In the absence of noise, $H_i(j)$ is zero.[1]

In order to consider a specific case, let us go back to the example of Fig. 1-3. Suppose that part of the sequence of the weights found at the weighing room is

$$\ldots\ 126,\ 183,\ 174,\ 98,\ 222,\ \ldots$$

Let us suppose that this group of weights is received at the recording room as

$$\ldots\ 126,\ 183,\ 474,\ 98,\ 222,\ \ldots$$

Let us determine the amount of information received with each weight. Let us suppose that the probabilities p_A and p_B of the various weights above are

$$p_A(126) = 0.01 \tag{73}$$

$$p_A(183) = 0.009 \tag{74}$$

$$p_A(174) = 0.01 \tag{75}$$

$$p_A(98) = 0.002 \tag{76}$$

$$p_A(222) = 0.003 \tag{77}$$

$$p_B(126)_{126} = 0.99 \tag{78}$$

$$p_B(183)_{183} = 0.99 \tag{79}$$

$$p_B(174)_{474} = 0.0001 \tag{80}$$

$$p_B(98)_{98} = 0.99 \tag{81}$$

$$p_B(222)_{222} = 0.99 \tag{82}$$

[1] Conditional entropy is discussed more fully at the end of Appendix VIII.

These probabilities can be determined from a study of the system over a long period of time. Now

$$\log\left[\frac{p_B(126)_{126}}{p_A(126)}\right] = \log\frac{0.99}{0.01} = +6.62 \text{ binits} \tag{83}$$

$$\log\left[\frac{p_B(183)_{183}}{p_A(183)}\right] = \log\frac{0.99}{0.009} = +6.78 \text{ binits} \tag{84}$$

$$\log\left[\frac{p_B(174)_{474}}{p_A(174)}\right] = \log\frac{0.0001}{0.01} = -6.64 \text{ binits} \tag{85}$$

$$\log\left[\frac{p_B(98)_{98}}{p_A(98)}\right] = \log\frac{0.99}{0.002} = 8.93 \text{ binits} \tag{86}$$

$$\log\left[\frac{p_B(222)_{222}}{p_A(222)}\right] = \log\frac{0.99}{0.003} = 8.35 \text{ binits} \tag{87}$$

From these values, we note two things. In the first place, the amount of information received is not the same for each message. In the second place, when an erroneous message is received, the amount of information received is negative. The receiver never knows how much information is received with any particular message, since he does not know what message was sent. He can, however, determine the *average* information per message by using Eq. (70) provided that he knows the characteristic probabilities of the system.

Let us next consider a somewhat simpler example used by Shannon. This example is very useful in illustrating the effects of noise. Suppose we have a system with two possible symbols 0 and 1. The messages in this case are single symbols. Suppose

$$p_A(1) = 0.5 = P_1 \tag{88}$$

$$p_A(0) = 0.5 = P_0 \tag{89}$$

During transmission, noise introduces errors so that on the average 1 in 100 symbols is received incorrectly (a 0 as 1 or a 1 as 0), and the errors are equally likely to occur among the 0's and 1's. Thus

$$p_{11} = 0.99 \tag{90}$$

$$p_{00} = 0.99 \tag{91}$$

$$p_{10} = 0.01 \tag{92}$$

$$p_{01} = 0.01 \tag{93}$$

For the received signals, we then have, according to Eq. (63),

$$P'_1 = p_A(1)p_{11} + p_A(0)p_{01} = 0.5 \times 0.99 + 0.5 \times 0.01 = 0.5 \quad (94)$$

$$P'_0 = P_A(1)p_{10} + p_A(0)p_{00} = 0.5 \times 0.01 + 0.5 \times 0.99 = 0.5 \quad (95)$$

From Eq. (68), we then have

$$p_B(1)_1 = \frac{0.5}{0.5}(0.99) = 0.99 \quad (96)$$

$$p_B(1)_0 = \frac{0.5}{0.5}(0.01) = 0.01 \quad (97)$$

$$p_B(0)_1 = \frac{0.5}{0.5}(0.01) = 0.01 \quad (98)$$

$$p_B(0)_0 = \frac{0.5}{0.5}(0.99) = 0.99 \quad (99)$$

Thus if a 1 is transmitted and a 1 is received, the information received with the symbol is

$$\log\left[\frac{p_B(1)_1}{p_A(1)}\right] = \log\frac{0.99}{0.5} = 0.985 \text{ binit/symbol} \quad (100)$$

On the other hand, if a 1 is transmitted and a 0 is received, the information received with the symbol is

$$\log\left[\frac{p_B(1)_0}{p_A(1)}\right] = \log\frac{0.01}{0.5} = -5.64 \text{ binit/symbol} \quad (101)$$

The situation is analogous for 0's correctly and erroneously received. The receiver of course does not know whether the symbol has brought him +0.985 or −5.64 binits of information. From Eqs. (100) and (101), we have

$$\begin{aligned}\text{average information/symbol} &= 0.99 \times 0.985 + 0.01 \times (-5.64)\\ &= 0.919 \text{ binit/symbol} \quad (102)\end{aligned}$$

We could equally well have determined the average information per symbol by using Eqs. (70), (71), and (72). Thus

$$\begin{aligned}H(j) = -(P'_1 \log P'_1 + P'_0 \log P'_0) &= -(0.5 \log 0.5 + 0.5 \log 0.5)\\ &= 1.00 \text{ binit/symbol} \quad (103)\end{aligned}$$

$$
\begin{aligned}
H_i(j) &= -[P_1(p_{11} \log p_{11} + p_{10} \log p_{10}) \\
&\qquad\qquad + P_0(p_{01} \log p_{01} + p_{00} \log p_{00})] \\
&= -[0.5(0.99 \log 0.99 + 0.01 \log 0.01) \\
&\qquad\qquad + 0.5(0.01 \log 0.01 + 0.99 \log 0.99)] \\
&= 0.081 \text{ binit/symbol} \qquad (104)
\end{aligned}
$$

$$
\begin{aligned}
\text{average information/symbol} &= H(j) - H_i(j) \\
&= 1.00 - 0.081 \\
&= 0.919 \text{ binit/symbol} \qquad (105)
\end{aligned}
$$

in agreement with Eq. (102).

It is interesting to note that if the percentage of errors is increased from 1 to 50% the value of $H(j)$ is unchanged, but $H_i(j)$ rises to 1.00 binit/symbol. Thus the average rate of transmission of information is zero. The result that no information is received in this case is quite reasonable, since an equal percentage of correctly received symbols could be obtained by tossing a coin at the receiver to determine whether the received symbol is a 0 or a 1.

On the other hand, if the percentage of errors is 100%, $H(j)$ is still unchanged but $H_i(j)$ falls to zero. The latter must be true because

$$p_{10} = 1 \qquad (106)$$

$$p_{01} = 1 \qquad (107)$$

$$p_{00} = 0 \qquad (108)$$

$$p_{11} = 0 \qquad (109)$$

Thus either p_{ij} or $\log p_{ij}$ will vanish in every term of the expression

$$H_i(j) = -\sum_i P_i \sum_j p_{ij} \log p_{ij} \qquad (110)$$

Consequently, the average rate of transmission of information is 1.00 binit/symbol just as in the noiseless case. This is reasonable, since there are only two symbols, and the fact that it is certain that the received symbol is not the transmitted symbol thus allows us to identify the transmitted symbol with certainty by observing the received symbol.

PROBLEMS[1]

1. An alphabet consists of 8 consonants and 8 vowels. Suppose that all letters of the alphabet are equally probable and that there is no intersymbol

[1] A complete solution of Prob. 1 is given in Appendix XII.

influence. If consonants are always understood correctly, but vowels are understood correctly only half the time, being mistaken for other vowels the other half of the time, all vowels being involved in errors the same percentage of the time, what is the average rate of information transmission per symbol?

Ans. $(3.75 - 0.25 \log_2 14)$ binits/symbol = 2.80 binits/symbol.

2. In an ergodic sequence having the alphabet A, B, C, and D, let all the symbols have equal frequencies of occurence and let there be no intersymbol influence. Suppose that half of the symbols are received erroneously, the errors being equally spread among the other symbols. Find the average information per symbol.

Ans. $(1 - \frac{1}{2} \log_2 3)$ binit/symbol = 0.21 binit/symbol.

3. (a) A news reporter in a busy area sends back 10 reports per day, each of which on the average would contain 20 binits of information if they were true. However, half of them are not true, and for the purposes of this problem are to be considered as having no effect on the probability estimates of actual events. How much information per day is there on the average in his reports? Assume that it is known that on the average only half of them are true. *Ans.* 95 binits per day.

(b) If this reporter took time to check his facts thoroughly, suppose that he could send only 40% as many reports, but that they would all be correct. How much information, on the average, would there be in his reports per day in this case? *Ans.* 80 binits per day.

(c) If you were a reader, would you prefer for the reporter to act as in (a) or (b)? Why?

1.10 On the reliability of information transmitted through a noisy channel

In Eq. (70) we have a formula for the average information per message for messages transmitted in a noisy transmission channel. We will now derive the rather remarkable result that it is possible by proper encoding to transmit this amount of information through the noisy channel with a vanishingly small percentage of errors. This means that by proper encoding, which may involve long time delays, it is theoretically possible to improve the signal-to-noise ratio in a transmission system until the noise vanishes.

To get an idea of how this may work, suppose we are transmitting English letters and suppose there are only 26 messages which we may ever care to send. Then each message can be represented by one English letter. Due to noise, a certain percentage of letters may be received incorrectly. We can, however, reduce the probability of erroneous reception of a message by repeating it. Thus our encoding process may simply be to repeat each message 10 times. Then if such a message as

AAMACAABAA

is received, it will be interpreted as message A. The probability of making an error when such a code is used is obviously much less than if single letters were used. On the other hand, the number of messages which can be sent in a given time is also obviously much less, so that the increased reliability is purchased at a price of decreased message-handling capacity.

If the number of possible messages which it may be desired to send is large, then multisymbol code groups must be used. For example, it may be desired to be able to send all possible English words. Then our method of encoding might be to repeat each word four times. Thus

BOI BOY TOY BIY

would be interpreted as BOY. Repetition, as just described, is not a particularly efficient method of encoding to reduce noise. More efficient methods[1] encode the possible messages into relatively long coded symbol groups. This enables a large number of intersymbol relationships to be used as a basis for eliminating erroneous messages. On the other hand, the use of these long coded symbol groups has the disadvantage that it causes a *delay*[2] in reception, since obviously the entire symbol group must be received before the message can be identified.

With these ideas in mind, let us now proceed with the derivation of the theorem for the amount of information which can be sent through a noisy channel with a vanishingly small percentage of errors. Suppose we have a noisy transmission channel whose characteristics are described by Eqs. (59), (60), (61), (71), and (72). Let us encode all messages into symbol sequences having a length

[1] By more efficient methods, we mean methods which obtain more error reduction for the same amount of redundancy.

[2] In Chap. VII we will show that an effective increase in message length may be obtained by increased bandwidth rather than increased time duration, so that time delays can be avoided when removing noise by using extra bandwidth.

of m symbols, where m is a large number. According to Eq. (16) of Appendix II and Eq. (71), the number of possible sequences which can appear at the receiver is, approximately,

$$n_1 = A_1 2^{mh(j)} \tag{111}$$

Likewise, the number of possible sequences which can be transmitted is approximately

$$n_2 = A_2 2^{mh(i)} \tag{112}$$

where
$$h(j) = -\frac{1}{q} \sum_j P'_j \log P'_j = \frac{1}{q} H(j) \tag{113}$$

$$h(i) = -\frac{1}{q} \sum_i P_i \log P_i = \frac{1}{q} H(i) \tag{114}$$

and q is the average number of symbols in a message.

By methods analogous to those used in the derivation of Eq. (16) of Appendix II, it may be shown that the number of different possible sequences due to noise which can appear at the receiver for any particular single transmitted sequence is, on the average,

$$n_3 = A_3 2^{mh_i(j)} \tag{115}$$

Now suppose we wish to be able to transmit any one of

$$n_4 = \frac{n_1}{n_3} 2^{-\eta m} \tag{116}$$

messages from the transmitter. We will choose η as a small quantity, in particular we will choose it as small in comparison with $[h(j) - h_i(j)]$. Nevertheless, if m is chosen sufficiently large,

$$2^{-\eta m} \tag{117}$$

will be very small in comparison with unity.

If m is a very large number, the n_1 different possible received sequences are ergodic and are therefore equally probable. Likewise, the n_2 different possible transmitted sequences are ergodic and they also are equally probable. Now suppose that the n_4 desired transmitted messages are, respectively, coded[1] to n_4, chosen

[1] By this we mean that each of the n_4 chosen transmitted sequences is interpreted as having the same meaning as a different one of the n_4 messages which it may be desired to transmit.

at random,[1] of the n_2 possible transmitted sequences (see Fig. 1-6). Let us consider the possible resulting received sequences which can arise, because of noise, from these n_4 transmitted sequences. In particular, we first wish to know how much duplication to expect among the possible received sequences. Since the n_2 possible transmitted sequences are ergodic and thus equally probable among

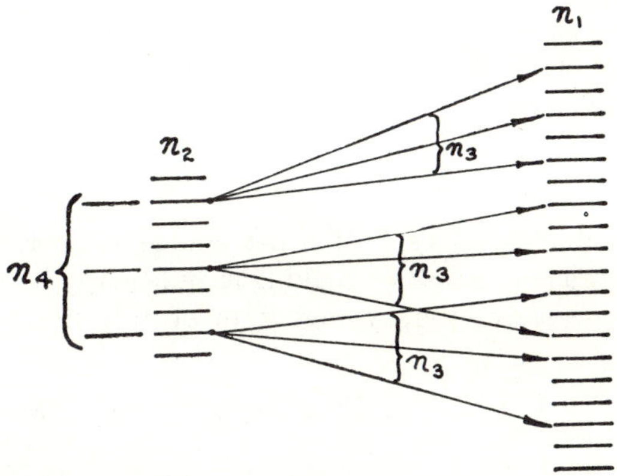

Fig. 1-6. Significance of symbols in noise reducing analysis:

$n_1 \equiv$ different possible received messages;

$n_2 \equiv$ different possible transmitted messages;

$n_3 \equiv$ different possible received messages due to noise superimposed on a given transmitted message;

$n_4 \equiv$ different transmitted messages available for use in the chosen noise reducing plan.

themselves and since the chosen transmitted sequences are chosen at random, the percentage of duplication among the possible resulting received sequences will be approximately equal to the percentage which they represent of all the possible received sequences. If there were no duplication, the number of different possible resulting received sequences would be a maximum, namely,

$$n_5 = n_4 \cdot n_3 = n_1 2^{-\eta m} \tag{118}$$

The percentage of duplication will therefore be approximately

$$\% \text{ duplication} = 2^{-\eta m} \text{ times } 100\% \tag{119}$$

[1] Since each of the chosen sequences is ergodic, the particular selection made will not affect the symbol probabilities.

which is a very small percentage, if m is sufficiently large. If n_4 is given by Eq. (116), it is therefore possible to encode n_4 possible messages to be transmitted through the channel with a percentage of errors in reception equal to the percentage of duplication given by Eq. (119). We have just shown that this can be done[1] by choosing at random[2] the n_4 transmitted sequences to be used, and then interpreting each of the n_3 different received sequences which can arise from any particular transmitted sequence S_i as having the meaning of S_i. By choosing m sufficiently large, it is thus possible to transmit

$$n_4 = \frac{n_1}{n_3} 2^{-\eta m} = \frac{A_1}{A_3} 2^{[h(j)-h_i(j)-\eta]m} \tag{120}$$

different possible messages, with a percentage of errors that can be made as small as we please by making m sufficiently large. For this number of messages, the average rate of transmitting language information is

$$h(j) - h_i(j) - \eta = h(j) - h_i(j) \quad \text{(approximately)} \tag{121}$$

We thus arrive at the theorem (originally derived by Shannon):

THEOREM C: *By proper encoding, a noisy channel can be made to transmit language information at the average rate of*

$$[h(j) - h_i(j)] \quad \text{binits/symbol}$$

with a vanishingly small percentage of errors.

Comparing this result with Eq. (70), we see that this is all the *average language information* which the channel can transmit in any case.

[1] While the method of noise reduction here described is simple in theory, it is not easy to make simple equipment which does what the theory calls for in practical cases. Thus, if the information is transmitted as speech, it is not easy to build automatic equipment which encodes speech into long sequences, and then properly identifies the sequences after noise has acted upon them, and finally decodes them into the original speech. Nevertheless, wideband frequency modulation and many other noise-reducing schemes actually use encoding into long sequences for noise reduction (although their encoding is not random). A detailed analysis would show that these methods of noise reduction have considerable similarity to the method described above, but they are less effective.

[2] It is perfectly possible that a choice made according to some planned pattern may do even better than a random choice, especially for short sequences.

The probability p_B is on the average

$$p_B = \left[1 - \frac{\text{percentage of errors in reception}}{100\%}\right] \tag{122}$$

This is an excellent measure of the *reliability* of transmission. Our analysis above indicates that the reliability of transmission of language information through a channel can be improved up to an essentially perfect value by proper encoding, without thereby decreasing the average rate of transmission of language information. On the other hand, it should also be realized that since this encoding requires the reception of long symbol sequences before the messages can be interpreted, the above described encoding to eliminate noise may involve considerable *time delays* in reception.[1,2]

In case we should similarly use sequences of m symbols duration in an attempt to transmit language information through a channel at an average rate per symbol greater than $h(j) - h_i(j)$, we could express this by writing

$$n_4 = A2^{[h(j)-h(j)+\eta]m} \tag{123}$$

where η is a positive quantity. In this case only about one symbol out of each $A2^{+\eta m}$ would be transmitted without error, so that the transmission would be useless. We thus see that if language information in the form of long sequences is sent into a channel at a rate greater than $[h(j) - h_i(j)]$ per symbol only an unconsequential fraction of it is likely to be received correctly.

Since $h_i(j)$ is a description of the noise properties of the channel, it would appear that $h(j) - h_i(j)$ is closely related to the noise-reduction threshold[3] of the system. The fact that the errors decrease exponentially with m when the attempted rate of language information transmission is less than $h(j) - h_i(j)$ and increase exponen-

[1] It should perhaps be pointed out that this type of encoding which reduces the percentage of errors to zero without reducing the rate at which information is received still does reduce the rate at which messages can be handled by the transmitter. Each received message now carries more information because its correctness is certain, but since the average rate of receiving information has been unchanged, it follows that fewer messages can be handled.

[2] The usefulness of a transmission system depends upon the rate of transmission of information, the reliability of the transmission, and the delay, if any, in reception. All three of these items must be considered when judging the system.

[3] See Section 7.10.

tially with m when the attempted rate of language information transmission is greater than $h(j) - h_i(j)$ is closely related to the mechanism of noise reduction. This will be considered in more detail later on when we study continuous information and deal with practical noise-reduction systems.

In the case of a noiseless channel, we have for the language transmission capacity [see Eq. (46)], if the output message set is the same as the input,

$$R = \frac{1}{b_{io}} H(i) = -\frac{1}{b_{io}} \sum_i P_i \log P_i = \frac{1}{b_{jo}} H(j)$$
$$= -\frac{1}{b_{jo}} \sum_j P'_j \log P'_j \tag{124}$$

where $b_{io} = \sum_i P_i b_i$ = average duration of input messages (125)

$b_{jo} = \sum_j P'_j b'_j$ = average duration of output messages (126)

If the channel is noisy,

$$\frac{1}{b_{io}} H(i) \quad \text{and} \quad \frac{1}{b_{jo}} H(j)$$

need no longer be equal, and in any case they will not, in general, represent the rate at which language information is transmitted through the channel. In the noisy case we will therefore distinguish between the two interpretations of R by means of definitions given by the following formulas:

$$R_1 = \frac{H(i)}{b_{io}} = -\frac{1}{b_{io}} \sum_i P_i \log P_i = \begin{bmatrix}\text{average rate at which language} \\ \text{information can be coded into} \\ \text{the transmission channel}\end{bmatrix} \tag{127}$$

$$R_0 = \frac{1}{b_{jo}} [H(j) - H_i(j)] = \begin{bmatrix}\text{average rate at which language infor-} \\ \text{mation can be sent through the channel}\end{bmatrix} \tag{128}$$

Ordinarily $R_1 > R_0$ in a noisy channel. We have shown, however, that by proper encoding language information can be sent through the channel at the rate R_0 with a vanishingly small percentage of errors. When this is done, again assuming that the output message set is the same as the input,

$$R_1' = -\frac{1}{b_{i'o}} \sum_{i'} P_{i'} \log P_{i'}$$

$$= -\frac{1}{b_{j'o}} \sum_{j'} P_{j'}' \log P_{j'}' = R_0 \qquad (129)$$

where i' and j' refer to messages in the new code of the symbol groups which eliminate errors. In other words, in the new errorless

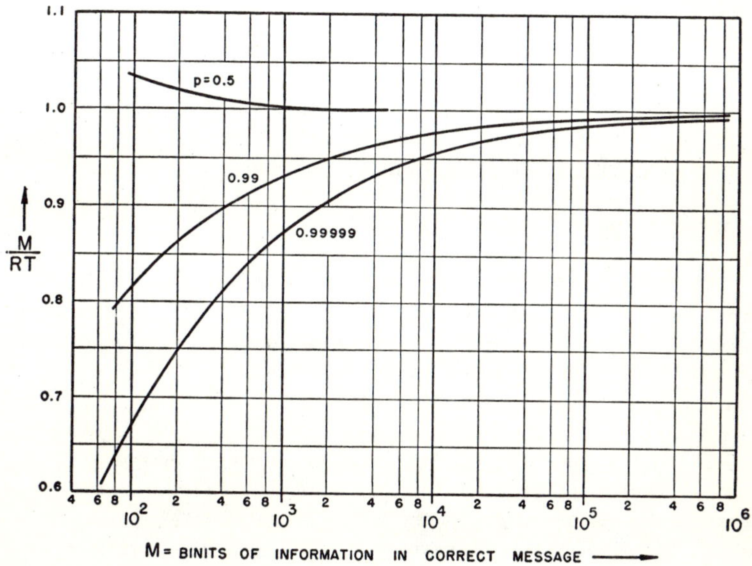

Fig. 1-7.

p = probability of no error;
R = language transmission capacity;
T = message duration;
0.1 = noise-to-signal power ratio.

(*From S. O. Rice,* Bell System Tech. Jour., *vol. 29, January, 1950, pp. 60–93.*)

code, language information cannot be coded into the channel at a greater rate than R_0.

The discussion in this section has dealt with the limiting case as the messages become infinitely long. When messages are of finite length, the analysis becomes very involved. In this connection, Fig. 1-7 is of interest. It gives an example of how the information per unit time in messages of finite length approaches the limiting value as the messages become longer. These curves were calculated by S. O. Rice for a particular method of encoding and a particular value of noise-to-signal ratio. Each curve represents a different value of p, where p in this case stands for the probability of no error in a message.

In concluding this section, we wish once more to emphasize that, while we have shown that by proper encoding it is possible to reduce the percentage of errors to zero without decreasing the rate of transmission of information, this type of encoding may involve enormous delays. Figure 1-7 gives some idea of the relation between the information transmission rate, the delay and the probability of error. In particular, it shows that very long messages are required in order to approach the maximum information transmission rate with a high probability of no error.

Problems

1. Suppose a transmitter of discrete quantized signals can transmit any signal whose value is an integer between -10 and $+10$ including zero. Suppose that there is noise in the transmission such that, for any given input, the output has the following probabilities:

$$p_{i\to i-1} = 0.2 \qquad p_{i\to i} = 0.6 \qquad p_{i\to i+1} = 0.2$$

(a) What is the entropy of the transmitted signal, $H(i)$?
(b) What is the entropy of the noise, $H_i(j)$?
(c) What is the entropy of the output signal, $H(j)$?
(d) What is the maximum average rate of transmission of information?

2. (a) In Prob. 1, assume that only the integers 9, 6, 3, 0, -3, -6, and -9 are used as transmitted signals. Assume that the noise in transmission is the same as stated in Prob. 1. If each received signal is interpreted as that integer of the used transmitted signals which is nearest to it in value, what is the percentage of errors in interpretation of the received signals?

(b) What is the maximum average rate of transmission of information in this case? Compare this with the value in (d) of Prob. 1.

3. (a) In Prob. 1, suppose that the noise probabilities are changed to

$$p_{i\to i-1} = \tfrac{1}{3} \qquad p_{i\to i} = \tfrac{1}{3} \qquad p_{i\to i+1} = \tfrac{1}{3}$$

and that the other conditions of the problem are unchanged. What is the maximum average rate of transmission of information?

(b) Assume the noise conditions of part (a) of Prob. 3, but assume that only those transmitted signals used in part (a) of Prob. 2 are used and that the received signals are interpreted as indicated there. What is the maximum average rate of transmission of information?

(c) Can the encoding of signals into long sequences be used in part (a) of Prob. 3 to obtain a larger rate of transmission of practically errorless information than can be obtained by the method of choice and interpretation of part (b)?

1.11 Concerning the value of information

The quantitative theory of information which we have been developing may appear incomplete and perhaps disappointing to the reader because it does not treat the *value* of information. There are just as many binits in the information which tells whether John Smith's wife had a boy or a girl, as in the information which tells whether your own wife had a boy or a girl. The latter is of course of much greater value and importance to you. It might be possible to develop a useful mathematical theory which deals with the relations between information and the observer and thereby to treat the value as well as the quantity of information. Such a theory has not yet been developed, as far as the writer is aware. The quantity of information which we have talked about deals only with the relative probabilities of events. Human thinking is done and decisions are made partly on the basis of these probabilities and partly on the basis of the relation of the events to the person doing the thinking and deciding. Information theory as we have developed it thus deals with only one aspect of the situation. Although it is thus incomplete, information theory does deal with an important part of the situation, and it is therefore of great interest and utility. In particular, it treats all the aspects of the communication of information in systems which may be treated as ergodic.

Another matter related to the value of information is the distinction between quantitative and qualitative information. New quantitative data of three-place accuracy has about 10 binits (*i.e.*, $\log_2 1000$) of information in each independent datum. Ten binits of qualitative information is numerically just as large from the point of view of information theory. However, it is believed in some circles that quantitative information is intrinsically more reliable than qualitative information. This belief is often based upon mistrust of the reliability of observers who report qualitative data. On the other hand, it also has an objective basis. The alphabet of possible values of quantitative data is arranged in numerical order. Therefore, all except the last few digits in a datum of quantitative information are pretty sure to be reliable. This is not so likely to be true for qualitative information. To offset this, it is also likely that at least some of the digits in quantitative information are pretty well known ahead of time, so that they carry very little information.[1] It is of interest to note that in the sciences, highly accurate quantitative information often brings to light important qualitative information. For example, highly accurate measurements of the masses of isotopes brought to light the mass defects, which were the starting point in the study of nuclear energies.

[1] Further discussion related to this section will be found in Chap. IX, especially Sec. 9.2.

CHAPTER II

SOME PROPERTIES OF CONTINUOUS SIGNALS

2.0 Some matters of terminology

Such variables as current, voltage, sound pressure, magnetic field strength, etc., can be superimposed by algebraic addition in linear systems. These are to be distinguished from current squared, voltage squared, etc., which are usually proportional to power and which cannot be superimposed by algebraic addition, since there may be additional interaction terms. Suppose that, at a given time, one signal is described as having a value of 3.6 units and another signal as having a value of 9.2 units. In order to know what happens when these signals are combined, it is usually important to know whether they can be added algebraically when superimposed in a linear system. For the sake of precision in dealing with such questions, certain new words have been introduced in the present book. We shall use the term *liniva*[1] for the instantaneous value of a variable which is subject to algebraic addition when superimposed with a variable of the same type in a linear system. In case the independent variable is some variable other than time, say distance or frequency, we shall still use the term liniva for values of the dependent variable which are subject to algebraic addition when superimposed in a linear system. We shall use the term *quadiva*[2] for the square of the instantaneous value of a variable of the foregoing type.[3,4] The time integral of the liniva of a signal we shall

[1] Liniva is a contraction of "linear instantaneous value."

[2] Quadiva is a contraction of "quadratic instantaneous value."

[3] The term "instantaneous amplitude" has often been used for what we call "liniva." The objection to using amplitude in this sense is that amplitude usually means a peak value, in particular the peak value of a sine wave.

[4] The word "power" has sometimes been used for what we call "quadiva." Although this use of power is usually clear, we sometimes will prefer to use the more precise term "quadiva."

call its *linear content.* The time integral of the quadiva we shall call the *quadratic content.* The quadiva of a signal is proportional to its power, while the quadratic content is proportional to its energy. Figure 2-1 illustrates the new terminology.[1] Generally

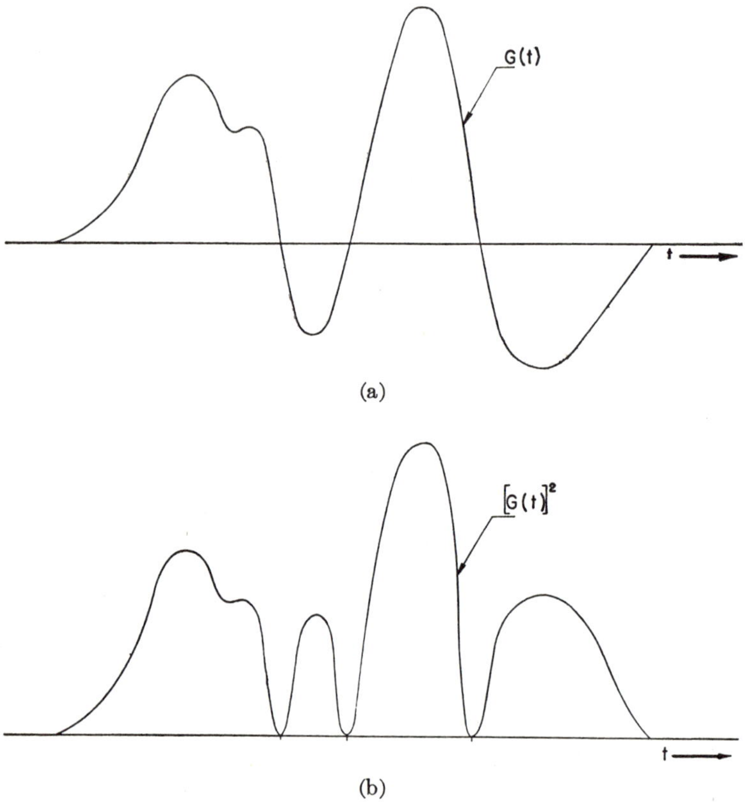

Fig. 2-1. (a) Liniva of the signal $G(t)$ plotted as a function of time. (b) Quadiva of the signal $G(t)$ plotted as a function of time.

speaking, the terms "liniva" and "quadiva" will be used only when the more common terms "value" and/or "power" are ambiguous.

[1] The terms "linear content" and "quadratic content" have already been used with the same meaning in Goldman, *Frequency Analysis, Modulation, and Noise.*

2.1 The sampling theorem in the time domain[1]

As a preliminary to the study of the information theory of continuous signals, we wish to derive certain general theorems. Consider a signal, in the form of a continuous function, which has passed through a transmission system having a finite bandwidth. For signals of this type, we shall prove the following theorem:

THEOREM 1: *If a function $G(t)$ contains no frequencies higher than W cycles per second, it is completely determined by giving its ordinates at a series of points spaced $1/2W$ seconds apart, the series extending throughout the time domain.*

We shall call this the *sampling theorem in the time domain.* The fact, described by the theorem, that a signal cannot have independent values closer together on the average than $1/2W$ has been known, for practical purposes, for a generation. We now wish to prove the theorem and thereby specify the conditions which hold.[2]

Let $\Omega(\omega)$ be the frequency spectrum of $G(t)$. Then

$$G(t) = \frac{1}{2\pi} \int_{-\infty}^{+\infty} \epsilon^{j\omega t} \Omega(\omega) d\omega \qquad (1)^{3}$$

$$= \frac{1}{2\pi} \int_{-2\pi W}^{+2\pi W} \epsilon^{j\omega t} \Omega(\omega) d\omega \qquad (2)$$

since $\Omega(\omega)$ is assumed to be zero outside the band W. Suppose we now let

$$t = \frac{n}{2W} \qquad (3)$$

[1] The proof of the theorem given here is due to C. E. Shannon, *Proc. I.R.E.* vol. 37, January, 1949, pp. 10–21. Our method of derivation of Eq. (10) is due to C. R. Cahn.

[2] The proof of the theorem will be based upon Fourier analysis. For purposes of simple reference, the notation used in Goldman, *Frequency Analysis, Modulation, and Noise,* will be followed, and sections or pages of the book itself will be abbreviated as *F.A.,M., and N.,* followed by the appropriate designation.

The only difference in notation of which the author is aware is that $a(\omega)$ and $b(\omega)$ in the present book are each equal to π times the values of the corresponding terms in *F.A.,M., and N.*

[3] *F.A.,M., and N.,* p. 60 or 66.

[4] Throughout the present book, the symbol ϵ will be used for the number 2.71828. . . .

in Eq. (2), where n is any positive or negative integer. Then

$$G\left(\frac{n}{2W}\right) = \frac{1}{2\pi}\int_{-2\pi W}^{+2\pi W} \Omega(\omega)\epsilon^{j\omega\frac{n}{2W}}\,d\omega \tag{4}$$

The right side of Eq. (4) will be recognized as having the value $2WC_{-n}$ where C_{-n} is the $(-n)$th complex Fourier series coefficient[1] obtained by expanding the function $\Omega(\omega)$ in a Fourier series in the range from $\omega = -2\pi W$ to $\omega = 2\pi W$. The set of values of $G(n/2W)$ for all positive and negative integral values of n thus determine all the coefficients in the Fourier expansion of $\Omega(\omega)$. Consequently they completely determine $\Omega(\omega)$ itself in the range $(-2\pi W < \omega < 2\pi W)$. Since $\Omega(\omega)$ is assumed to be zero outside this range, the set of values of $G(n/2W)$ completely specify $\Omega(\omega)$, which in turn completely specifies $G(t)$. This proves the theorem. There is thus one and only one function whose spectrum $\Omega(\omega)$ is limited to the frequency bandwidth W and which passes through a set of given values at sample points $1/2W$ sec apart. Let us now find an analytical expression for this function in terms of its values at the sample points.

First of all let us expand $\Omega(\omega)$ in a Fourier series in the interval $(-2\pi W \leq \omega \leq 2\pi W)$. Thus[2]

[1] *F.A.,M., and N.*, p. 35, Eq. (14).

[2] *F.A.,M., and N.*, p. 35, Eq. (10). Strictly speaking, this reference derived only the equation for the case in which $\Omega(\omega)$ is a real function of ω. In case $\Omega(\omega)$ is a complex function of ω, which is the usual case, we can readily show that the equation still holds. Thus let $\Omega(\omega) = a(\omega) - jb(\omega)$, where $a(\omega)$ and $b(\omega)$ are real. Then

$$a(\omega) = \sum_{n=-\infty}^{+\infty} A_n\epsilon^{-j\frac{2\pi n\omega}{4\pi W}} \quad\text{and}\quad b(\omega) = \sum_{n=-\infty}^{+\infty} B_n\epsilon^{-j\frac{2\pi n\omega}{4\pi W}}$$

where

$$A_n = \frac{1}{4\pi W}\int_{-2\pi W}^{+2\pi W} a(\omega)\epsilon^{j\frac{2\pi n\omega}{4\pi W}}d\omega, \text{ and } B_n = \frac{1}{4\pi W}\int_{-2\pi W}^{+2\pi W} b(\omega)\epsilon^{j\frac{2\pi n\omega}{4\pi W}}d\omega$$

Therefore

$$a(\omega) - jb(\omega) = \Omega(\omega) = \sum_{n=-\infty}^{+\infty} (A_n - jB_n)\epsilon^{-j\frac{2\pi n\omega}{4\pi W}} = \sum_{n=-\infty}^{+\infty} \bar{C}_n\epsilon^{-j\frac{2\pi n\omega}{4\pi W}}$$

where

$$\begin{aligned}\bar{C}_n = A_n - jB_n &= \frac{1}{4\pi W}\int_{-2\pi W}^{+2\pi W} [a(\omega) - jb(\omega)]\epsilon^{j\frac{2\pi n\omega}{4\pi W}}d\omega \\ &= \frac{1}{4\pi W}\int_{-2\pi W}^{+2\pi W} \Omega(\omega)\epsilon^{j\frac{2\pi n\omega}{4\pi W}}d\omega\end{aligned}$$

$$\Omega(\omega) = \sum_{n=-\infty}^{+\infty} \bar{C}_n \epsilon^{-j\frac{2\pi n\omega}{4\pi W}} = \sum_{n=-\infty}^{+\infty} \bar{C}_n \epsilon^{-j\omega\frac{n}{2W}} \tag{5}$$

where $$\bar{C}_n = \frac{1}{4\pi W} \int_{-2\pi W}^{+2\pi W} \Omega(\omega) \epsilon^{j\omega\frac{n}{2W}} d\omega = \frac{1}{2W} G\left(\frac{n}{2W}\right) \tag{6}$$

(Note that $\bar{C}_n$ is identical with C_{-n} in the usual complex Fourier series. See *F.A.,M. and N.*, p. 35.)

Therefore $$\Omega(\omega) = \frac{1}{2W} \sum_{n=-\infty}^{+\infty} G\left(\frac{n}{2W}\right) \epsilon^{-j\omega\frac{n}{2W}} \tag{7}$$

Consequently, from Eqs. (1) and (7),

$$\begin{aligned} G(t) &= \frac{1}{2\pi} \int_{-2\pi W}^{2\pi W} \frac{1}{2W} \left[\left(\sum_{n=-\infty}^{+\infty} G\left(\frac{n}{2W}\right) \epsilon^{-j\omega\frac{n}{2W}}\right)\right] \epsilon^{j\omega t} d\omega \\ &= \frac{1}{4\pi W} \sum_{n=-\infty}^{+\infty} G\left(\frac{n}{2W}\right) \int_{-2\pi W}^{2\pi W} \epsilon^{j\omega\left(t-\frac{n}{2W}\right)} d\omega \end{aligned} \tag{8}$$

Now

$$\begin{aligned} \int_{-2\pi W}^{2\pi W} \epsilon^{j\omega\left(t-\frac{n}{2W}\right)} d\omega &= \left[\frac{\epsilon^{j\omega\left(t-\frac{n}{2W}\right)}}{j\left(t-\frac{n}{2W}\right)}\right]_{-2\pi W}^{2\pi W} \\ &= \frac{\epsilon^{j2\pi W\left(t-\frac{n}{2W}\right)} - \epsilon^{-j2\pi W\left(t-\frac{n}{2W}\right)}}{j\left(t-\frac{n}{2W}\right)} = \frac{2\sin(2\pi Wt - n\pi)}{t-\frac{n}{2W}} \end{aligned} \tag{9}$$

Thus $$\begin{aligned} G(t) &= \frac{2}{4\pi W} \sum_{n=-\infty}^{+\infty} G\left(\frac{n}{2W}\right) \frac{\sin(2\pi Wt - n\pi)}{\left(t-\frac{n}{2W}\right)} \\ &= \sum_{n=-\infty}^{+\infty} G\left(\frac{n}{2W}\right) \frac{\sin(2\pi Wt - n\pi)}{2\pi Wt - n\pi} \end{aligned} \tag{10}$$

In the form of Eq. (10), $G(t)$ may thus be represented everywhere in terms of its values at the sample points.

Figure 2-2 shows a graph of the function

$$Sa(2\pi Wt) = \frac{\sin 2\pi Wt}{2\pi Wt} \tag{11}$$

This function, which we shall call the *sampling function*, has some simple and useful properties. Its value is unity at $t = 0$ and is zero at $t = n/2W$, *i.e.*, at all the sample points excepting $t = 0$. Furthermore, its spectrum is constant in the frequency band W and is zero outside.[1]

Let us next consider an arbitrary function of time whose spectrum is limited to the bandwidth W. Let us sample this function at points

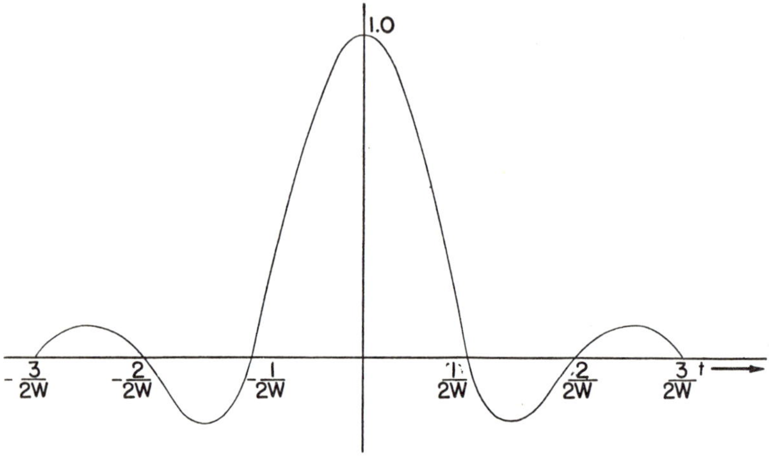

Fig. 2-2. The sampling function

$$Sa(2\pi Wt) = \frac{\sin(2\pi Wt)}{2\pi Wt}.$$

$1/2W$ apart. Suppose that the value of the function is very small outside the range from T_1 to T_2, as shown in Fig. 2-3, and let

$$T = T_2 - T_1 \tag{12}$$

There are thus $2TW$ sampling points[2] in the range where the function has appreciable value. The values of the function at these $2TW$ points then account for all the terms in Eq. (10) which make a substantial contribution to $G(t)$. It follows from the rapid attenuation of the sampling function $Sa(2\pi Wt)$ that the effects of any term in Eq. (10) will be of consequence only within a relatively

[1] *F.A.,M., and N.*, p. 133 (footnote 1).

[2] More accurately, if the sampling points are at the centers of the intervals, there are $2TW$ of them. If the sampling points are at the edges of the intervals, there are $2TW + 1$ sampling points.

small number of intervals in the neighborhood of the corresponding sampling point. Therefore, even if the function $G(t)$ is not small outside the range from T_1 to T_2, the values of $G(t)$ at the $2TW$ sampling points determine the value of $G(t)$ within the range to a very high degree of approximation excepting very near to T_1 and T_2.

It is well known that, if a function has a spectrum which is strictly limited to a finite bandwidth, the time representation of the func-

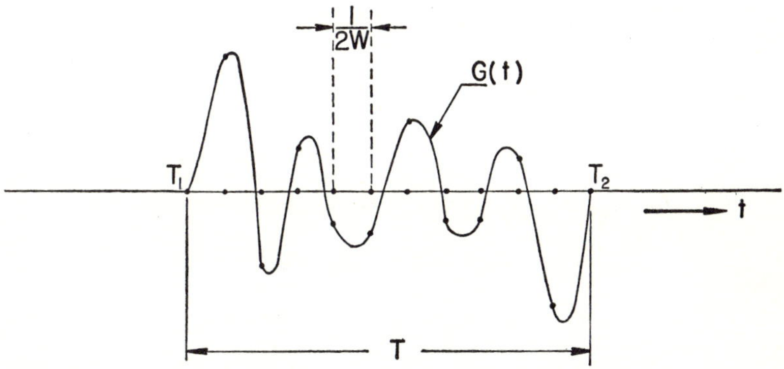

Fig. 2-3. A function of bandwidth W, having values differing appreciably from zero only in the range ($T_1 < t < T_2$).

tion will trail on indefinitely, even though it may become small. It is therefore not possible to have a function of strictly finite time duration and finite spectral bandwidth. However, it is possible to have a function which has *almost* all its quadratic content[1] within a finite time duration and within a finite bandwidth. In fact, most communication signals are functions of this type. It therefore follows from the foregoing discussion that if T is the approximate duration of a signal and if W is its approximate spectral bandwidth and if

$$2TW \gg 1 \tag{13}$$

the function is determined everywhere to a high degree of accuracy by its values at $2TW$ sampling points spaced at time locations $1/2W$ apart.

2.2 The sampling theorem in the frequency domain

We will now derive a sampling theorem in the frequency domain which is the analogue of the one already derived for the time

[1] This is proportional to its energy.

domain. Suppose we have a function of time $G(t)$ which is zero everywhere outside the range from T_1 to T_2 (see Fig. 2-4).

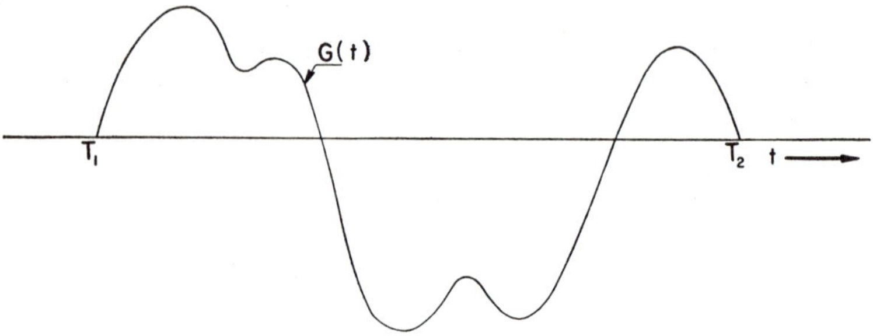

Fig. 2-4. A function which is zero outside the range ($T_1 \leq t \leq T_2$).

The frequency representation of $G(t)$ is then

$$\Omega(\omega) = \int_{T_1}^{T_2} G(t)\epsilon^{-j\omega t}dt \tag{14}$$

Now we can also express $G(t)$ as a Fourier series, everywhere in the range from T_1 to T_2, as

$$G(t) = \sum_{n=-\infty}^{+\infty} D_n\epsilon^{j\frac{2\pi nt}{T_2-T_1}} \tag{15}$$

where

$$D_n = \frac{1}{T_2 - T_1}\int_{T_1}^{T_2} G(t)\epsilon^{-j\frac{2\pi nt}{T_2-T_1}}dt$$

$$= \frac{1}{T_2 - T_1}\,\Omega\left(\frac{2\pi n}{T_2 - T_1}\right) \tag{16}$$

Since $G(t)$ is completely determined by its Fourier series expansion in the range $T_1 < t < T_2$, the values of the spectrum function $\Omega(\omega)$ at the set of points

$$\omega = n\left(\frac{2\pi}{T_2 - T_1}\right) = n\omega_0 \tag{17}$$

where

$$\omega_0 = \frac{2\pi}{T_2 - T_1} \tag{18}$$

and n takes on all positive and negative integral values, including zero, will completely determine $G(t)$. It then follows from Eq. (14) that they will also completely determine $\Omega(\omega)$ for all values of ω. We thus arrive at the *sampling theorem in the frequency domain:*

THEOREM: *If $\Omega(\omega)$ represents the spectrum of a function $G(t)$ which is zero everywhere except in the range $T_1 < t < T_2$, then $\Omega(\omega)$ is exactly determined for all values of ω by giving its values at a series of points $1/(T_2 - T_1)$ cycles per second apart in frequency,*[1] *the series extending throughout the frequency domain.*

Now

$$\Omega(\omega) = \int_{-\infty}^{+\infty} G(t)\epsilon^{-j\omega t}dt$$

$$= \int_{-\infty}^{+\infty} G(t) \cos \omega t \, dt - j \int_{-\infty}^{+\infty} G(t) \sin \omega t \, dt \qquad (19)$$

It follows that $\Omega(n\omega_0)$ is, in general, a complex quantity. Accordingly, two items of information are necessary to specify $\Omega(n\omega_0)$, namely, its real and imaginary parts. On the other hand $\Omega(-n\omega_0)$ is the complex conjugate of $\Omega(+n\omega_0)$, as may be seen from Eq. (19). The values of $\Omega(n\omega_0)$ in the negative frequency range are thus redundant, since they are completely determined by its values in the positive frequency range. It therefore follows from the foregoing that if $T = T_2 - T_1$ is the approximate duration of a signal and if W is its approximate spectral bandwidth, and if $2TW \gg 1$, then the function is determined everywhere to a high degree of precision by the values of its spectrum at the $(TW + 1)$ sampling points[2] spaced at frequency intervals $1/T$ apart. Since the real and the imaginary part of the spectrum are required at every sampling point, other than zero frequency, the number of separate items of information necessary is

$$2TW + 1 = 2TW \quad \text{approximately} \qquad (20)$$

just as in the time domain case.

Corresponding to Eq. (10), we have as an analytical expression for $\Omega(\omega)$ in terms of its values at the sampling points

$$\Omega(\omega) = \sum_{n=-\infty}^{+\infty} \Omega\left(\frac{2\pi n}{T}\right) \frac{\sin\left(\frac{\omega T}{2} - \pi n\right)}{\left(\frac{\omega T}{2} - \pi n\right)} \qquad (21)^3$$

[1] $2\pi/(T_2 - T_1)$ radians/sec apart in ω.

[2] The 1 in $(TW + 1)$ is required in order to include the zero frequency term $n = 0$.

[3] This may also be expressed as

provided that $$T_2 = \frac{T}{2} \quad \text{and} \quad T_1 = -\frac{T}{2} \tag{22}$$

It is of interest to consider the relation between the sampling theorem in the frequency domain and a conventional Fourier series expansion. If we have a function $G(t)$, which is zero everywhere except in the range from T_1 to T_2, then in the interval from T_1 to T_2, $G(t)$ may be expressed either as a Fourier integral in the form

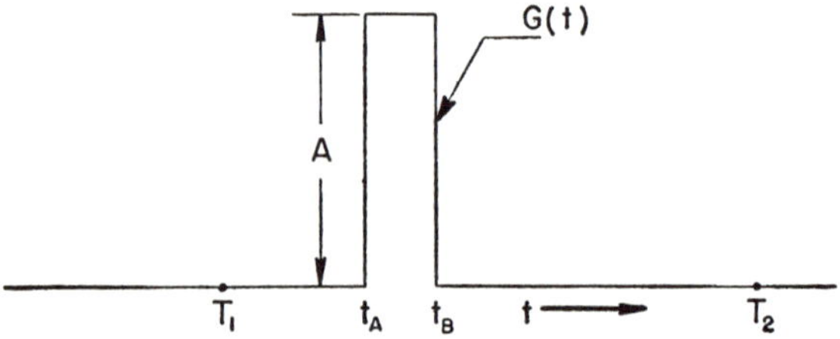

Fig. 2-5. A function of time which is zero everywhere except in the range $(T_1 < t < T_2)$. This particular function happens also to be zero except in the more restricted range $(t_A < t < t_B)$, but that is unimportant in the present argument.

$$G(t) = \frac{1}{2\pi} \int_{-\infty}^{+\infty} \Omega(\omega)\epsilon^{j\omega t} d\omega \tag{22a}$$

or from Eqs. (15) and (16) as a Fourier series in the form

$$G(t) = \sum_{n=-\infty}^{+\infty} \frac{1}{T_2 - T_1} \Omega\left(\frac{2\pi n}{T_2 - T_1}\right) \epsilon^{j\frac{2\pi n t}{T_2 - T_1}} \tag{22b}$$

The Fourier series expression is valid only in the interval from T_1 to T_2, whereas the Fourier integral expression is valid everywhere. Writing $G(t)$ in the form of Eq. (22b) shows that *a Fourier series expansion of $G(t)$ is an expression of $G(t)$ in terms of its sampling values in the frequency domain.*

$$\Omega(\omega) = \Omega(0)\left[\frac{\sin\left(\frac{\omega T}{2}\right)}{\frac{\omega T}{2}}\right] + \sum_{n=1}^{\infty}\left\{a_n\left[\frac{\sin\left(\frac{\omega T}{2} - \pi n\right)}{\frac{\omega T}{2} - \pi n} + \frac{\sin\left(\frac{\omega T}{2} + \pi n\right)}{\frac{\omega T}{2} + \pi n}\right] - jb_n\left[\frac{\sin\left(\frac{\omega T}{2} - \pi n\right)}{\frac{\omega T}{2} - \pi n} - \frac{\sin\left(\frac{\omega T}{2} + \pi n\right)}{\frac{\omega T}{2} + \pi n}\right]\right\}$$

where $\Omega(n\omega_0) = a_n - jb_n$.

By way of illustration we show a pulse of magnitude A and duration $(t_B - t_A)$ in Fig. 2-5 and we show its spectrum in Fig. 2-6. The sample-point values are indicated by the heavy dots in Fig. 2-6. These will give both the magnitudes [except for a factor $1/(T_2 - T_1)$]

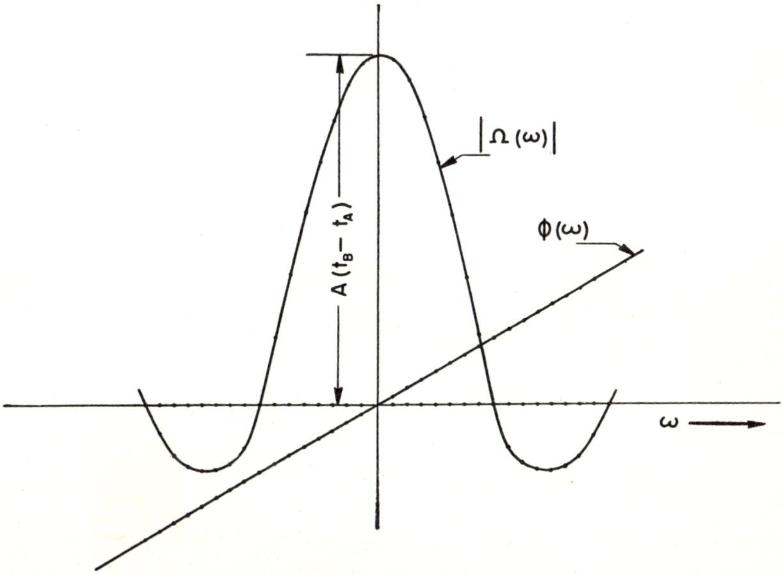

Fig. 2-6. Spectrum of the function in Fig. 2-5 showing sample points corresponding to the temporal range $(T_1 < t < T_2)$.

and the phases of the complex Fourier series coefficients, if $G(t)$ is expanded in a Fourier series in the interval $(T_1 < t < T_2)$.

2.3 The sampling theorem in the time domain when the frequency band does not start at zero[1]

We next wish to determine the form of the sampling theorem in the time domain for a function $G(t)$ whose spectrum is limited to the frequency range between f_1 and f_2. The reader may recognize that the methods used in the following analysis are somewhat similar to those used in studying unsymmetrical sideband distributions.[2]

[1] This section can be omitted in a first reading of the book.
[2] *F.A.,M., and N.*, Secs. 4.9 and 5.6.

Let

$$\omega_1 = 2\pi f_1 \tag{23}$$

and

$$\omega_2 = 2\pi f_2 \tag{24}$$

In order to express $G(t)$ in terms of values at sampling points, we now wish to obtain a continuous frequency range to be used for a Fourier series expansion. The elimination of the frequency range around $\omega = 0$ makes this difficult, unless we can limit our interval to either the range from ω_1 to ω_2 or the range from $-\omega_2$ to $-\omega_1$. Let us try the former interval. Thus, let us define the complex function of time

$$g(t) = \frac{1}{\pi}\int_{\omega_1}^{\omega_2} \Omega(\omega)\epsilon^{j\omega t}d\omega \tag{25}$$

$$= \frac{1}{\pi}\int_{\omega_1}^{\omega_2} [a(\omega) - jb(\omega)](\cos \omega t + j \sin \omega t)d\omega \tag{26}$$

$$= \frac{1}{\pi}\int_{\omega_1}^{\omega_2} a(\omega) \cos \omega t \, d\omega + \frac{1}{\pi}\int_{\omega_1}^{\omega_2} b(\omega) \sin \omega t \, d\omega$$
$$- \frac{j}{\pi}\int_{\omega_1}^{\omega_2} b(\omega) \cos \omega t \, d\omega + \frac{j}{\pi}\int_{\omega_1}^{\omega_2} a(\omega) \sin \omega t \, d\omega \tag{27}$$

where, as before, $\Omega(\omega)$ is the spectrum of $G(t)$.

The real part of Eq. (27) is nothing other than $G(t)$ itself expressed as a Fourier integral in real form. Let us now introduce the function $G_1(t)$, defined as

$$G_1(t) = \frac{1}{\pi}\int_{\omega_1}^{\omega_2} b(\omega) \cos \omega t \, d\omega - \frac{1}{\pi}\int_{\omega_1}^{\omega_2} a(\omega) \sin \omega t \, d\omega \tag{28}$$

$G_1(t)$ may therefore be called the quadrature function of $G(t)$ since it may also be written

$$G_1(t) = \frac{1}{\pi}\int_{\omega_1}^{\omega_2} b(\omega) \sin \left(\omega t + \frac{\pi}{2}\right) d\omega + \frac{1}{\pi}\int_{\omega_1}^{\omega_2} a(\omega) \cos \left(\omega t + \frac{\pi}{2}\right) d\omega \tag{29}$$

In mathematical literature it is called the Hilbert transform[1] of $G(t)$.

[1] A short discussion of Hilbert transforms is given in Appendix V. See also E. A. Guillemin, *The Mathematics of Circuit Analysis*, Wiley, New York (1949), p. 338–349.

From Eqs. (27) and (28) we see that

$$g(t) = G(t) - jG_1(t) \tag{30}$$

Let us now expand $\Omega(\omega)$ in a Fourier series in the range ω_1 to ω_2. Accordingly

$$\Omega(\omega) = \sum_{n=-\infty}^{+\infty} \bar{C}_n \epsilon^{-j\frac{2\pi n}{\omega_2-\omega_1}\omega} \tag{31}$$

where
$$\bar{C}_n = \frac{1}{\omega_2 - \omega_1} \int_{\omega_1}^{\omega_2} \Omega(\omega) \epsilon^{j\frac{2\pi n}{\omega_2-\omega_1}\omega} d\omega$$

$$= \frac{\pi}{\omega_2 - \omega_1} g\left(\frac{2\pi n}{\omega_2 - \omega_1}\right) \tag{32}$$

The last equality in Eq. (32) is obtained by the substitution $t = 2\pi n/(\omega_2 - \omega_1)$ in Eq. (25). From Eqs. (31), (32), and (25)

$$g(t) = \frac{1}{\pi} \int_{\omega_1}^{\omega_2} \sum_{n=-\infty}^{\infty} \frac{\pi}{\omega_2 - \omega_1} g\left(\frac{2\pi n}{\omega_2 - \omega_1}\right) \epsilon^{-j\frac{2\pi n}{\omega_2-\omega_1}\omega} \epsilon^{j\omega t} d\omega$$

$$= \frac{1}{\omega_2 - \omega_1} \sum_{n=-\infty}^{\infty} g\left(\frac{2\pi n}{\omega_2 - \omega_1}\right) \int_{\omega_1}^{\omega_2} \epsilon^{j\omega\left(t - \frac{2\pi n}{\omega_2-\omega_1}\right)} d\omega \tag{33}$$

Performing the indicated integration, we obtain

$$g(t) = \frac{1}{\omega_2 - \omega_1} \sum_{n=-\infty}^{\infty} g\left(\frac{2\pi n}{\omega_2 - \omega_1}\right) \left[\frac{\epsilon^{j\omega\left(t - \frac{2\pi n}{\omega_2-\omega_1}\right)}}{j\left(t - \frac{2\pi n}{\omega_2 - \omega_1}\right)}\right]_{\omega_1}^{\omega_2} \tag{34}$$

It is desired to express $G(t)$ in terms of the above. From Eq. (30), $G(t)$ is the real part of $g(t)$. Now[1]

$$\text{Re}\left\{g\left(\frac{2\pi n}{\omega_2 - \omega_1}\right)\left[\frac{\epsilon^{j\omega_2\left(t - \frac{2\pi n}{\omega_2-\omega_1}\right)} - \epsilon^{j\omega_1\left(t - \frac{2\pi n}{\omega_2-\omega_1}\right)}}{j\left(t - \frac{2\pi n}{\omega_2 - \omega_1}\right)}\right]\right\}$$

$$= \text{Re}\left\{\left[G\left(\frac{2\pi n}{\omega_2 - \omega_1}\right) - jG_1\left(\frac{2\pi n}{\omega_2 - \omega_1}\right)\right]\right.$$

[1] The notation "Re" stands for "the real part of."

$$\left.\left[\frac{\cos \omega_2\left(t-\frac{2\pi n}{\omega_2-\omega_1}\right)+j\sin \omega_2\left(t-\frac{2\pi n}{\omega_2-\omega_1}\right) - \cos \omega_1\left(t-\frac{2\pi n}{\omega_2-\omega_1}\right)-j\sin \omega_1\left(t-\frac{2\pi n}{\omega_2-\omega_1}\right)}{j\left(t-\frac{2\pi n}{\omega_2-\omega_1}\right)}\right]\right\}$$

$$= G\left(\frac{2\pi n}{\omega_2-\omega_1}\right)\left[\sin \omega_2\left(t-\frac{2\pi n}{\omega_2-\omega_1}\right) - \sin \omega_1\left(t-\frac{2\pi n}{\omega_2-\omega_1}\right)\right]\left(\frac{1}{t-\frac{2\pi n}{\omega_2-\omega_1}}\right)$$
$$- G_1\left(\frac{2\pi n}{\omega_2-\omega_1}\right)\left[\cos \omega_2\left(t-\frac{2\pi n}{\omega_2-\omega_1}\right) - \cos \omega_1\left(t-\frac{2\pi n}{\omega_2-\omega_1}\right)\right]\left(\frac{1}{t-\frac{2\pi n}{\omega_2-\omega_1}}\right) \tag{35}$$

Therefore applying the appropriate trigonometric identities and substituting in Eq. (34),

$$G(t) = \frac{2}{\omega_2-\omega_1}\sum_{n=-\infty}^{\infty}\left[G\left(\frac{2\pi n}{\omega_2-\omega_1}\right)\frac{\sin\left(\frac{\omega_2-\omega_1}{2}\right)\left(t-\frac{2\pi n}{\omega_2-\omega_1}\right)}{t-\frac{2\pi n}{\omega_2-\omega_1}}\cos\frac{\omega_2+\omega_1}{2}\left(t-\frac{2\pi n}{\omega_2-\omega_1}\right) + G_1\left(\frac{2\pi n}{\omega_2-\omega_1}\right)\frac{\sin\left(\frac{\omega_2-\omega_1}{2}\right)\left(t-\frac{2\pi n}{\omega_2-\omega_1}\right)}{t-\frac{2\pi n}{\omega_2-\omega_1}}\sin\frac{\omega_2+\omega_1}{2}\left(t-\frac{2\pi n}{\omega_2-\omega_1}\right)\right] \tag{36}$$

$$= \frac{2}{\omega_2-\omega_1}\sum_{n=-\infty}^{\infty}\sqrt{G_n^2+(G_{1n})^2}\,\frac{\sin\left(\frac{\omega_2-\omega_1}{2}\right)\left(t-\frac{2\pi n}{\omega_2-\omega_1}\right)}{t-\frac{2\pi n}{\omega_2-\omega_1}}\cos\left[\left(\frac{\omega_2+\omega_1}{2}\right)\left(t-\frac{2\pi n}{\omega_2-\omega_1}\right)-\phi_n\right] \tag{37}$$

where

$$G_n = G\left(\frac{2\pi n}{\omega_2 - \omega_1}\right) \tag{38}$$

$$G_{1n} = G_1\left(\frac{2\pi n}{\omega_2 - \omega_1}\right) \tag{39}$$

$$\phi_n = \tan^{-1}\left(\frac{G_{1n}}{G_n}\right) \tag{40}$$

It follows from Eq. (37) that a knowledge of $G(t)$ and of $G_1(t)$ at all the sample points, *i.e.*, all points where

$$t = \frac{2\pi n}{\omega_2 - \omega_1} = \frac{n}{f_2 - f_1} = \frac{n}{W} \tag{41}$$

completely determines $G(t)$ everywhere. In Eq. (41), we use the symbol

$$W = f_2 - f_1 \tag{42}$$

for the bandwidth, as has been our custom. We note that the sample points determined by Eq. (41) are twice as far apart as those determined by Eq. (3), which is used when the frequency band starts at zero frequency. However, since we must know the values of both $G(t)$ and $G_1(t)$ at these new sample points, there are still $2TW$ separate items of information necessary to specify completely a function of duration T and bandwidth W.[1]

The form of Eq. (36) and (37) allows us to regard $G(t)$ as the superposition of a set of modulated carrier sampling functions, one at each sample point in the time domain, with the carrier frequency at the center of the spectral band. These modulated carrier sampling functions are of the form

$$A_n \frac{\sin\left(\frac{\omega_2 - \omega_1}{2}\right)\left(t - \frac{2\pi n}{\omega_2 - \omega_1}\right)}{t - \frac{2\pi n}{\omega_2 - \omega_1}} \cos\left[\left(\frac{\omega_2 + \omega_1}{2}\right)\left(t - \frac{2\pi n}{\omega_2 - \omega_1}\right) - \phi_n\right]$$

[1] This statement assumes that if $G(t)$ differs appreciably from zero only in a range of duration T, $G_1(t)$ will do the same. This can be proved for signals of zero average value and of long duration T which are typical functions of ergodic ensembles. The latter are discussed in the next chapter. They are the principal signals of interest in information theory.

and have both an amplitude A_n and a phase ϕ_n. A knowledge of the amplitude and phase of the modulated carrier sampling functions at every sample point determines G_n and G_{1n} at every sample point, and thus completely determines $G(t)$.

PROBLEM

1. What is the analytical expression for the sampling theorem in the frequency domain for a function whose spectrum is limited to the range between f_1 and f_2 and whose duration extends from $-(T/2)$ to $+(T/2)$?

2.4 Expression of the quadratic content of a signal in terms of its values at the sampling points

The *quadratic content* of the signal $G(t)$ is the term which we shall use for

$$\int [G(t)]^2 dt \tag{43}$$

If $G(t)$ is a voltage or current, then the quadratic content will usually be proportional to an energy. If we substitute the value of $G(t)$ from Eq. (10) into Formula (43), we find

$$\begin{aligned}\int_{-\infty}^{+\infty} [G(t)]^2 dt &= \int_{-\infty}^{+\infty} \left[\sum_{n=-\infty}^{+\infty} G\left(\frac{n}{2W}\right) \frac{\sin (2\pi Wt - n\pi)}{2\pi Wt - n\pi} \right]^2 dt \\ &= \frac{1}{2W} \sum_{n=-\infty}^{n=+\infty} \left[G\left(\frac{n}{2W}\right)\right]^2 \end{aligned} \tag{44}$$

The final step in Eq. (44) is arrived at by using the known formulas of integration

$$\int_{-\infty}^{+\infty} \frac{\sin^2 x}{x^2}\, dx = \pi \tag{45}$$

and
$$\int_{-\infty}^{+\infty} \frac{\sin x}{x} \cdot \frac{\sin (x + n\pi)}{x + n\pi}\, dx = 0 \qquad \text{if } n \neq 0 \tag{46}$$

from which it follows that

$$\int_{-\infty}^{+\infty} \frac{\sin (2\pi Wt - n\pi)}{2\pi Wt - n\pi} \cdot \frac{\sin (2\pi Wt - m\pi)}{2\pi Wt - m\pi}\, dt = \begin{cases} 0 & \text{if } m \neq n \\ \dfrac{1}{2W} & \text{if } m = n \end{cases} \tag{47}$$

For a signal of approximate duration T and bandwidth W, $G(n/2W)$ is different from zero only for $2TW$ sampling points. The summation on the right of Eq. (44) need then be taken only over the $2TW$

values corresponding to these sampling points. In particular, if the signal is essentially limited to the range between 0 and T, we may write

$$\int_0^T [G(t)]^2 dt = \frac{1}{2W} \sum_{n=0}^{n=2TW} \left[G\left(\frac{n}{2W}\right)\right]^2 \tag{48}$$

Equations (44) and (48) are formulas for the quadratic content of a signal in terms of its values at the sampling points.

We can also readily derive formulas for the quadratic content in terms of the values of the spectrum $\Omega(\omega)$ at the sampling points in the frequency domain. From the Fourier integral energy theorem,[1] we have

$$\int_{-\infty}^{+\infty} [G(t)]^2 dt = \frac{1}{2\pi} \int_{-\infty}^{+\infty} \Omega(\omega) \cdot \Omega^*(\omega) d\omega \tag{49}$$

where $\Omega^*(\omega)$ is the complex conjugate of $\Omega(\omega)$. From Eqs. (21) and (47), it follows that

$$\int_{-\infty}^{+\infty} \Omega(\omega) \cdot \Omega^*(\omega) d\omega = \frac{2\pi}{T} \sum_{n=-\infty}^{+\infty} \Omega\left(\frac{2\pi n}{T}\right) \cdot \Omega^*\left(\frac{2\pi n}{T}\right) \tag{50}$$

From Eqs. (49) and (50) we then have the alternative formula for the quadratic content of a signal which differs from zero only between 0 and T

$$\begin{aligned} \int_0^T [G(t)]^2 dt &= \frac{1}{T} \sum_{n=-\infty}^{+\infty} \Omega\left(\frac{2\pi n}{T}\right) \cdot \Omega^*\left(\frac{2\pi n}{T}\right) \\ &= \frac{a_0^2}{T} + \frac{2}{T} \sum_{n=1}^{+\infty} (a_n^2 + b_n^2) \end{aligned} \tag{51}$$

where a_n and b_n are defined in footnote 1, page 74. If the signal is also approximately limited to the bandwidth W, we then have

$$\begin{aligned} \int_0^T [G(t)]^2 dt &= \frac{1}{T} \sum_{n=-TW}^{n=+TW} \Omega\left(\frac{2\pi n}{T}\right) \Omega^*\left(\frac{2\pi n}{T}\right) \\ &= \frac{1}{T} a_0^2 + \frac{2}{T} \sum_{n=1}^{n=TW} (a_n^2 + b_n^2) = \frac{1}{2W} \sum_{n=0}^{n=2TW} \left[G\left(\frac{n}{2W}\right)\right]^2 \end{aligned} \tag{52}$$

[1] *F.A.,M., and N.*, p. 110.

2.5 Transformations between time and frequency sampling values, considered as independent variables[1]

In the analysis of succeeding chapters, the $2TW$ values of a function at the sampling points in the time domain or, alternatively, at the sampling points in the frequency domain will often be considered the independent variables in functions of interest, such as entropy. We will therefore find it of interest to be able to transform from one set of variables to the other. Accordingly, we will now develop the formulas of transformation.

From Eq. (7) we have

$$\Omega(\omega) = a(\omega) - jb(\omega) = \frac{1}{2W} \sum_{n=-\infty}^{+\infty} G\left(\frac{n}{2W}\right) \epsilon^{-j\omega\frac{n}{2W}} \tag{53}$$

while from Eqs. (15) and (16)

$$G(t) = \frac{1}{T_2 - T_1} \sum_{q=-\infty}^{+\infty} \Omega\left(\frac{2\pi q}{T_2 - T_1}\right) \epsilon^{j\frac{2\pi q t}{T_2 - T_1}} \tag{54}$$

Let us set our origin of time so that

$$T_1 = -\frac{T}{2} \tag{55}$$

and

$$T_2 = +\frac{T}{2} \tag{56}$$

where

$$T = T_2 - T_1 \tag{57}$$

This will simplify our formulas without fundamentally restricting their generality. The limits of $(+\infty)$ and $(-\infty)$ in the summations of Eqs. (53) and (54) can now also be changed to $+TW$ and $-TW$, respectively, since the terms outside this range are, by hypothesis, negligible. Then letting ω equal $2\pi m/T$ in Eq. (53), we obtain

$$\begin{aligned}\Omega\left(\frac{2\pi m}{T}\right) &= \frac{1}{2W} \sum_{n=-TW}^{+TW} G\left(\frac{n}{2W}\right) \epsilon^{-j\frac{\pi mn}{WT}} \\ &= \frac{1}{2W} \sum_{n=-TW}^{+TW} G\left(\frac{n}{2W}\right) \cos\left(\frac{\pi mn}{WT}\right) \\ &- \frac{j}{2W} \sum_{n=-TW}^{+TW} G\left(\frac{n}{2W}\right) \sin\left(\frac{\pi mn}{WT}\right) = a\left(\frac{2\pi m}{T}\right) - jb\left(\frac{2\pi m}{T}\right)\end{aligned} \tag{58}$$

[1]This derivation and that of Appendix VI are due to C. R. Cahn

Likewise letting $t = p/2W$ in Eq. (54), we obtain

$$G\left(\frac{p}{2W}\right) = \frac{1}{T}\sum_{q=-TW}^{+TW} \Omega\left(\frac{2\pi q}{T}\right)\epsilon^{j\frac{\pi q p}{WT}}$$

$$= \frac{1}{T}a(0) + \frac{2}{T}\sum_{q=1}^{TW} a\left(\frac{2\pi q}{T}\right)\cos\left(\frac{\pi q p}{WT}\right)$$

$$+ \frac{2}{T}\sum_{q=1}^{TW} b\left(\frac{2\pi q}{T}\right)\sin\left(\frac{\pi q p}{WT}\right) \tag{59}$$

Equations (58) and (59) are the desired equations of transformation between the values at the sample points in the time domain and the values at the sample points in the frequency domain. We see that they are linear algebraic equations with constant coefficients. On the basis of these equations, it is shown in Appendix VI that the transformation between time and frequency sampling values is orthogonal.

PROBLEMS[1]

1. *The Sampling Theorem for Periodic Functions.* (Due to C. R. Cahn)[2] Making use of the identities

$$\sum_{q=0}^{2N} \epsilon^{j\frac{2\pi n q}{2N+1}} = \sum_{q=0}^{2N} \epsilon^{j\frac{2\pi n}{T}\cdot\frac{qT}{(2N+1)}} \begin{cases} = \left[\begin{array}{l}2N+1 \text{ if } n \text{ is an integral multiple}\\ \text{of } (2N+1)\text{, including zero}\end{array}\right] \\ = [0 \text{ for other values of } n] \end{cases}$$

and
$$\sum_{n=-N}^{+N} \epsilon^{j\frac{2\pi n}{T}x} = 1 + 2\sum_{n=1}^{N}\cos\left(\frac{2\pi n}{T}\right)x = \frac{\sin\left[\frac{(2N+1)\pi}{T}x\right]}{\sin\left(\frac{\pi}{T}x\right)}$$

show that if $f(t)$ is a periodic function of period T, and that if all the Fourier coefficients vanish above the Nth harmonic, then

$$f(t) = \sum_{q=0}^{2N} f\left(\frac{qT}{2N+1}\right)\frac{\sin\left[(2N+1)\frac{\pi}{T}\left(t - \frac{qT}{2N+1}\right)\right]}{(2N+1)\sin\left[\frac{\pi}{T}\left(t - \frac{qT}{2N+1}\right)\right]}$$

Therefore $f(t)$ has only $(2N + 1)$ sampling values and these have a spacing of $T/(2N + 1)$ between them.

2. *A Modified Form of the Sampling Theorem for Periodic Functions.* (Due to C. R. Cahn)

[1] A solution to Prob. 1 is given in Appendix XII.

[2] This theorem has also been derived independently by Wesley Peterson.

If the sine component of the Nth harmonic in Prob. 1 vanishes, show that

$$f(t) = \sum_{q=0}^{2N-1} f\left(\frac{qT}{2N}\right) \frac{\sin\left[2N\frac{\pi}{T}\left(t - \frac{qT}{2N}\right)\right]}{2N\tan\left[\frac{\pi}{T}\left(t - \frac{qT}{2N}\right)\right]}$$

3. *Alternative Derivation of the Sampling Theorem in the Time Domain When the Frequency Band Does Not Start at Zero.* (Due to Sidney Applebaum)

Suppose that $\Omega(\omega)$ is the Fourier transform of $G(t)$, a real function of t. Also suppose that

$$\Omega(\omega) = 0 \quad \text{except when } (\omega_1^2 < \omega^2 < \omega_2^2)$$

Let

$$2\pi W_m = \frac{\omega_2 - \omega_1}{2} \quad \text{and} \quad \omega_0 = \frac{\omega_1 + \omega_2}{2}$$

Show that

$$\begin{aligned} 2\pi G(t) &= \cos \omega_0 t \int_{-2\pi W_m}^{+2\pi W_m} \{\Omega(\omega_0 + u) + \Omega(-\omega_0 + u)\}\epsilon^{jut}du \\ &\quad + \sin \omega_0 t \int_{-2\pi W_m}^{2\pi W_m} j\{\Omega(\omega_0 + u) - \Omega(-\omega_0 + u)\}\epsilon^{jut}du \\ &= f_1(t) \cos \omega_0 t + f_2(t) \sin \omega_0 t \end{aligned}$$

Show that $f_1(t)$ and $f_2(t)$ are both real functions of time of bandwidth W_m.

CHAPTER III

ERGODIC ENSEMBLES AND RANDOM NOISE

3.1 Special characteristics of the continuous case in information theory

In this and succeeding chapters, we shall extend information theory to deal with continuous signals. In order to do this, the elementary question regarding a continuous signal to which we wish the answer is, "What is the probability of occurrence of this

T R M Z W P O L Y A R Q B Z T L M P R S

(a)

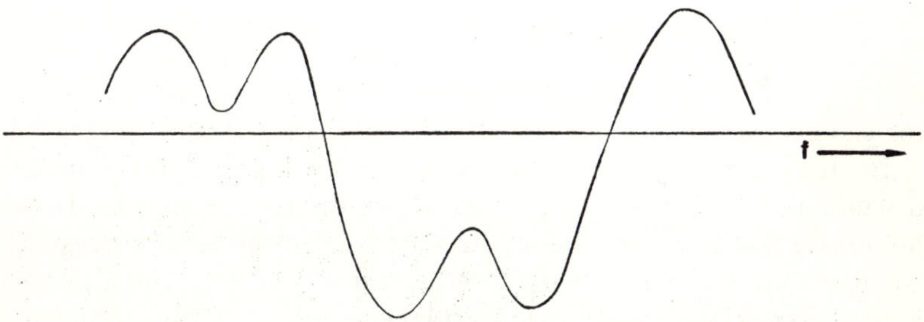

(b)

Fig. 3-1. (a) A discrete signal. (b) A continuous signal.

particular signal?" This is the same question to which we needed an answer in the discrete case. We shall try to develop the continuous theory as a limiting case of the discrete theory, but, as will be pointed out, there are important differences between the two.

Consider the discrete signal in Fig. 3-1(a) and the continuous signal in Fig. 3-1(b). In any finite length of the discrete signal in Fig. 3-1(a), there is only a finite number of symbols each chosen from a finite alphabet. On the other hand, in any finite length of the continuous signal in Fig. 3-1(b), there is an infinite number of points. The value or characteristic of each of these points is its ordinate, and there is an infinite number of possible values of the ordinate. It might, therefore, appear that a continuous signal of finite length has the properties of an infinite sequence of symbols each drawn from an infinite alphabet. From our experience with discrete signals, this would lead us to expect that continuous signals, even of finite length, would carry infinite amounts of information. We know that this is not really the case, and it is not hard to tell why. In the first place, all actual signals have spectra of finite bandwidth. The sampling theorem in the time domain then tells us that only the values of the signal at the sampling points can be chosen independently. This reduces the continuous signal to the equivalent of a finite sequence of symbols. In the second place, all actual signals have some noise superimposed upon them. We shall show later that this has the effect of reducing the number of distinguishable amplitude levels to a finite value. Thus, because of the presence of noise, a continuous signal has the equivalent of only a finite alphabet.

The foregoing discussion indicates that continuous signals may be expected to have the same order of magnitude of quantity of information as discrete signals. We mention, in passing, that no useful information theory of continuous signals has been developed for the noiseless case, and it is unlikely that one will be developed.

In dealing with continuous signals an item will enter into our discussions which was not treated in the discrete case. This is the quadratic content, or the related items, power and energy. In the discrete case, while it is true that the symbols in the alphabet have an order, no importance is attached to the order of symbols in the alphabet. In practical continuous signals, however, the equivalent of the order of symbols in the alphabet has important physical significance, for upon it is based the signal power. We will find that, in the continuous case, the size of the usable alphabet will be determined by the signal power and, in turn, more accurately by the signal-to-noise ratio.

Finally, one of the special characteristics of the continuous case is the possibility of using frequency analysis. This, as might be expected, is often a powerful method of investigation because the properties of linear transmission systems are often expressed as frequency characteristics.

3.2 Probability distributions for continuous ranges of values

If x is a continuous stochastic (chance) variable, the probability of obtaining a value of x which lies in the range between x and $x + dx$ in any particular trial is,

$$P(x)dx \tag{1}$$

The foregoing statement is, in effect, the definition of $P(x)$. We shall call $P(x)$ *a probability distribution function.*[1] Since the value of x must lie somewhere between $-\infty$ and $+\infty$, we also know that

$$\int_{-\infty}^{+\infty} P(x)dx = 1 \tag{2}$$

The probability of a function of n variables $x_1,x_2, \ldots ,x_n$ is defined in an analogous manner. The normalization condition for n variables is, of course,

$$\int \cdots \int P(x_1,x_2, \ldots ,x_n)dx_1dx_2 \ldots dx_n = 1 \tag{3}$$

The quantity $P(x_1,x_2, \ldots ,x_n)$ in Eq. (3) is the probability distribution function of the simultaneous occurrence of $x_1,x_2, \ldots ,x_n$. Thus suppose

$$P(x_1,x_2) = \begin{cases} b\epsilon^{-x_1/A} & \begin{cases}\text{from } x_2 = C \text{ to } x_2 = D \\ \text{and } x_1 = 0 \text{ to } x_1 = \infty\end{cases} \\ 0 & \begin{cases}\text{for all other values of } x_1 \\ \text{and } x_2\end{cases} \end{cases} \tag{4}$$

Then by Eq. (3)

$$\begin{aligned} \int_{-\infty}^{+\infty} \int_{-\infty}^{+\infty} P(x_1,x_2)dx_1dx_2 &= 1 \\ = \int_{x_1=0}^{x_1=+\infty} \int_{x_2=C}^{x_2=D} b\epsilon^{-\frac{x_1}{A}}dx_1dx_2 & \\ = (D - C) \int_0^{+\infty} b\epsilon^{-\frac{x_1}{A}} dx_1 &= (D - C)bA \end{aligned} \tag{5}$$

Equation (3) thus requires that

$$bA(D - C) = 1 \quad \text{or} \quad b = \frac{1}{A(D - C)} \tag{6}$$

[1] Some authors have called $P(x)$ a *probability density function* and have called its integral the *distribution function.*

Let us suppose that in Eq. (4), in a particular case,

$$x_1 \equiv \text{distance that a school child has to walk to school} \tag{7}$$

$$x_2 = \text{age of child} \tag{8}$$

$$D = 14 \text{ years} \tag{9}$$

$$C = 6 \text{ years} \tag{10}$$

$$A = \tfrac{1}{4} \text{ mile} \tag{11}$$

$$b = \tfrac{1}{2} \text{ year}^{-1} \text{ mile}^{-1} \tag{12}$$

Then we may expect that the fraction of the children at the school who are under 8 years of age and who must walk more than a mile to school is

$$\begin{aligned} \text{fraction} &= \int_{x_1=1}^{x_1=\infty} \int_{x_2=6}^{x_2=8} \tfrac{1}{2}\epsilon^{-4x_1} dx_1 dx_2 \\ &= \frac{2}{2} \int_1^\infty \epsilon^{-4x_1} dx_1 = \frac{\epsilon^{-4}}{4} = 0.00458 \end{aligned} \tag{13}$$

Thus, if there are 2000 children at the school we can expect, on the average, that since

$$0.00458 \times 2000 = 9 \quad \text{(approximately)} \tag{14}$$

approximately 9 children under 8 years of age will have to walk more than a mile to school.

The foregoing example was particularly simple since the distribution in x_1 was independent of the distribution in x_2. In this case, it is shown in books on probability theory that

$$P(x_1,x_2) = P(x_1)P(x_2) \tag{15}$$

where $P(x_1)$ is the distribution in x_1 and $P(x_2)$ is the distribution in x_2. In addition to Eq. (3) we then also have the added normalization conditions that

$$\int_{-\infty}^{+\infty} P(x_1)dx_1 = 1 = \int_{-\infty}^{+\infty} P(x_2)dx_2 \tag{16}$$

In the foregoing case, the reader can readily show that

$$P(x_1) = \begin{cases} \dfrac{1}{A}\epsilon^{-x_1/A} & \text{from } x_1 = 0 \text{ to } x_1 = \infty \\ 0 & \text{for negative values of } x_1 \end{cases} \tag{17}$$

$$P(x_2) = \begin{cases} \dfrac{1}{D - C} & \text{from } x_2 = C \text{ to } x_2 = D \\ 0 & \text{elsewhere} \end{cases} \tag{18}$$

In general, the distribution in x_1 need not be independent of the distribution in x_2. In such a case we would say that there are *constraints* between x_1 and x_2. A distribution function

$$P(x_1,x_2) = b\epsilon^{-ax_1/x_2} \tag{19}$$

would be a case in which there are constraints between the distributions in x_1 and x_2.

With these preliminary ideas about continuous probability distribution functions, we are now ready to start the study of ergodic ensembles.

3.3 Ergodic ensembles of functions

A number of functions all of which possess one or more characteristic properties is called a *set* of functions. The characteristic properties identify those functions which are members of the set. Thus all sinusoidal functions of time form a set, their identifying characteristic being that they are sinusoidal. In fact, this entire particular set of functions can be expressed explicitly by the formula

$$f(t) = A \sin (\omega t + \phi) \tag{20}$$

In other words, every sinusoidal function of time is of the form of the right side of Eq. (20).

Another example of a set of functions is the set of all functions of time containing no frequency components above W cycles/sec.

If a set of functions has associated with it a probability distribution from which we may determine the probability of occurrence of each function in the set, then the set is called an *ensemble* of functions. Thus the set of functions

$$f(t) = \sin \omega t \tag{21}$$

is an ensemble of functions if the probability distribution of ω, namely, $P(\omega)$, is given. In this case, ω is called the *stochastic variable* (*i.e.*, chance variable) of the ensemble. When the value of ω is specified, a particular function of the ensemble is thereby specified. The relative probability of occurrence of any particular $\sin \omega t$ is specified by $P(\omega)$.

The set of functions

$$f(t) = A \sin (\omega t + \phi) \tag{22}$$

is an ensemble of functions if the probability distribution $P(A,\omega,\phi)$ of the simultaneous occurrence of A, ω, and ϕ is given. The stochastic variables of this ensemble are A, ω, and ϕ. In case the value of A (or ω or ϕ) is the same for all functions in the ensemble, then it is not a stochastic variable of the ensemble. The stochastic variables of an ensemble are also called *degrees of freedom* of the ensemble. Thus, in Eq. (22), if A, ω, and ϕ are all stochastic variables, then the ensemble has 3 degrees of freedom, namely, A, ω, and ϕ. If A is constant, and only ω and ϕ are stochastic variables, then the ensemble has only 2 degrees of freedom. In case the value of A is determined by the values of ω and ϕ, then also A is not an independent degree of freedom of the ensemble.

The types of ensembles which are of interest in information theory usually have large numbers of degrees of freedom. Thus consider the ensemble of functions of bandwidth W and approximate duration T. Any particular function of this ensemble is specified by the $2TW$ values of the function at its sample points. Therefore the ensemble has $2TW$ degrees of freedom. The probability distribution which is necessary in order to define such an ensemble is the probability distribution

$$P(y_1,y_2, \ . \ . \ . \ ,y_{2TW})$$

where $y_1, \ . \ . \ . \ ,y_{2TW}$ are the values of a particular function at the sample points. Here $P(y_1,y_2, \ . \ . \ . \ ,y_{2TW})$ is the relative probability of occurrence of that particular function in the ensemble whose values at the sample points are $y_1,y_2, \ . \ . \ . \ ,y_{2TW}$.

We now recall that the information theory of discrete systems was built around ergodic systems of symbols. In a similar manner, the information theory of continuous systems is built around *ergodic ensembles* of functions, a term which will be defined below. If the functions of the ensemble may be considered to be limited to some bandwidth W, and practically all functions of interest in information theory may be considered to be so limited, then the correspondence between the continuous case and the discrete case is very close, since a band-limited function is completely specified by the sequence of values of the function at its sample points. In

Chap. I, we defined an ergodic system of symbols as one in which the occurrence of symbols is controlled by probability but no appreciable intersymbol influence extends over more than a finite number of symbols. We found that all except a vanishingly small fraction of the sequences occurring in ergodic systems of symbols were "typical" sequences of the ergodic system,[1] and for simplicity we called them *ergodic sequences*. These sequences had the distinguishing characteristic that any sufficiently long sequence had a percentage composition of symbols, digrams, trigrams (and all the other combinations which comprise the intersymbol influence) which corresponded exactly to the system probabilities. We have previously shown that a band-limited function of time can be completely determined by its values at the sample points. We can therefore specify a band-limited function of time in terms of the sequence of the values which it has at its sample points. With this background, we now define an *ergodic ensemble* of band-limited functions of time as an ensemble of functions whose sequences of values at the sample points form an ergodic system of sequences. Thus, in an ergodic ensemble of functions the sum total of the probabilities of occurrence of those functions whose values at the sample points form typical sequences of an ergodic system is unity. The sum total of the probabilities of occurrence of those functions of the ensemble whose values at the sample points do not form typical sequences of this ergodic system is vanishingly small.

Since the number ($2TW$) of sample points of a typical function of an ergodic ensemble must thus be large, our definition of an ergodic ensemble requires that the ensemble shall have a large number of degrees of freedom.[2]

On the basis of the foregoing definition, the properties of ergodic systems of sequences can now be carried over to ergodic ensembles of functions. Thus, the intersymbol influence must be of finite range, the system probabilities must be independent of time, and,

[1] By this we mean that the sum total of the probabilities of occurrence of non-typical sequences approached zero.

[2] It is possible to define an ergodic ensemble in such a way as to include certain ensembles with small numbers of degrees of freedom. (This is done by C. E. Shannon, "A Mathematical Theory of Communication," *Bell System Tech. Jour.*, vol. 27, p. 625, October, 1948.) It appears to the author, however, that such a definition would lead to confusion at the present state of our analysis.

as a consequence, all probabilities must be independent of the choice of the origin of time.

We shall find that a study of ergodic ensembles of band-limited functions will serve as a foundation for the information theory of continuous systems.

3.4 Examples of ergodic ensembles

The output over a long period of time of practically any source of communication signals is a good approximation to a typical function of an ergodic ensemble. A radio loudspeaker, a telephone receiver, and the human mouth would be examples of sources of communication signals. Consider, for example, the output of a loudspeaker. It will produce sound pressures as a function of time

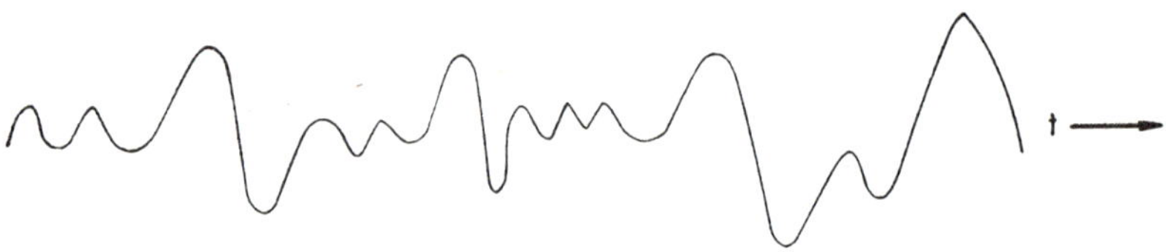

Fig. 3-2. Sound pressure at a point as a function of time.

at some listening point (see Fig. 3-2). This function will have a maximum bandwidth W of say 15 kc (kilocycles). If a length of time T of a week's duration is chosen, the resulting function of duration T will consist of a large number of messages in which there is a certain amount of coherence, *i.e.*, the quarter-hour broadcasting periods, together with certain intervals of silence. The maximum range of the intersymbol influence will be a matter of seconds, or at most a matter of minutes, if we confine our attention to amplitude as a function of time. As mentioned earlier,[1] if we are interested in the semantic content (meaning) of the messages, the range of the intersymbol influence, *i.e.*, the coherence of the messages, may be very long and the message may not be ergodic at all. From the standpoint of apparatus design, we will be interested primarily in the amplitude as a function of time in the messages which the apparatus is likely to be called upon to handle. A large group of these messages having a combined duration of a week will have symbol (sampling-point value) probabilities and

[1] Chapter I, p. 37.

intersymbol influences which will be practically the same for any reasonably likely week of programs. The function consisting of a particular week's programs coming out of the loudspeaker will thus be a member of an ergodic ensemble of functions. The same group of messages taken in a different order is another typical function of the ensemble. Other functions of the ensemble will be the programs of other weeks. Still other functions of the ensemble will be other weeks of programs which might reasonably be expected to occur, but which happen not to occur. It was first emphasized by Wiener[1] that the design of communication equipment should properly be based upon the ensemble of signals which the equipment is likely to be called upon to handle.

A somewhat simpler example of an ergodic ensemble than the foregoing could be obtained from a high-gain amplifier whose output is due entirely to thermal noise and tube noise of its early stages. If the amplifier has an effective bandwidth of 1 megacycle, the oscillograms of 1 sec duration of this amplifier's output would be members of an ergodic ensemble of functions of time. If the transmission of the amplifier is uniform in its passband, it is actually possible to express this ensemble analytically. This will be shown in Sec. 3.12.

PROBLEMS

1. In Fig. 3-A is a sample of a signal consisting of a group of pulses each of unit height and 1 microsecond duration and occurring at random times.

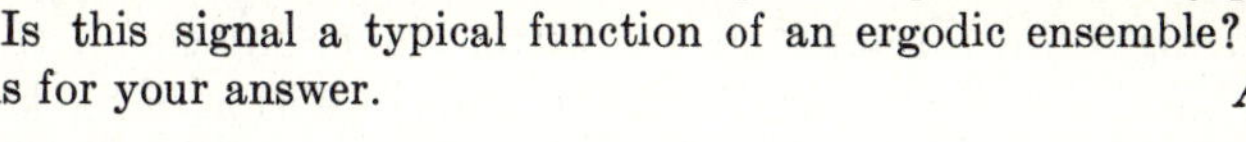

Fig. 3-A.

Suppose that the probability of the presence of a pulse at any particular time is $(\frac{1}{10})$ and that not more than one pulse is present at any particular time. Is this signal a typical function of an ergodic ensemble? Give the reasons for your answer. *Ans.* Yes.

2. In Fig. 3-B is a sample of a signal consisting of a group of pulses of unit height and random duration starting 1 sec apart. Is this signal a typical function of an ergodic ensemble? Give reasons for your answer. *Ans.* No.

[1] Norbert Wiener, *Extrapolation, Interpolation and Smoothing of Stationary Time Series*. Wiley, New York (1949).

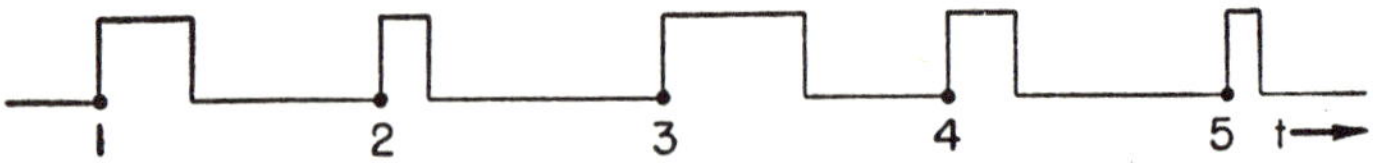

Fig. 3-B.

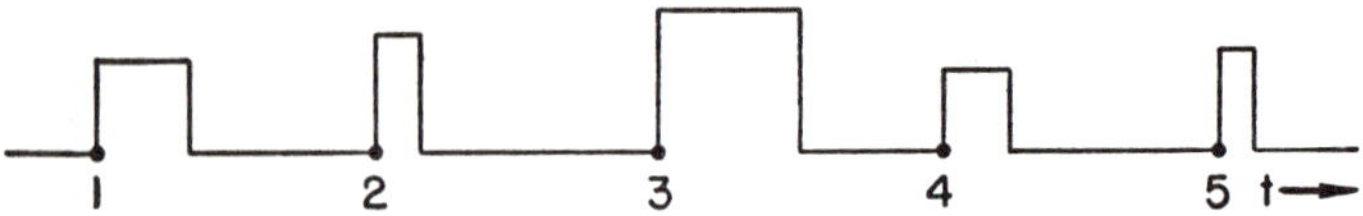

Fig. 3-C.

3. In Fig. 3-C is a sample of a signal consisting of a group of pulses of random height and random duration starting 1 sec apart. Is this signal a typical function of an ergodic ensemble? Give reasons for your answer.

Ans. No.

3.5 The problem of the random walk in one dimension

Suppose a wanderer starts from the point 0 and walks for a distance d in a straight line. After walking this distance d, he abruptly changes his direction of travel, as in Fig. 3-3, to a new direction chosen at random. If he makes a large number N of such random changes in direction, each time after walking a distance d, it is a matter of interest in many physical problems to know how far, on the average, he will be from the origin after these N displacements. The foregoing problem is called the problem of the *random walk*.

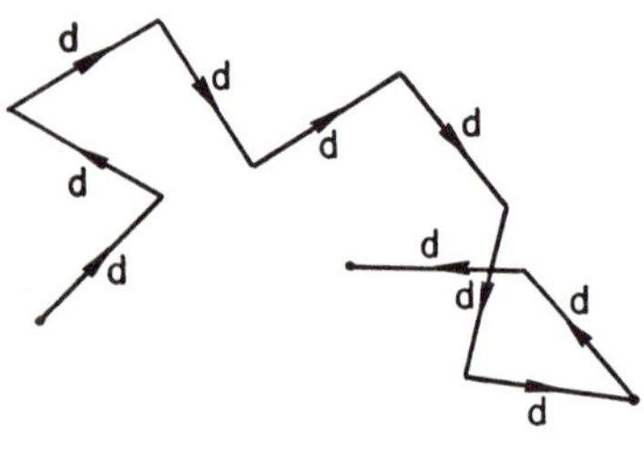

Fig. 3-3. A section of the path of a typical random walk in two dimensions.

The case which will be of particular interest to us is that of the random walk in one dimension. In this case the wanderer walks along a straight line for a distance d. After traversing the distance d, he either continues for the next d in the same direction or backtracks the distance d. The decision of which he will do is chosen at random and is as likely to be one as the other. He might, for example, flip a coin to decide. He continues in this manner making a random choice of whether to go forward or back after each distance d. After walking N such distances, he could be anywhere between a distance Nd in one direction and Nd in the other. However, it is possible to get a fair idea of how far from the starting point he is likely to be. Thus, it is shown in Appendix VII that, if a large number of wanderers take such walks, we have the formula

$$\begin{bmatrix}\text{root-mean-square value of total}\\ \text{displacements from starting point}\end{bmatrix} = \sqrt{N}\,d \tag{23}$$

One might at first have supposed that, if N is a large number, the wanderer will in general return essentially to the origin, because of

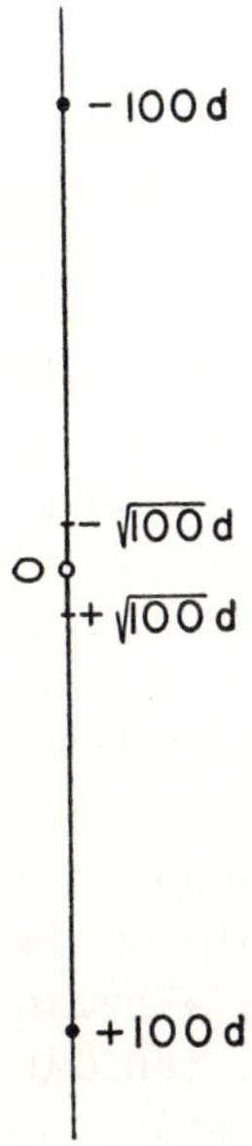

Fig. 3-4. Illustration of the relative sizes of Nd and $\sqrt{N}\,d$ when $N = 100$.

cancellation of equal walks in opposite directions. It is certainly true, as illustrated in Fig. 3-4, that the root-mean-square distance of wanderers from the origin, namely, $\sqrt{N}\,d$, will be very small in comparison with their maximum possible distance Nd. However, if N is a large number, the magnitude of $\sqrt{N}\,d$ may still be quite large; and it may be important in practical problems, especially when no term of the size of Nd is present.

Let us now use Eq. (23) to investigate the root-mean-square value of the sum of sequences of N terms, each of unit magnitude, having signs chosen at random. Such a sequence might be

$$1 + 1 - 1 - 1 + 1 - 1 - 1 + 1 - \ldots + 1 - 1 \quad (N \text{ terms})$$

The sum of such sequences will lie anywhere between $(-N)$ and $(+N)$. However, according to Eq. (23), the root-mean-square value of a large group of such sums will be $\sqrt{N}$. The average value of the algebraic values of the sums will of course be zero. In other words, if $S_1, S_2, \ldots, S_m$ represent the respective values of m such sums chosen at random, then

$$\lim_{m\to\infty} \frac{1}{m}(S_1^2 + S_2^2 + \ldots + S_m^2) = \overline{S^2} = N \tag{24}$$

or

$$\text{root mean square value of } S = \sqrt{N} \tag{24a}$$

and

$$\lim_{n\to\infty} \frac{1}{m}(S_1 + S_2 + \ldots + S_m) = \bar{S} = 0 \tag{25}$$

We will find these formulas to be very useful in studying the fluctuations which arise owing to the superposition of a large number of small random components.

3.6 The transformation of constraints from the time to the frequency domain[1]

(a) *Constraints on the Amplitude.* In Chap. I we discussed the constraints which affect the probabilities of occurrence of the sequences of symbols which may be used in information transmission systems. In a similar manner, we will now discuss the constraints which affect the probabilities of occurrence of the sequences of values at sample points in band-limited ensembles of functions. Consider an ensemble of functions of bandwidth W, and let $G_0(t)$ in Fig. 3-5 be a typical function of this ensemble. Let the sample points be

$$\ldots t_{-n}, t_{-(n-1)}, \ldots, t_0, t_1, \ldots, t_k, \ldots, t_n \ldots$$

If the probability distribution of $G_0(k/2W)$ at any t_k depends on the values of $G_0(t)$ at previous sample points or future sample points,

[1] The relations between the power spectrum and the correlation function are closely related to this section. They will be discussed in Chap. VIII.

then the ensemble has constraints[1] between the values at sample points.

In the following discussion we will be dealing only with ergodic ensembles. The probabilities involved can therefore be obtained either from averages over the ensemble of functions or from averages over a given typical function of the ensemble. The answers in either case will be the same.

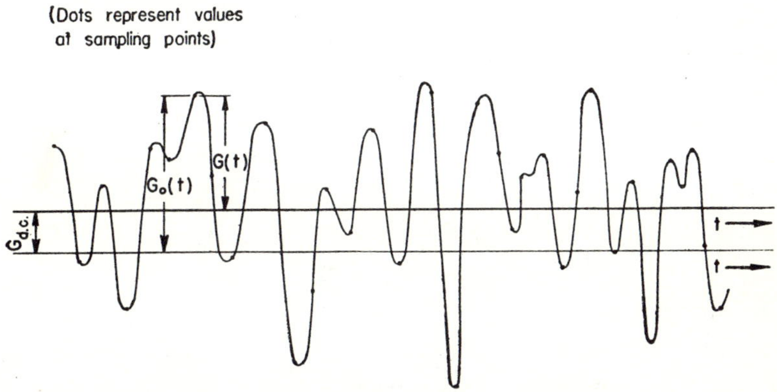

Fig. 3-5. A typical function of the ensemble.

Strictly speaking, Fourier integral analysis, upon which the sample-point theory is based, applies only to functions with finite quadratic content. For this reason, among others, in order to find the properties of $G_0(t)$, we will study the property of a typical section of $G_0(t)$ of some long duration T, and we will then let T approach infinity.

Let G_{dc} be the average value of $G_0(t)$ in the interval T. Let us now define the function $G(t)$ as

$$G(t) = G_0(t) - G_{dc} \tag{26}$$

The value of $G_0(t)$ at any sample point can then be obtained from $G(t)$, merely, by adding G_{dc}. As T approaches infinity, $G(t)$ will have the same spectrum as $G_0(t)$, except at zero frequency. We will find $G(t)$ particularly valuable in the analysis which follows.

If the probability distribution of $G(k/2W)$ at any t_k is inde-

[1] If the constraints are fixed constraints, then there are probabilities of zero or unity involved.

pendent[1] of the values of $G(t)$ at other sample points, then there is no intersymbol influence and the values of $G(t)$ at its sample points are said to be *uncorrelated*. If the distribution of $G(k/2W)$ is dependent in any way upon the values of $G(t)$ at other sample points, then there is intersymbol influence and the values of $G(t)$ at its sample points are said to be *correlated*. According to the definition of $G(t)$ in Eq. (26), we have for the values of $G(t)$ at its sample points the equation

$$\frac{1}{2TW} \sum_{n=-TW}^{+TW} G\left(\frac{n}{2W}\right) = 0 \tag{27}$$

Let us now find the power spectrum of a typical function $G(t)$ of an ergodic ensemble and see how it is related to the constraints. Starting with Eq. (7) of Chap. II, we have

$$\Omega(\omega) = \frac{1}{2W} \sum_{n=-\infty}^{+\infty} G\left(\frac{n}{2W}\right) \epsilon^{-j\omega\frac{n}{2W}} \tag{28}$$

The function

$$\lim_{T\to\infty} \left[\frac{1}{T} |\Omega(\omega)|^2\right] = \lim_{T\to\infty} \left[\frac{1}{T} \Omega(\omega) \cdot \Omega^*(\omega)\right] \tag{29}$$

is called the *power spectrum* of $G(t)$. From Eqs. (28) and (29) we have

$$\begin{aligned}
\frac{1}{T} |\Omega(\omega)|^2 &= \frac{1}{4TW^2} \left[\sum_{n=-TW}^{+TW} G\left(\frac{n}{2W}\right) \epsilon^{-j\omega\frac{n}{2W}}\right] \left[\sum_{m=-TW}^{+TW} G\left(\frac{m}{2W}\right) \epsilon^{+j\omega\frac{m}{2W}}\right] \\
&= \frac{1}{4W^2T} \sum_n \sum_m G\left(\frac{n}{2W}\right) G\left(\frac{m}{2W}\right) \epsilon^{j\frac{\omega(m-n)}{2W}} \\
&= \frac{1}{4W^2T} \sum_n \left[G\left(\frac{n}{2W}\right)\right]^2 \\
&\qquad + \frac{1}{4W^2T} \underbrace{\sum_n \sum_m}_{m \neq n} G\left(\frac{n}{2W}\right) G\left(\frac{m}{2W}\right) \epsilon^{j\frac{\omega(m-n)}{2W}}
\end{aligned} \tag{30}$$

[1] Since we are dealing with an ergodic ensemble, the probabilities must be independent of the origin of time. The probability distributions for all sample points must therefore be the same, in the absence of intersymbol influence.

Of course,

$$\epsilon^{j\frac{\omega}{2W}(m-n)} = \cos\left[\frac{\omega}{2W}(m-n)\right] + j\sin\left[\frac{\omega}{2W}(m-n)\right] \tag{31}$$

Now

$$\underbrace{\sum_n \sum_m}_{m \neq n} G\left(\frac{n}{2W}\right) G\left(\frac{m}{2W}\right) \cos\left[\frac{\omega}{2W}(m-n)\right]$$

consists of pairs of equal terms such as

$$G\left(\frac{3}{2W}\right) G\left(\frac{5}{2W}\right) \cos\left[\frac{\omega}{2W}(5-3)\right]$$

and

$$G\left(\frac{5}{2W}\right) G\left(\frac{3}{2W}\right) \cos\left[\frac{\omega}{2W}(3-5)\right]$$

Similarly,

$$\underbrace{\sum_n \sum_m}_{m \neq n} G\left(\frac{n}{2W}\right) G\left(\frac{m}{2W}\right) \sin\left[\frac{\omega}{2W}(m-n)\right]$$

consists of pairs of canceling terms. Thus Eq. (30) reduces to

$$\frac{1}{T}|\Omega(\omega)|^2 = \frac{1}{4W^2T}\sum_n \left[G\left(\frac{n}{2W}\right)\right]^2 + \frac{1}{2W^2T} \sum_n \sum_m G\left(\frac{n}{2W}\right) G\left(\frac{m}{2W}\right) \cos\left[\frac{\omega}{2W}(m-n)\right] \tag{32}$$

where

$$m - n > 0 \tag{33}$$

Let us next find the value of $|\Omega(\omega)|^2$ at the sampling point in the frequency domain where

$$\omega = \frac{2\pi q}{T} \tag{34}$$

Here

$$\frac{1}{T}\left|\Omega\left(\frac{2\pi q}{T}\right)\right|^2 = \frac{1}{4W^2T}\sum_n \left[G\left(\frac{n}{2W}\right)\right]^2 + \frac{1}{2W^2T}\sum_n \sum_m G\left(\frac{n}{2W}\right) G\left(\frac{m}{2W}\right) \cos\left[\frac{2\pi q(m-n)}{2TW}\right] \tag{35}$$

or $$\frac{1}{T}\left|\Omega\left(\frac{2\pi q}{T}\right)\right|^2 = \frac{1}{4W^2T}\sum_n\left[G\left(\frac{n}{2W}\right)\right]^2$$
$$+ \frac{1}{2W^2T}\sum_p \cos\left(\frac{\pi qp}{TW}\right)\left[\sum_n G\left(\frac{n}{2W}\right)G\left(\frac{n+p}{2W}\right)\right] \tag{36}$$
where $$p = m - n \tag{37}$$

Let us now consider the last term on the right side of Eq. (36). For a value of p which is within the range of the intersymbol influence, the quantity

$$\frac{1}{2W^2T}\cos\left(\frac{\pi qp}{TW}\right)\sum_n G\left(\frac{n}{2W}\right)G\left(\frac{n+p}{2W}\right) \tag{38}$$

may well have a value which is comparable with the constant term

$$\frac{1}{4W^2T}\sum_n\left[G\left(\frac{n}{2W}\right)\right]^2 \tag{39}$$

Let us next choose any particular value of p which is larger than the range of the intersymbol influence. Then the values of $G(n/2W)$ and $G(n + p/2W)$ are unrelated. The product terms $G(n/2W) \cdot G(n + p/2W)$ may be either positive or negative, and since the ensemble is ergodic, the average value of the positive terms will be equal to that of the negative terms. The average value of the cross-product terms in Eq. (36), for a given value of p, namely,

$$\frac{1}{2TW}\sum_n G\left(\frac{n}{2W}\right)G\left(\frac{n+p}{2W}\right) \tag{40}$$

will consequently approach zero for large values of $2TW$. It would, however, be a mistake to neglect them, since there are approximately $2TW$ values of p which must be used, so that there are $TW \cdot 2TW$ cross-product terms, whereas the divisor in Eq. (36) is only $2W^2T$.

To get an idea of the order of magnitude of

$$\frac{1}{2W^2T}\sum_p \cos\left(\frac{\pi qp}{TW}\right)\left[\sum_n G\left(\frac{n}{2W}\right)G\left(\frac{n+p}{2W}\right)\right] \tag{41}$$

where the summation is taken over values of p beyond the range of the intersymbol influence, we can use the random-walk formula, since the situations are analogous. We wish to find the probable magnitude of the sum of $TW \cdot 2TW = 2T^2W^2$ terms of average value

$$\frac{1}{2W^2T}\overline{\left|G\left(\frac{n}{2W}\right)\right|\left|G\left(\frac{n+p}{2W}\right)\right|} \tag{42}$$

According to the random-walk formula, this probable magnitude will be of the order of

$$\sqrt{2}\,TW \cdot \frac{1}{2W^2T}\overline{\left|G\left(\frac{n}{2W}\right)\right|\left|G\left(\frac{n+p}{2W}\right)\right|} \tag{43}$$

The quantity (43) is of the same order of magnitude as (39) since there are $2TW$ terms in $\sum\limits_n$. The actual value of (41) may, of course, be either negative or positive. The average value over the ensemble of (41) will be zero, but since (43) is of the same order of magnitude as (39), we cannot neglect (41) when dealing with a particular function of the ensemble.

With the aid of the foregoing analysis we can now discuss the effects of the intersymbol influence on the power spectrum. Let us rewrite Eq. (36) in the form

$$\begin{aligned}\frac{1}{T}\left|\Omega\left(\frac{2\pi q}{T}\right)\right|^2 &= \frac{1}{4W^2T}\sum_n\left[G\left(\frac{n}{2W}\right)\right]^2 \\ &+ \frac{1}{2W^2T}\sum_p \cos\left(\frac{\pi qp}{TW}\right)\left[\sum_n G\left(\frac{n}{2W}\right)G\left(\frac{n+p}{2W}\right)\right] \\ &\qquad\text{(over the range of the intersymbol influence)} \\ &+ \frac{1}{2W^2T}\sum_p \cos\left(\frac{\pi qp}{TW}\right)\left[\sum_n G\left(\frac{n}{2W}\right)G\left(\frac{n+p}{2W}\right)\right] \\ &\qquad\text{(outside the range of intersymbol influence)}\end{aligned} \tag{44}$$

In Fig. 3-6, a curve is illustrated of $(1/T)|\Omega(\omega)|^2$ for a typical function of an ergodic ensemble. The values of $\frac{1}{T}\left|\Omega\left(\frac{2\pi q}{T}\right)\right|^2$ at the sample points are also shown. The figure is of necessity diagrammatic since it is impractical to show more than a few of the $2TW$ sample points and of the oscillations which go through them. The general features of $(1/T)|\Omega(\omega)|^2$ in Fig. 3-6 can be predicted from

Eq. (44). Thus there is a constant component (independent of q and therefore of ω) given by

$$\frac{1}{4W^2T}\sum_n \left[G\left(\frac{n}{2W}\right)\right]^2 \tag{45}$$

Then there is a component that varies relatively slowly with q but varies hardly at all from one sampling point to a neighboring

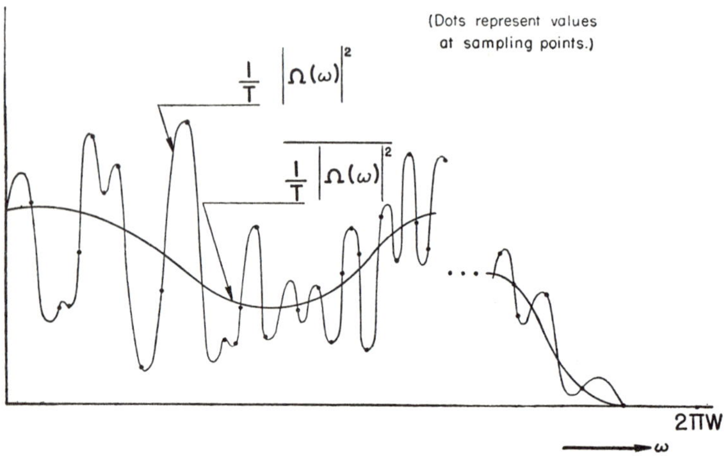

Fig. 3-6. Power spectrum of a particular $G(t)$ and the ensemble average of the power spectrum.

one. This component is

$$\frac{1}{2W^2T}\sum_p \cos\left(\frac{\pi qp}{TW}\right)\left[\sum_n G\left(\frac{n}{2W}\right)G\left(\frac{n+p}{2W}\right)\right] \tag{46}$$

(over the range of intersymbol influence)

The value of this term is determined by the values of

$$\frac{1}{2TW}\sum_n G\left(\frac{n}{2W}\right)G\left(\frac{n+p}{2W}\right) \quad \text{for } p < \text{range of intersymbol influence} \tag{47}$$

for the different values of p. In contradistinction to (40), the values of (47) will not in general vanish, since p is inside the range of the intersymbol influence. Finally, there is the component

$$\frac{1}{2W^2T}\sum_p \cos\left(\frac{\pi qp}{TW}\right)\left[\sum_n G\left(\frac{n}{2W}\right)G\left(\frac{n+p}{2W}\right)\right] \tag{48}$$

(outside range of intersymbol influence,
i.e., $p > p_0$ = max range of intersymbol influence)

This component varies markedly from one sampling point to another, and is quite likely to vary any amount up to its peak-to-peak value between neighboring sampling points.

If we take an average over the ensemble of $\frac{1}{T}\left|\Omega\left(\frac{2\pi q}{T}\right)\right|^2$, the values of (45) and (46) will not change, but the average value of (48) will be zero. In fact the average value of (48) over even a large number[1] of neighboring sample points in a particular function of an ergodic ensemble will tend toward zero.

We may thus draw the following conclusions:

(1) If there is no intersymbol influence, *i.e.*, if there are no constraints, the ensemble average of the power spectrum will be constant throughout the frequency range W.

(2) Intersymbol influence in the time domain will tend to cause variation with frequency in the ensemble average of the power spectrum. This variation represents a constraint on the amplitude in the frequency domain.

(3) In the case of any particular function $G(t)$ of an ensemble, the individual values of $|\Omega(\omega)|^2$ at the sample points are not determined by the constraints. These individual values at the sample points represent the amplitude information in the frequency representation of $G(t)$. It is these fluctuations which represent the differences between the different $G(t)$'s of an ensemble and allow them to be distinguished and to have meanings assigned to them and thus carry information.

In the practical measurement of power spectra, an individual measurement in a particular frequency range will usually include a large number of neighboring sampling points. Conclusions (1) and (2) above will therefore apply to the practical measurement of power spectra of an individual typical function as well as to the ensemble average.

We note that, in Fig. 3-6, the power spectrum approaches zero

[1] It is reasonable to speak of a large number of neighboring sample points, since if $2TW = 10^6$, then one could speak of 1000 neighboring sample points.

well before the frequency W is reached. This is shown to illustrate the fact that, if a value of W is chosen which is higher than the actual spectral range of the ensemble, the intersymbol influence (particularly for small values of p) will make (46) cancel a large part of (45) in the high-frequency range.

Before leaving the subject of constraints on the amplitude, it should be noted that we have not mentioned any constraints upon the amplitude at the frequency sampling points due to the probability distributions of values at sampling points in the time domain, as distinguished from the intersymbol influence. This does not mean that such constraints do not exist, but only that we have nothing to report about them here.

(b) *Constraints on the Phase.* The foregoing discussion has indicated how the intersymbol influence in the time domain can be transformed into constraints on the amplitude of the spectrum in the frequency domain. We next wish to find out what we can about constraints on the phase of the spectrum.

Let us consider a typical signal $G(t)$ of an ergodic ensemble. Let us now transform $G(t)$ into the frequency domain and consider the phase of $\Omega(\omega)$ at the qth sample point. This phase is ϕ_q where

$$\Omega\left(\frac{2\pi q}{T}\right) = \left|\Omega\left(\frac{2\pi q}{T}\right)\right| \epsilon^{j\phi_q} \tag{49}$$

In the first place, since all properties of the ensemble are independent of the choice of the origin of time, it follows that the phases ϕ_q, say ϕ_{768}, of the different $G(t)$'s of the ensemble will be distributed uniformly between 0 and 2π.

We next will show that, as $T \to \infty$, the phases ϕ_n, where n ranges over all integral values from $-TW$ to $+TW$, at the n sample points of a *particular* $\Omega(\omega)$, *i.e.*, a *particular* $G(t)$, are also uniformly distributed from 0 to 2π.[1] We first point out that this is true for every narrow frequency range, which is still large enough so that there is a very large number of sample points in the range.[2,3] If it were not true, there would be some phase angle ϕ_m where the dis-

[1] We here assume that the effective bandwidth of $\Omega(\omega)$ has a finite value W.

[2] We here refer only to frequency ranges in which $\Omega(\omega)$ does not vanish.

[3] Since $T \to \infty$, a narrow frequency range can still have a large number of frequency sampling points.

tribution had a maximum. Then near the time

$$t = \frac{\phi_m}{\omega}$$

if $\omega/2\pi$ is the average frequency of the narrow frequency range, there would be a concentration of an appreciable part of the energy due to that frequency range in a narrow interval of time. If this were true, the function $G(t)$ would not be a typical function of an ergodic ensemble. Since by hypothesis $G(t)$ is a typical function of an ergodic ensemble, it follows that the phase distribution in every narrow frequency range must be uniform. Since the property of uniform phase distribution thus holds for every narrow frequency range, it also holds for the entire spectrum.

Next let us write

$$\Omega(\omega) = a(\omega) - jb(\omega) \tag{50}$$

Then, since all phases are equally probable, we have for the average over an ergodic ensemble

$$\overline{\left[a\left(\frac{2\pi q}{T}\right)\right]^2} = \overline{\left[b\left(\frac{2\pi q}{T}\right)\right]^2} = \frac{1}{2}\overline{\left[\Omega\left(\frac{2\pi q}{T}\right)\right]^2} \tag{51}$$

In ergodic time ensembles the intersymbol influence between frequency components is of such a nature that the correct probability distribution of the values at the sampling points and groups of sampling points in time is obtained. These intersymbol influences, as we have shown in Sec. 3.6(a), can depend upon the absolute frequencies of the components. Therefore, any phase constraints which exist are likely to depend upon the absolute frequencies of the components whose phases are under consideration.

It is the opinion of the author that an investigation of phase constraints and a general investigation of constraints due to probability distributions may well lead to important results in information theory.

PROBLEMS

1. Show that for positive ω

$$j\Omega(\omega) = \frac{1}{2W}\sum_{n=-\infty}^{+\infty} G_1\left(\frac{n}{2W}\right)\epsilon^{j\frac{\omega n}{2W}}$$

where $G_1(t)$ is the Hilbert transform of $G(t)$.

2. (a) What is the function whose spectrum is $\Omega^*(\omega)$, where $\Omega^*(\omega)$ is the complex conjugate of $\Omega(\omega)$, and $\Omega(\omega)$ is the Fourier transform of $G(t)$?

Ans. $G(-t)$.

3. Show that

$$|\Omega(\omega)|^2 = [a(\omega)]^2 + [b(\omega)]^2 = \frac{1}{4W^2} \sum_{p=1}^{\infty} \cos \frac{\omega p}{2W} \sum_{n=-\infty}^{+\infty} G_1\left(\frac{n}{2W}\right) G_1\left(\frac{n+p}{2W}\right)$$

where $G_1(t)$ is the Hilbert transform of $G(t)$.

4. Show that

$$a(\omega) = \frac{1}{2W} \sum_{n=-\infty}^{+\infty} G\left(\frac{n}{2W}\right) \cos \frac{\omega n}{2W} = \frac{-1}{2W} \sum_{n=-\infty}^{+\infty} G_1\left(\frac{n}{2W}\right) \sin \left(\frac{\omega n}{2W}\right)$$

and

$$b(\omega) = \frac{1}{2W} \sum_{n=-\infty}^{+\infty} G\left(\frac{n}{2W}\right) \sin \left(\frac{\omega n}{2W}\right)$$

$$= \frac{1}{2W} \sum_{=n-\infty}^{+\infty} G_1\left(\frac{n}{2W}\right) \cos \left(\frac{\omega n}{2W}\right)$$

where $G_1(t)$ is the Hilbert transform of $G(t)$.

5. (a) In an ergodic ensemble of bandwidth W, the range of the intersymbol influence of the sample point values is finite. Does it follow that the range of the intersymbol influence in the ensemble of functions consisting of the Hilbert transforms of the original ergodic ensemble is also finite?

(b) Give an example to illustrate your answer.

6. If a function $G(t)$, which is a typical function of an ergodic ensemble, is sent through a linear network having a phase characteristic that varies with frequency, show that the phases of the frequency sampling values in the output of the network are still uniformly distributed from zero to 2π.

7. Show that in an ergodic ensemble

(a) A complete set of symbol probabilities is derivable from a complete set of digram probabilities.

(b) A complete set of n-gram probabilities is derivable from a complete set of pair-separation probabilities, *i.e.*, a complete set of $p(x_k,y_{k+j})$ for all values of x and y and all integral values of j. Since the ensemble is ergodic, $p(x_k,y_{k+j})$ is independent of k.

3.7 Coherence

Suppose we have two signals[1] $G_1(t)$ and $G_2(t)$. If we can get some of the information in $G_2(t)$ by knowing $G_1(t)$, then the signals are

[1] In this section, $G_1(t)$ is just a function of time and does not particularly represent a quadrature function as it did in the preceding problems.

at least partially dependent. This dependence between the detailed values (at the sample points) which carry the information of two signals is described as *coherence.* Two signals (or two parts of the same signal) are said to be *coherent* when there is a dependence between their detailed values. Two signals (or two parts of the same signal) are said to be *incoherent* if they are independent of one another.

We can also express the foregoing ideas in a somewhat more precise manner. Given two signals having sample-point sequences

$$\ldots\ x_{-1},x_0,x_1,x_2,\ \ldots$$

and

$$\ldots\ y_{-1},y_0,y_1,y_2,\ \ldots$$

Then, if the probability distribution at any x_k depends on the values of any of the y's, or on the values of any combinations of the y's, then we may say that the y signal constrains the x signal. Ordinarily, in practice, if the y signal constrains the x signal, then the x signal will also constrain the y signal. It would be difficult to prove that this is necessarily true in all cases.

If the y signal constrains the x signal, then the x signal may be said to be partially dependent upon the y signal. This may also be expressed by saying that the x signal is *partially coherent* with the y signal.[1] If the x signal is completely determined in all its detailed properties, when the values of the y signal are known, then the x signal is said to be *completely coherent* with the y signal.

When two identical signals are superimposed "in phase" (*i.e.*, algebraically added), it is well known that the quadratic content (energy) of the resultant is equal to four times that of either signal taken separately. On the other hand, if they are superimposed 180 deg out of phase (*i.e.*, algebraically subtracted), the resultant has zero quadratic content. In general, if two signals $G_1(t)$ and $G_2(t)$ are algebraically added, the resultant quadratic content is

$$\int[G_1(t) + G_2(t)]^2dt = \int\{[G_1(t)]^2 + [G_2(t)^2\}dt + \int 2G_1(t)\cdot G_2(t)dt \tag{52}$$

The first integral on the right of Eq. (52) represents the quadratic contents of the two signals taken separately, and the second integral

[1] Intersymbol influence in a signal represents partial coherence between parts of the same signal.

represents the interaction quadratic content. If the interaction quadratic content of two signals is zero, we will call the signals *orthogonal*. The ratio

$$\frac{\int 2G_1(t)G_2(t)dt}{\int 2|G_1(t)||G_2(t)|dt} = S_{12} \tag{53}$$

will be called the *interaction ratio* between the signals $G_1(t)$ and $G_2(t)$. The value of the interaction ratio thus must lie between -1 and $+1$.

It is of interest to find the value of the interaction quadratic content between two signals in terms of their frequency-sampling values. We will assume that the d-c components have been removed from the signals (or at least from one of them) since that will be true in all cases of interest. Thus

$$G_1(t) = \sum_{n=1}^{+\infty} \frac{2}{T}\left[a_1\left(\frac{2\pi n}{T}\right)\cos\left(\frac{2\pi nt}{T}\right) + b_1\left(\frac{2\pi n}{T}\right)\sin\left(\frac{2\pi nt}{T}\right)\right] \tag{54}$$

$$G_2(t) = \sum_{m=1}^{+\infty} \frac{2}{T}\left[a_2\left(\frac{2\pi m}{T}\right)\cos\left(\frac{2\pi mt}{T}\right) + b_2\left(\frac{2\pi m}{T}\right)\sin\left(\frac{2\pi mt}{T}\right)\right] \tag{55}$$

Since as shown in Sec. 2.2, the a's and b's are identical in value with Fourier series coefficients (except for a factor $2/T$), it follows[1] that

$$\int_0^T G_1(t)G_2(t)dt = \frac{2}{T}\sum_{n=1}^{\infty}\left[a_1\left(\frac{2\pi n}{T}\right)\cdot a_2\left(\frac{2\pi n}{T}\right) + b_1\left(\frac{2\pi n}{T}\right)\cdot b_2\left(\frac{2\pi n}{T}\right)\right] \tag{56}$$

In case the signals are members of ergodic ensembles which are independent of one another, then

$$\frac{2}{T}\sum_{n=1}^{\infty} a_1\left(\frac{2\pi n}{T}\right)\cdot a_2\left(\frac{2\pi n}{T}\right) = 0 = \frac{2}{T}\sum_{n=1}^{\infty} b_1\left(\frac{2\pi n}{T}\right)\cdot b_2\left(\frac{2\pi n}{T}\right) \tag{57}$$

so that the interaction quadratic content is zero.

We can also express the interaction quadratic content in terms of the values at the sampling points in the time domain. This gives us

[1] *F.A.,M., and N.*, Sec. 1.11, Eq. (99).

$$\int G_1(t) \cdot G_2(t) dt$$

$$= \int \sum_n \sum_m G_1\left(\frac{n}{2W}\right) G_2\left(\frac{m}{2W}\right) \frac{\sin (2\pi Wt - n\pi)}{2\pi Wt - n\pi} \frac{\sin (2\pi Wt - m\pi)}{2\pi Wt - m\pi} dt$$

$$= \frac{1}{2W} \sum_n G_1\left(\frac{n}{2W}\right) G_2\left(\frac{n}{2W}\right) \tag{58}$$

If $G_1(t)$ and $G_2(t)$ are members of ergodic ensembles which are independent of one another, the right side of Eq. (58) is negligible, so that in this way also we know the interaction quadratic content is zero. We thus see that incoherent ergodic signals are orthogonal.[1] We thus obtain the useful result that, *when two incoherent signals which are typical functions of ergodic ensembles are superimposed in a linear system, their quadratic contents are additive.* Since the interaction quadratic content of incoherent signals (which are members of ergodic ensembles) vanishes, non-vanishing interaction quadratic content is an indication of dependence of the signals, *i.e.*, coherence between the signals. In all the foregoing, it has been assumed that any d-c components have been removed.

We next wish to consider whether two functions which are members of ergodic ensembles and which have a vanishing interaction quadratic content will necessarily be incoherent. The answer is certainly "no," for let

$$G_2(t) = G_1(t + t_1) \tag{59}$$

where t_1 is greater than the range of the intersymbol influence. In this case $G_2(t)$ and $G_1(t)$ are completely coherent, but their inter-

[1] In cases where the signals involved are not typical functions of ergodic ensembles, it is a general practical fact that any two signals arising from independent sources are almost certain to be orthogonal, if the signals are sufficiently long and if they do not both have d-c components. Even in the case of signals of presumably the same frequency, if the signals are independent there will always be a wandering of the phase between them so that over a long time the average value of the interaction quadratic content will be zero. In other words, the right sides of Eqs. (56) and (58) will almost always vanish for independent signals of the type just described.

action quadratic content vanishes. Next, however, suppose that

$$\lim_{T\to\infty}\left[\frac{1}{2T}\int_{-T}^{+T} G_1(t)G_2(t+\tau)\,dt\right] = 0 \quad \text{(for all values of } \tau) \tag{60}$$

Then also

$$\frac{1}{4TW}\sum_{n=-TW}^{+TW} G_1\left(\frac{n}{2W}\right) G_2\left(\frac{n+p}{W}\right) = 0 \quad \text{(for all values of } p) \tag{61}$$

Even this strong condition does not ensure that the signals are not coherent. Thus let $G_1(t)$ and $G_2(t)$ be the upper and lower sidebands of an amplitude-modulated signal, modulated by a cosine wave.

$$G_1(t) = \frac{Am}{2}\sin\left[(\omega+\alpha)t+\phi\right] \tag{62}$$

$$G_2(t) = \frac{Am}{2}\sin\left[(\omega-\alpha)t+\phi\right] \tag{63}$$

Then

$$\lim_{T\to\infty}\left\{\frac{1}{2T}\int_{-T}^{+T}\frac{A^2m^2}{4}\sin\left[(\omega+\alpha)t+\phi\right]\sin\left[(\omega-\alpha)(t+\tau)+\phi\right]dt\right\} = 0 \quad \text{(for all values of } \tau) \tag{64}$$

Nevertheless $G_1(t)$ and $G_2(t)$ are not incoherent since the values of m, α, and ϕ, which are the degrees of freedom in $G_1(t)$, have identical values with those in $G_2(t)$ and the properties of amplitude modulation depend upon the fact that these values are identical. However $G_1(t)$ and $G_2(t)$ are not members of ergodic ensembles. On the other hand, let us suppose that the modulating signal were a typical member of an ergodic ensemble instead of being a pure sine wave. Then the signals made up of the upper and lower sidebands, respectively, would also appear to be typical members of ergodic ensembles, and it appears that they would still satisfy Eq. (60) even though they are completely coherent, *i.e.*, all the detailed properties of one of them could be determined from the other.

We will find that coherence is an important concept in our work, as it is in other branches of communication engineering. Thus coherence between the feedback signal and the input signal is necessary for feedback systems to be effective. Coherence is necessary in the radiation of the different elements of an antenna array in order to achieve directivity. Coherence between the sidebands of a frequency-modulated signal is necessary in order to achieve noise reduction. On the other hand, it is the incoherence between noise and the desired signal which is responsible for the noise reduction obtainable by various correlation techniques and for the noise reduction possible with various types of modulation.[1]

3.8 The properties of random noise in the time domain[2]

Random noise is an idealization of the type of disturbances present in modern communication systems due to the atomicity of matter and electricity. Several alternative and equivalent definitions may be given for random noise. Starting from any one of these definitions, the other definitions may be shown to be equivalent to it. Some of these definitions will be tabulated below.

A random-noise signal is a stochastic function of time (and frequency). By this we mean that the amplitude of a random-noise signal for any particular value of time is determined by *chance.* It cannot be predicted. However, there are certain average properties of a random-noise signal over long periods of time whose future values can be predicted from the past. Thus in the case of any particular process whose properties remain unchanged and which gives rise to random noise, the quadratic content per unit of time of the noise remains constant and is a property of the process.[3] Furthermore, the liniva distribution of random noise as a function of time is *normal* (gaussian). Thus the probability that the liniva I will have a value between I and $I + dI$ is at any time equal to

$$P(I)dI = \frac{1}{(\sqrt{2\pi\sigma_I^2})}\epsilon^{-I^2/2\sigma_I^2}dI \tag{65}$$

where $$\sigma_I = \text{root mean-square value of } I = \sqrt{\overline{I^2}} \tag{66}$$

[1] When two ergodic ensembles are coherent, either wholly or in part, the coherent components of the ensembles invariably arise from the same ultimate source.

[2] For an extensive background on the subject of random noise, the reader is referred to the classical paper of S. O. Rice, "The Mathematical Theory of Random Noise," *Bell System Tech. Jour.*, vol. 23, p. 282, July, 1944, and vol. 25, p. 45, January, 1945, and to J. L. Lawson and G. E. Uhlenbeck, *Threshold Signals*, McGraw-Hill, New York, (1950), and Goldman, *Frequency Analysis, Modulation, and Noise.*

[3] For example, the mean-square thermal noise voltage generated in a resistor is $\overline{E^2} = 4kTR\,\Delta F$. See *F.A.,M., and N.*, Chaps. VI through IX.

In Fig. 3-7 is a diagram of a sample of random noise of duration T. This random noise is assumed to be taken from a system of bandwidth W. The probability distribution for I in Eq. (65) above then also approximates the actual distribution of the linivas of I at the sample points of Fig. 3-7, if T is sufficiently long.

Finally, in the case of random noise (of bandwidth W) there is no intersymbol influence between the values of the noise function at the sample points in the time domain.

From the foregoing properties of random noise, we can derive corresponding properties for random noise in the frequency domain. Before doing so, however, we must acquaint the reader with the central-limit theorem.

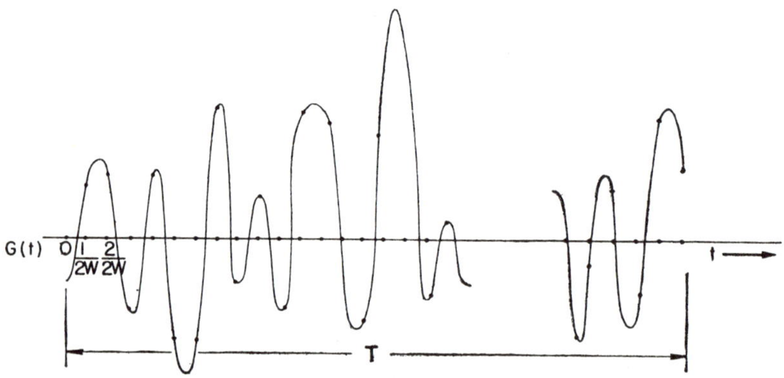

Fig. 3-7. A sample of random noise of bandwidth W and duration T in the time domain.

3.9 The central-limit theorem

There is a very general mathematical phenomenon of probability theory which we shall find useful in the study of random noise. This phenomenon is known as the *central-limit theorem*. The central-limit theorem concerns the form of the probability distribution which is obtained by the superposition of a large number of independent random contributions. A special form of the theorem which we shall find adequate for our purposes is as follows:

(1) If $x_1, x_2, \ldots, x_p$ are independent random variables

(2) The average value of each random variable is zero, *i.e.*,

$$\sum_a x_{sa} P(x_{sa}) = 0 \tag{67}$$

where x_{sa} is any possible value of x_s and $P(x_{sa})$ is the probability of obtaining this value.

(3) In case the random variables are multidimensional (such as frequency components, which are two dimensional, having an in-phase and a quadrature component), then the values in the different dimensions are independent of one another or, have a vanishing cross correlation, or, at least, X, Y, etc., defined below have a vanishing cross correlation.

(4) The value of every particular x_s is negligible in comparison with the ensemble average of the absolute value of the resultant of all the others, *i.e.*,

$$\operatorname*{Lim}_{p \to \infty} \frac{x_s}{\overline{\left| \sum_{s=1}^{p} x_s \right|}} = 0 \tag{68}$$

then if the dimensions are x, y, z, etc., the central-limit theorem states that the probability that the resultant of the superposition will lie in the element of volume $dX\,dY\,dZ$ . . . is

$$P(X,Y,Z)dX\,dY\,dZ = \frac{\alpha}{\sqrt{\pi}} \epsilon^{-\alpha^2 X^2} dX \cdot \frac{\beta}{\sqrt{\pi}} \epsilon^{-\beta^2 Y^2} dY \cdot \frac{\gamma}{\sqrt{\pi}} \epsilon^{-\gamma^2 Z^2}\, dZ \ . \ . \ . \tag{69}$$

where

$$X = \sum_{s=1}^{\infty} x_s \tag{70}$$

$$Y = \sum_{s=1}^{\infty} y_s \tag{71}$$

$$Z = \sum_{s=1}^{\infty} z_s \tag{72}$$

$$\frac{1}{\alpha^2} = 2 \int_{-\infty}^{+\infty} X^2 P(X) dX \tag{73}$$

$$\frac{1}{\beta^2} = 2 \int_{-\infty}^{+\infty} Y^2 P(Y) dY \tag{74}$$

$$\frac{1}{\gamma^2} = 2 \int_{-\infty}^{+\infty} Z^2 P(Z) dZ \tag{75}$$

A distribution of the form of Eq. (69) is called a *normal distribution*.

Derivations of the central-limit theorem may be found in the literature.[1] We will now illustrate the meaning of the theorem by applying it directly to the study of random noise.

[1] H. Cramer, *Mathematical Methods of Statistics*, Princeton University Press, Princeton, N.J., (1946), p. 213.

3.10 The properties of random noise in the frequency domain

According to Eq. (7) of Chap. II the spectrum function for any value of ω may be expressed in terms of the values of a function at the sample points in the time domain as follows:

$$\begin{aligned}\Omega(\omega) = a(\omega) - jb(\omega) &= \frac{1}{2W} \sum_{s=-TW}^{+TW} G\left(\frac{s}{2W}\right) \epsilon^{-j\omega\frac{s}{2W}} \\ &= \frac{1}{2W} \sum_{s=-TW}^{+TW} G\left(\frac{s}{2W}\right) \cos\left(\frac{\omega s}{2W}\right) \\ &\qquad - \frac{j}{2W} \sum_{s=-TW}^{+TW} G\left(\frac{s}{2W}\right) \sin\left(\frac{\omega s}{2W}\right)\end{aligned} \tag{76}$$

Thus

$$a(\omega) = \frac{1}{2W} \sum_{s=-TW}^{+TW} G\left(\frac{s}{2W}\right) \cos\left(\frac{\omega s}{2W}\right) \tag{77}$$

$$b(\omega) = \frac{1}{2W} \sum_{s=-TW}^{+TW} G\left(\frac{s}{2W}\right) \sin\left(\frac{\omega s}{2W}\right) \tag{78}$$

If $G(t)$ is a random-noise function, then the individual terms in the summations of Eqs. (77) and (78) satisfy all the requirements necessary for the central-limit theorem. Thus, let

$$x_s = \frac{1}{2W} G\left(\frac{s}{2W}\right) \cos\left(\frac{\omega s}{2W}\right) \tag{79}$$ [1]

$$y_s = \frac{1}{2W} G\left(\frac{s}{2W}\right) \sin\left(\frac{\omega s}{2W}\right) \tag{80}$$

Since, according to Sec. 3.8, pure random noise is a stochastic function with no intersymbol influence, the different x_s variables are independent random variables and so are the y_s variables. Thus requirement (1) is satisfied.

Furthermore the average value (for fixed s) of each x_s and each y_s is zero, so that condition (2) is satisfied.[2] Condition (3) is also

[1] x_s is thus the value of the right side of Eq. (79) at the sth sample point, with a corresponding value for y_s. An ensemble of samples of random noise of duration T from the source in question will generally give different values of x_s and y_s in the different functions of the ensemble. However, we know that the distribution function for $G(s/2W)$ over the ensemble is given by Eq. (65), and this then determines the distribution functions of x_s and y_s.

[2] The average value of x_s for all the different values of s in a given function of the ensemble is also zero. However, that is not what condition (2) is about.

satisfied since $a(\omega)$ and $b(\omega)$ as expressed in Eqs. (77) and (78) have zero cross-correlation. This is true because the average value of

$$\cos\left(\frac{\omega s}{2W}\right)\cdot\sin\left(\frac{\omega s}{2W}\right)$$

over all values of s vanishes and the ensemble average value of $G(s/2W)$ is independent of s. Finally condition (4) is satisfied, since each individual x_s or y_s is negligibly small because $2TW \gg 1$.

Since all the conditions required for the central-limit theorem to be applicable are satisfied, it follows from Eqs. (69) through (75) that the probability distribution of $a(\omega)$ and $b(\omega)$ is given by the equation

$$P(a,b)da\,db = \frac{1}{\rho\sqrt{\pi}}\,\epsilon^{-a^2/\rho^2}\cdot\frac{1}{\rho\sqrt{\pi}}\,\epsilon^{-b^2/\rho^2}da\,db$$

$$= \frac{1}{\rho^2\pi}\,\epsilon^{-(a^2+b^2)/\rho^2}da\,db \tag{81}$$

where
$$\rho^2 = \overline{2a^2} = \overline{2b^2} = \overline{|\Omega|^2} \tag{82}$$

The averages in Eq. (82) are ensemble averages of a^2, b^2, and $|\Omega|^2$ for a given value of ω. However, since there is no intersymbol influence in the time domain, it follows from Sec. 3.6 that $\overline{a^2}$,$\overline{b^2}$, and $\overline{|\Omega|^2}$ do not vary with frequency and that they may also be considered as averages taken over all frequency sampling points for a given $G(t)$, *i.e.*, a given $\Omega(\omega)$. Furthermore, by Eq. (52) of Chap. II, since the ensemble of noise signals is ergodic

$$\overline{a_n^2} + \overline{b_n^2} = \frac{T}{2W}\overline{\left|G\left(\frac{s}{2W}\right)\right|^2} \tag{83}$$

and by Sec. 3.6
$$\overline{a_n^2} = \overline{b_n^2} \tag{84}$$

From Eq. (82), (83), and (84) it then follows that

$$\rho^2 = \frac{T}{2W}\overline{\left|G\left(\frac{s}{2W}\right)\right|^2} = \frac{T}{2W}\,\sigma_I^2 \tag{85}$$

Equation (81) is the two-dimensional normal distribution function for random noise in the frequency domain, corresponding to Eq. (65) which is a one-dimensional normal distribution for the same random noise expressed in the time domain. If we transform

Eq. (81) from the variables a and b to the variables $|\Omega|$ and ϕ, where

$$|\Omega|\epsilon^{j\phi} = a - jb \tag{86}$$

we can make use of the equations

$$da\,db = |\Omega|d|\Omega|d\phi \tag{87}$$

and

$$\int_{-\infty}^{+\infty}\int_{-\infty}^{+\infty} P(a,b)da\,db = \int_0^{+\infty}\int_0^{2\pi} \frac{1}{\rho^2\pi}\,\epsilon^{-|\Omega|^2/\rho^2}|\Omega|d|\Omega|d\phi$$

$$= \int_0^{\infty} \frac{2}{\rho^2}\,|\Omega|\epsilon^{-|\Omega|^2/\rho^2}d|\Omega| \tag{88}$$

Thus the probability distribution for $|\Omega|$ is

$$P(|\Omega|)d|\Omega| = \frac{2}{\rho^2}\,|\Omega|\epsilon^{-|\Omega|^2/\rho^2}d|\Omega| \tag{89}$$

The phase ϕ of the spectrum function is, of course, uniformly distributed between 0 and 2π as pointed out in Sec. 3.6 and as is evident from the fact that ϕ does not appear in the integrand of Eq. (88). Thus

$$P(\phi)d\phi = \frac{d\phi}{2\pi} \tag{90}$$

On the basis of the foregoing analysis we will be able to give two alternative but remarkably analogous definitions of random noise, one based on its properties in the time domain, the other based on its properties in the frequency domain.

3.11 Definitions and descriptions of various kinds of noise

The definitions of random noise which we are about to give will refer to random noise limited to bandwidth W and duration T. Later we shall see what happens to random noise when either the bandwidth or the duration becomes infinite. As stated earlier, random noise is a stochastic function, *i.e.*, its value at any point (in particular, any sampling point) is controlled by probability. On the other hand, since its values are controlled by probability, the law of large numbers tells us that if

$$2TW \gg 1$$

the distribution of sampling values in a given typical function of an ergodic ensemble will correspond approximately to the probability distributions. With these facts in mind, we now give the following equivalent definitions based upon the analysis of the foregoing sections.

DEFINITION I:

An ergodic ensemble of functions of time of duration T and bandwidth W (starting at zero frequency),[1] whose values at the sample points in time are independent of one another, but each of which have a normal ensemble distribution of the same mean-square value, will be called a *random-noise ensemble.* Any typical function of this ensemble will be called a *random-noise function. Random noise* is the output of a source which generates an uninterrupted and uncorrelated succession of random-noise functions from the same ensemble.

The fact that the ensemble is ergodic requires that the number of sample points $2TW$ be very large and that the values at the sampling points be determined by a stochastic process and not be determined ahead of time. The definition in addition requires that there be no intersymbol influence and that the values at the sample points shall have a normal distribution.

On the basis of the results derived in Sec. 2.10, we can also define a random-noise ensemble in terms of its properties in the frequency domain as follows:

DEFINITION II:

An ensemble of stochastic functions of frequency of bandwidth W (starting at zero frequency) and duration T, whose values at the sample points in the frequency domain are independent of one another,[2] but each of which have a two-dimensional normal distribution having the same mean-square value for both dimensions

[1] One reason we have required that the bandwidth start at zero frequency for random noise is that we have not shown that the values of $2TW$ sampling points of $G(t)$ specify $G(t)$ except when the bandwidth starts at zero frequency. Furthermore, with this definition, all ensembles of random noise are geometrically similar or at least can be derived from each other by linear stretching.

[2] The fact that there are no constraints between the frequency-sampling values of random noise is a consequence of the fact, to be derived in Chap. IV, that the entropy of random noise comes out the same, whether calculated from its frequency or its time components, when the assumption is made that there are no constraints between the frequency components.

and for all frequencies, will be called a random-noise ensemble if $2TW$ is very large.[1]

Definitions I and II tell us under what circumstances a particular function of time, *i.e.*, a particular signal or a particular oscillogram or the like, may be described as random noise. The requirements are

(a) $2TW \gg 1$.

(b) There is no apparent relationship between the linivas at different sampling points.

(c) The linivas have a normal distribution.

If requirements (b) and (c) hold for either the time or the frequency domain, they hold for the other also.

In Figs. 3-7 and 3-8 are diagrams in the time and frequency domains of random noise of duration T and bandwidth W. If we increase the bandwidth W, for random noise, the general waveshape characteristics remain the same. The effect of increased bandwidth in random noise is like an accordion compression of the noise wave in the time domain. There is an increase in average height and a finer structure, but the complete lack of regularity in the details, which is characteristic of random noise, is still present. If the bandwidth is made infinite, the average power is infinite and the separation of sampling points in the time domain approaches zero. The noise wave oscillates back and forth with infinite excursions. The effect on the representation in the frequency domain of keeping the bandwidth finite and allowing the duration T to become infinite is completely analogous.

This is probably a good place to emphasize once more the fact that, in the case of any particular sample of random noise, the frequency characteristic is not flat in detail but varies back and forth as shown in Fig. 3-8. In equal bandwidths Δf which include a large number of frequency domain sample points, the mean-square value of $|\Omega|$ will be approximately the same. Therefore, as the duration is made longer and longer, so that the frequency-

[1] If the values at the sampling points are independent and each has a two-dimensional normal distribution, the phase distribution will be random. Furthermore there cannot be an appreciable d-c component, since no single frequency component (the d-c component is the zero frequency component) can have an appreciable percentage of the total energy if $2TW \gg 1$ and the distribution is normal.

domain sampling points come closer and closer together, there are more and more sampling points in any bandwidth Δf, and the mean-square value of $|\Omega|$ in Δf will approach a constant value.

In addition to random noise, there are also other special types of noise which are of interest.[1] For example ergodic noise whose frequency components have amplitudes whose ensemble average values are uniformly distributed with respect to frequency in the

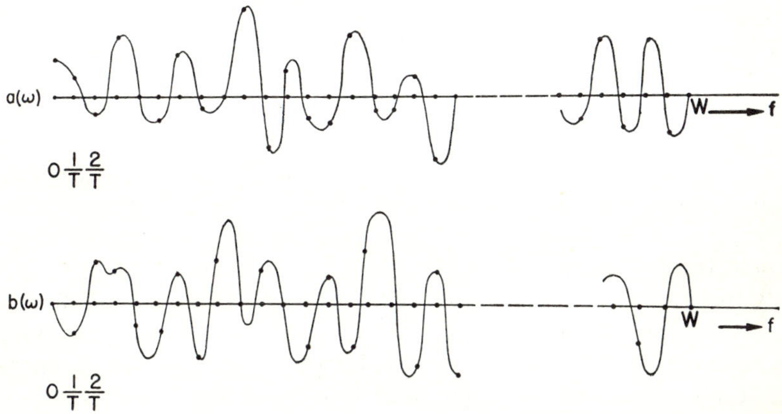

Fig. 3-8. A sample of random noise of bandwidth W and duration T in the frequency domain.

frequency range of interest is called *white noise*. Thus random noise is white noise, but white noise is not necessarily random noise. According to the analysis in Sec. 3.6, if there is no intersymbol influence between the values at the sample points of an ergodic ensemble of functions of time, then the ensemble average of the amplitudes of the spectrum will be uniformly distributed (*i.e.*, will be flat) with respect to frequency. Such an ensemble will therefore have the properties of white noise.

If random noise is sent through a linear selective network, then (1) the output will still be an ergodic ensemble of functions of time having a gaussian distribution of the amplitudes at its frequency sampling points and (2) there will still be no intersymbol influence in the frequency domain. However, the ensemble average of the amplitudes at the sampling points will not be flat with frequency.

[1] Certain noise definitions are also included in the table in Sec. 3.14.

Any noise which has characteristics (1) and (2) above is called *gaussian noise.* Random noise is then gaussian noise which has a flat spectrum.[1] In other words, random noise is white gaussian noise whose flat spectrum starts at zero frequency.

With the aid of the central-limit theorem it can readily be shown that gaussian noise has a gaussian (normal) distribution of values at the sample points in the time domain. However, when the spectrum is not flat, there is intersymbol influence between the values at the sample points in the time domain.[2]

Let us next point out certain general ways in which the types of noise discussed above may arise. First suppose we have the superposition of a large number of impulsive-type functions (each having a flat spectrum in the frequency range of interest). Let the location of these functions be determined by a random process. These functions need not be similar, but no one function can have an appreciable percentage of the total quadratic content. Then, if these functions are sufficiently numerous so that requirement (4) for the central-limit theorem is satisfied at all sample points in the time domain, the result of this superposition will be random noise. Random noise, in practice, almost invariably arises from a superposition process of this type. This is true of shot noise,[3] thermal noise,[4] the background noise of a phonograph record, and numerous other cases.

[1] In *F.A.,M., and N.*, the term "random noise" was used for what we here call "gaussian noise." However, the foregoing information-theory analysis has led to the new definitions used in the present book.

[2] It is of interest to note that, when there is no intersymbol influence between the values at the sample points in time, the central-limit theorem may be applied to show that the amplitude distribution of the frequency components is gaussian (normal). The frequency components for white non-random noise are, as shown below, not independent of one another, so that requirement (1) for the central-limit theorem is not satisfied and the central-limit theorem cannot be applied to the superposition of frequency components in order to get the distributions at the sampling points in time. If the frequency components were independent (unconstrained), the central-limit theorem could be applied to their superposition to show that the distribution of the linivas in time were gaussian (normal). This would be true only if the white noise in question were gaussian noise. Thus we know that there must be constraints between the frequency components of white, non-gaussian, noise.

[3] *F.A.,M., and N.*, Sec. 7.11.

[4] *F.A.,M., and N.*, Sec. 9.6.

If, in the foregoing paragraph, the functions are not necessarily impulsive, but all the functions to be superimposed are of identical size and shape, then the resultant will be gaussian noise. Certain types of wideband shot noise are of this type. As already mentioned, random noise sent through a linear selective transmission system will give rise to gaussian noise.

PROBLEMS

1. Using the central-limit theorem, give a formal proof to show that Definition II of random noise is a consequence of Definition I.

2. Using the central-limit theorem, give a formal proof to show that Definition I of random noise is a consequence of Definition II.

3. Prove that the superposition process described in the second last paragraph of Sec. 3.11 will give rise to random noise.

4. Prove that the superposition process described in the last paragraph of Sec. 3.12 will give rise to gaussian noise.

3.12 Analytical formulas for random noise

With the aid of Eqs. (10) and (21) of Chap. II we can give analytical formulas for random noise. If $I(t)$ is random noise of bandwidth W expressed as a function of time, we can write

$$I(t) = \sum_{n=-\infty}^{+\infty} g_n \frac{\sin (2\pi Wt - n\pi)}{2\pi Wt - n\pi} \tag{91}$$

where g_n is the value of the random noise at the $(+n)$th sampling point. According to our previous discussion in Sec. 3.8, each g_n is then an independent stochastic function having a normal distribution whose mean-square value is independent of n and is equal to

$$\overline{g_n^2} = \sigma_I^2 \tag{92}$$

Equation (91) describes every random-noise function of bandwidth W and mean-square value σ_I^2. The individual values of g_n cannot be predicted ahead of time for any individual signal, but if $2TW$ is large, the distribution of values of g_n will be the same as the probability distribution, namely, a normal distribution of mean square value σ_I^2.

If $\Omega(\omega)$ is random noise of bandwidth W and duration T expressed as a function of ω, we can write

$$\Omega(\omega) = \sum_{n=-TW}^{+TW} (a_n - jb_n) \frac{\sin\left(\frac{\omega T}{2} - \pi n\right)}{\frac{\omega T}{2} - \pi n} \tag{93}$$

where a_n is the value of the real component of the spectrum function at the $+n$th sampling point and $-b_n$ is the value of the imaginary component. Both a_n and b_n are independent stochastic functions having normal distributions whose mean-square value is independent of n and is given by the equation

$$\overline{a_n^2} = \frac{\rho^2}{2} = \overline{b_n^2} \tag{94}$$ [1]

The properties of a_n and b_n are analogous to those of g_n in Eq. (91).

3.13 Superposition laws for random noise

In Sec. 3.7 we showed that the interaction quadratic content vanishes for two incoherent signals which are typical functions of ergodic ensembles. It is also shown in books on probability theory that, if two normal distributions are superimposed, the resultant is another normal distribution. From these facts, we arrive at the following superposition laws for random noise. In these laws we can substitute the word "gaussian" or the word "white" for "random" and the laws will still hold.

I. If two or more independent random-noise signals are linearly superimposed (algebraically added), the resultant is another random-noise signal. If σ^2 is the mean-square value of the resultant, then

$$\sigma^2 = \sigma_1^2 + \sigma_2^2 + \cdots + \sigma_q^2 \tag{95}$$

where σ_1^2, σ_2^2, etc., are the respective mean-square values of the random noise signals which are superimposed. As a corollary we note that the quadratic contents of superimposed independent random-noise signals are additive, *i.e.*,

$$\int I^2 dt = \int I_1^2 dt + \int I_2^2 dt + \cdots + \int I_q^2 dt \tag{96}$$

II. If any other signal, whether random noise or not, is linearly superimposed on a random-noise signal, the quadratic contents are

[1] If ρ is not constant, but changes (slowly) with n, the noise is not random but is gaussian.

additive, *i.e.*,

$$\int I^2 dt = \int I_S^2 dt + I_N^2 dt \tag{97}$$

where I is the resultant signal, and I_S and I_N are the original signal and noise, respectively. We here assume that I_S and I_N are incoherent.

3.14 The classification of undesired signals

Based in part upon the ideas developed in these chapters, we will now attempt to give a systematic classification of undesired signals.

CLASSIFICATION OF UNDESIRED SIGNALS

I. *Noise:* Signals which are not coherent with any signals to which meaning is assigned in any transmission system.[1]
   A. *Ergodic Noise:* Consists of ergodic ensembles of functions. Predictable on the average, but not in detail.
   B. *Non-ergodic Noise*
      1. *Regular:* Shows characteristic waveshapes. Usually partially periodic.
         Examples:
         "Hum." Noise of the power-line frequency and its harmonics.
         "Howl." Noise at an audio frequency due to instability, usually involving an acoustic feedback path.
         "Motor-boating." Regular noise of low frequency caused by instability of the power supply.
         "Ignition noise." Impulse noise from automobile ignition systems.
      2. *Irregular:* Sporadic waveshapes.
         Examples:
         Atmospherics.
         Microphonics due to mechanical vibrations of tube elements.
         Stray transients.
         Effects of unpredictable backlash.

[1] Tube noise and thermal noise in a receiver will cause random-noise amplitude modulation and random-noise angle (phase) modulation of an RF carrier, provided that the carrier is large in comparison with the noise and that it is located at the center of the passband, and provided that the receiver has a uniform transmission characteristic in the passband. Since the usual situation in a radio receiver does not satisfy all these requirements, the actual acoustic output of a receiver in the absence of a desired signal modulation of the carrier is rarely random noise even if it is ergodic noise. Receiver noise output in the absence of a carrier (assuming that the noise is generated in the RF circuits) is, of course, also not random noise. In the absence of hum and cross talk, however, it is usually ergodic noise.

II. *Cross Talk:* Undesired signals which are desired signals in some other transmission system or which are coherent with desired signals of some other transmission system.

Examples:

Cross talk between telephone lines.

Common channel interference in broadcast programs.

Adjacent channel interference in broadcast programs.

Inverted speech ("monkey chatter") from adjacent channels.

III. *Distortion:* Undesired signals which are coherent with the desired signal.

A. Reversible Distortion (This can be eliminated, in theory at least, by sending the signal through a correcting network.)

Examples:

1. Effects of non-flat amplitude vs. frequency characteristics or of non-linear phase vs. frequency characteristics.
2. Effects of unsymmetrical sideband distributions.
3. Effects of non-linearity of circuit elements when these do not involve cutoff or limiting.
4. Effects of predictable backlash and hysteresis.

B. Irreversible Distortion (The undesired component of the signal may be predictable at the input side of the transducer, but is controlled by probability from the point of view of the output side. The probabilities concerned, however, vary with the signal values.)

Examples:

1. Effects of cutoff or limiting in circuit elements (special case of Example 2.)
2. Loss of information due to loss of accuracy in an element in the transmission system (This is also called alphabet deficit, see Sec. 7.8.)

PROBLEMS

In answering the problems below, disregard any d-c (average) component which may be present.

1. A current is made up of the superposition of rectangular pulses of unit height of 1 microsecond duration which occur at random times.

(a) If there are 100,000 pulses/sec and the signal is passed through a system having a bandwidth of 10 megacycles, is the output essentially random noise? If the bandwidth is 1 kc, is the output essentially random noise? Draw a diagram of the output in each case.

(b) Answer the same questions as in (a) when the number of pulses per second is increased to 100,000,000,000.

2. A current is made up of the superposition of rectangular pulses of 1 microsecond duration which occur at equally spaced intervals. The heights of the pulses are distributed at random, and the distribution is normal with a mean-square value of unity. Answer all the same questions as in 1(a) and 1(b).

3. In Prob. 2, suppose that all the pulses are of unit height, instead of having a normal distribution. Now answer the same questions again as in 1(a) and 1(b).

4. A current is made up of the superposition of rectangular pulses of 1 microsecond duration which occur at random times. The heights of the pulses are distributed at random, and the distribution is normal with a mean-square value of "7." If there are 100,000,000 pulses/sec, find the root-mean-square value of current in a bandwidth of 1 kc.

5. A resistor acts as a voltage generator of random-noise voltage of strength

$$\overline{e^2} = 4kTR\,\Delta F$$

with an internal resistance of the generator equal to R. Find the root-mean-square value of noise voltage across R_3 in Fig. 3-D. Is e_n random noise? Is it gaussian?

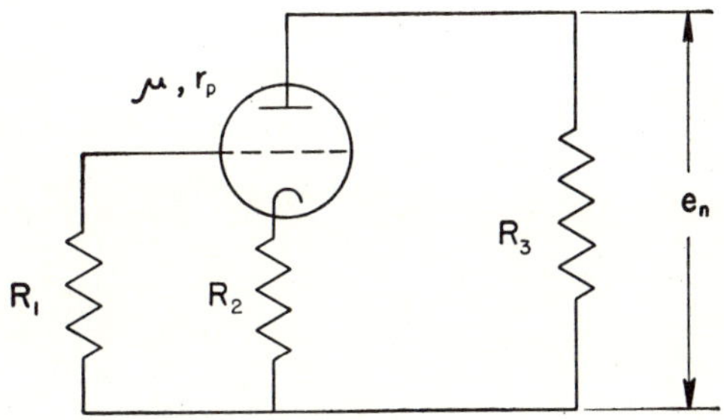

Fig. 3-D.

6. Random noise is sent into an amplifier with a square-law characteristic. Is the output random noise? Why?

7. Random noise is sent into a limiter. Is the output random noise? Why?

8. (a) A large audience is carrying on miscellaneous conversations among its members. Is the output of a microphone placed on the stage approxi-

mately random noise? Why? (Assume that no one on the stage is speaking.) Is it approximately gaussian noise? Why?

(b) A large audience is singing "For He's a Jolly Good Fellow." Is the output of a microphone placed on the stage approximately gaussian noise? (Assume that no one on the stage is singing.) Why?

9. In a stationary stochastic sequence of symbols MCANW----, assume that at any location, the probability values for the next symbol in the sequence are determined by the preceding symbol to the extent of the fraction $a/(a + b)$ and are determined at random by the fraction $b/(a + b)$. Show that the intersymbol influence ultimately falls off exponentially with distance.

CHAPTER IV

THE ENTROPY OF CONTINUOUS DISTRIBUTIONS[1]

4.1 The definition of entropy

It would be natural to try to define the entropy of a continuous distribution as a limiting case of the entropy of a discrete distribution, since the latter has already been defined in Chap. I. Let us see what happens when we do this. In the discrete case, entropy is defined as

$$H = -\sum_i p_i \log p_i \tag{1}$$

where p_i is the probability of obtaining the ith symbol (or ith set of symbols). Accordingly, in the continuous case we could try to define the entropy of the continuous distribution $p(x)$ of the single variable x as

$$H = \lim_{\Delta x_i \to 0} - \sum_i p(x_i)\Delta x_i \log [p(x_i)\Delta x_i] \tag{2}$$

where the range between $-\infty$ and $+\infty$ is divided into small increments Δx_i in the customary manner. Continuing Eq. (2), we can write

$$\begin{aligned} H &= \lim_{\Delta x_i \to 0} - \sum_i p(x_i)[\log p(x_i)]\Delta x_i + \lim_{\Delta x_i \to 0} - \sum_i p(x_i)[\log \Delta x_i]\Delta x_i \\ &= - \int p(x) \log p(x) dx - \lim_{\Delta x_i \to 0} \sum_i p(x_i)[\log \Delta x_i]\Delta x_i \end{aligned} \tag{3}$$

As $\Delta x_i \to 0$, the second term on the right side of Eq. (3) becomes infinite. To get an idea of the rate at which it becomes infinite, let us make all of the Δx_i's equal. Then the term reduces to

$$\begin{aligned} -\lim_{\Delta x_i \to 0} \sum_i p(x_i)[\log \Delta x_i]\Delta x_i \\ &= -\lim_{\Delta x_i \to 0} \log \Delta x_i \sum_i p(x)_i \Delta x_i \\ &= -\lim_{\Delta x_i \to 0} \log \Delta x_i \int p(x) dx = -\lim_{\Delta x_i \to 0} \log \Delta x_i \end{aligned} \tag{4}$$

[1] This chapter is based largely upon material by C. E. Shannon in *Bell System Tech. Jour.*, Vol. 27, October, 1948.

Because of the infinite term in this method of arriving at entropy in the continuous case, it might appear that entropy is not a useful concept for continuous distributions. This happens not to be true. It turns out, in effect, that when entropy is used in information studies in the continuous case the same infinite term invariably occurs twice in an expression, once with a (+) sign and once with a (−) sign, so that the term is canceled out. For the sake of simplicity we shall define the entropy of a continuous distribution directly as

$$H = -\int p(x) \log p(x) dx \qquad (5)^1$$

or, when n variables are involved, as

$$H = -\int \cdots \int p(x_1, \ldots, x_n) \log p(x_1, \ldots, x_n) dx_1 \ldots dx_n \qquad (6)$$

instead of defining entropy by a limiting process from the discrete case. It then remains to be shown whether entropy, with this definition, has analogous properties to those which we found for entropy as defined in the discrete case.

In the case of a band-limited ergodic ensemble of signals, it is useful to define the entropy per degree of freedom[2] as well as the entropy per unit of time. Accordingly, we write

$$\begin{aligned} H_n &= \text{entropy/degree of freedom} \\ &= -\lim_{n\to\infty} \frac{1}{n} \int \cdots \int p(x_1, \ldots, x_n) \log p(x_1, \ldots, x_n) dx_1 \ldots dx_n \end{aligned} \qquad (7)$$

where $x_1, x_2, \ldots, x_n$ are the values of a signal at the sample points. Since $n = 2TW$, it follows that

$$H_t = \text{entropy/unit of time} = 2WH_n \qquad (8)$$

Using the definitions in Eqs. (5) and (6) we will now investigate the properties of the entropy of continuous distributions.

[1] Unless otherwise stated, we will use $\log_\epsilon$ in calculating the entropy of continuous distributions. This will require a conversion factor ($\log_2 \epsilon = 1.443 \ldots$) when entropy is used to calculate binits of information.

[2] The number of independent variables in the signal, *i.e.*, the number of sample points, are here considered the number of degrees of freedom; even though, due to "intersymbol influence," the number of degrees of freedom may, strictly speaking, be less than this value.

4.2 A theorem from the calculus of variations

In the analyses which will follow in Sec. 4.3 we will want to find probability distributions which will give maximum entropies under certain conditions. Specifically, we will want to find the form of $p(x)$ so that the entropy

$$H = -\int p \log p \, dx \tag{9}$$

will be a maximum subject to certain constraints. This is a particular case of the so-called isoperimetric problem of the calculus of variations. In its general form, for our purposes, the problem is to find p as a function of x, so that the integral

$$I = \int_a^b F(x,p)dx \tag{10}$$

will yield a maximum (or minimum) value of I, subject to the conditions that

$$\int_a^b \phi_1(x,p)dx = K_1 \tag{11}$$

$$\int_a^b \phi_2(x,p)dx = K_2 \tag{12}$$

$$\cdots\cdots\cdots$$

$$\cdots\cdots\cdots$$

$$\int_a^b \phi_n(x,p)dx = K_n \tag{13}$$

K_1, K_2, etc., are preassigned constants, and the facts that the integrals on the left sides of Eqs. (11) to (13) must equal these preassigned constants are the constraints operating on the integral in Eq. (10).

According to the calculus of variations, the form of $p(x)$ which will make I a maximum (or a minimum) is obtained by solving the equation

$$\frac{\partial F}{\partial p} + \lambda_1 \frac{\partial \phi_1}{\partial p} + \lambda_2 \frac{\partial \phi_2}{\partial p} + \cdots + \lambda_n \frac{\partial \phi_n}{\partial p} = 0 \tag{14}$$

The quantities λ_1, λ_2, etc., are adjustable constants (called undetermined multipliers) whose values are then determined by substituting the value found for $p(x)$ from Eq. (14) into Eqs. (11) through (13).

The method will become clear in Sec. 4.3 when we solve some examples.

4.3 Some derivations of distributions having maximum entropy for given constraints

(a) *The Case of Limited Peak Quadiva.* Let us first consider the case in which the peak quadiva of an ensemble of values of x^2 is limited to the peak value S. Then the range of possible values of x lies between $-\sqrt{S}$ and $+\sqrt{S}$. Equations (10) through (13) may now be rewritten for this problem as

$$H = -\int_{-\sqrt{S}}^{+\sqrt{S}} p \log p \, dx \tag{15}$$

subject to the normalizing condition on the probability that

$$\int_{-\sqrt{S}}^{+\sqrt{S}} p \, dx = 1 \tag{16}$$

Thus

$$F(x,p) = -p \log p \tag{17}$$

$$\frac{\partial F}{\partial p} = -(1 + \log p) \tag{18}$$

$$\phi_1(x,p) = p \tag{19}$$

$$\frac{\partial \phi_1}{\partial p} = 1 \tag{20}$$

Substituting these values into Eq. (14), we obtain

$$-(1 + \log p) + \lambda_1 = 0 \tag{21}$$

or

$$p = \epsilon^{\lambda_1 - 1} \tag{22}$$

as the form of p for maximum H. Substituting this value of p into Eq. (16) to eliminate λ_1, we obtain

$$\int_{-\sqrt{S}}^{+\sqrt{S}} \epsilon^{\lambda_1 - 1} \, dx = \epsilon^{\lambda_1 - 1} \int_{-\sqrt{S}}^{+\sqrt{S}} dx = \epsilon^{\lambda_1 - 1} \cdot 2\sqrt{S} = 1 \tag{23}$$

Thus

$$\epsilon^{\lambda_1 - 1} = \frac{1}{2\sqrt{S}} \tag{24}$$

Substituting into Eq. (22), we obtain

$$p = \frac{1}{2\sqrt{S}} \tag{25}$$

We see that p has the constant value $1/2\sqrt{S}$ for maximum entropy in this case. The entropy itself is

$$H = -\int_{-\sqrt{S}}^{+\sqrt{S}} p \log p \, dx = -\frac{1}{2\sqrt{S}} \cdot \log \frac{1}{2\sqrt{S}} \cdot 2\sqrt{S}$$

$$= \log 2\sqrt{S} = \tfrac{1}{2} \log 4S \tag{26}$$

in this case of maximum entropy.

(b) *The Case of a Given Average Quadiva* (*Given Standard Deviation*). Let us next consider the case in which

$$\int_{-\infty}^{+\infty} x^2 p \, dx = \sigma^2 \tag{27}$$

where σ^2 is a given constant. We now wish to find the form of the probability distribution $p = p(x)$ which will give maximum entropy in this case.

Equation (10) now assumes the form

$$H = -\int_{-\infty}^{+\infty} p \log p \, dx \tag{28}$$

The constraints are expressed by the normalizing condition

$$\int_{-\infty}^{+\infty} p \, dx = 1 \tag{29}$$

and by Eq. (27). Thus

$$F(x,p) = -p \log p \tag{30}$$

$$\frac{\partial F}{\partial p} = -(1 + \log p) \tag{31}$$

$$\phi_1(x,p) = p \tag{32}$$

$$\frac{\partial \phi_1}{\partial p} = 1 \tag{33}$$

$$\phi_2(x,p) = x^2 p \tag{34}$$

$$\frac{\partial \phi_2}{\partial p} = x^2 \tag{35}$$

Substituting these values into Eq. (14), we obtain

$$-(1 + \log p) + \lambda_1 + \lambda_2 x^2 = 0 \tag{36}$$

$$p = \epsilon^{\lambda_2 x^2 + \lambda_1 - 1} = \epsilon^{\lambda_1 - 1} \cdot \epsilon^{\lambda_2 x^2} \tag{37}$$

In order to eliminate λ_1 and λ_2, we first substitute into Eq. (29). This gives

$$1 = \int_{-\infty}^{+\infty} p\,dx = \int_{-\infty}^{+\infty} \epsilon^{\lambda_1 - 1}\epsilon^{\lambda_2 x^2}\,dx = 2\epsilon^{\lambda_1 - 1}\int_0^\infty \epsilon^{\lambda_2 x^2} dx$$

$$= 2\epsilon^{\lambda_1 - 1} \cdot \frac{1}{2}\sqrt{\frac{\pi}{-\lambda_2}} \qquad (38)^1$$

or

$$\epsilon^{\lambda_1 - 1} = \sqrt{\frac{-\lambda_2}{\pi}} \qquad (39)$$

We note that λ_2 must be a negative quantity, for otherwise the integral in Eq. (38) would not converge.

We next substitute Eqs. (37) and (39) into Eq. (27). Thus

$$\sigma^2 = \int_{-\infty}^{+\infty} x^2 p\,dx = \int_{-\infty}^{+\infty} x^2 \sqrt{\frac{-\lambda_2}{\pi}}\,\epsilon^{\lambda_2 x^2} dx$$

$$= 2\sqrt{\frac{-\lambda_2}{\pi}} \int_0^\infty x^2 \epsilon^{\lambda_2 x^2}\,dx$$

$$= 2\sqrt{\frac{-\lambda_2}{\pi}} \cdot \frac{1}{4}\sqrt{\frac{\pi}{(-\lambda_2)^3}} = -\frac{1}{2\lambda_2} \qquad (40)^2$$

Thus

$$\lambda_2 = -\frac{1}{2\sigma^2} \qquad (41)$$

and

$$\epsilon^{\lambda_1 - 1} = \sqrt{\frac{-\lambda_2}{\pi}} = \sqrt{\frac{1}{2\pi\sigma^2}} \qquad (42)$$

Substituting into Eq. (37) we have finally

$$p = \epsilon^{\lambda_1 - 1}\epsilon^{\lambda_2 x^2} = \frac{1}{\sqrt{2\pi\sigma^2}}\,\epsilon^{-x^2/2\sigma^2} \qquad (43)$$

as the form of p which gives maximum entropy. Equation (43) will be recognized as a gaussian distribution. A gaussian distribution thus has maximum entropy for a given average power. To find H itself, we have from Eq. (43)

$$\log p = -\log\sqrt{2\pi\sigma^2} - \frac{x^2}{2\sigma^2} \qquad (44)$$

[1] The definite integral in Eq. (38) is a well-known form (No. 38, *F.A.,M., and N.*), p. 411. It is evaluated in Stanford Goldman, *Transformation Calculus and Electrical Transients*, Prentice-Hall, New York (1949), pp. 251–252.

[2] This integral is form No. 39, *F.A.,M., and N.*, p. 411.

Thus

$$H = -\int_{-\infty}^{+\infty} p \log p \, dx = \int_{-\infty}^{+\infty} p \log \sqrt{2\pi\sigma^2} \, dx + \int_{-\infty}^{+\infty} \frac{px^2 dx}{2\sigma^2}$$

$$= \log \sqrt{2\pi\sigma^2} + \tfrac{1}{2} \tag{45}$$

by direct application of Eqs. (29) and (27). Equation (45) may also be expressed as

$$H = \log \sqrt{2\pi\sigma^2} + \log \sqrt{\epsilon} = \log \sqrt{2\pi\epsilon\sigma^2} \tag{46}$$

This is an important equation, since we will often have occasion to use the entropy of a gaussian distribution.

Exercise

Show that, for a given entropy, the gaussian distribution

$$p(x) = \frac{1}{\sqrt{2\pi\sigma^2}} \epsilon^{-x^2/2\sigma^2}$$

has the smallest variance (mean-square deviation from the average value) of any one-dimensional probability distribution.

(c) *The Case of Given Average Positive x.* Let x be limited to positive values, and let the average value of x be a given constant, say A. The constraints are thus

$$A = \int_0^\infty xp \, dx \tag{47}$$

and

$$1 = \int_0^\infty p \, dx \tag{48}$$

and the equation for H is

$$H = -\int_0^\infty p \log p \, dx \tag{49}$$

To find the distribution function $p = p(x)$, which gives maximum H, we have

$$F(x,p) = p \log p \tag{50}$$

$$\frac{\partial F}{\partial p} = -(1 + \log p) \tag{51}$$

$$\phi_1(x,p) = p \tag{52}$$

$$\frac{\partial \phi_1}{\partial p} = 1 \tag{53}$$

$$\phi_2(x,p) = xp \tag{54}$$

$$\frac{\partial \phi_2}{\partial p} = x \tag{55}$$

Substituting these values into Eq. (14), we have

$$-(1 + \log p) + \lambda_1 + \lambda_2 x = 0 \tag{56}$$

or

$$p = \epsilon^{\lambda_2 x + \lambda_1 - 1} = \epsilon^{\lambda_1 - 1} \cdot \epsilon^{\lambda_2 x} \tag{57}$$

Proceeding as in the earlier cases to eliminate λ_1 and λ_2, we first substitute Eq. (57) into (48). Thus

$$1 = \int_0^\infty \epsilon^{\lambda_1 - 1}\epsilon^{\lambda_2 x} dx = \left.\frac{\epsilon^{\lambda_1 - 1}\epsilon^{\lambda_2 x}}{\lambda_2}\right|_0^\infty = -\frac{\epsilon^{\lambda_1 - 1}}{\lambda_2} \tag{58}$$

This evaluation has required that λ_2 be negative in order for the integral to converge. From Eq. (58) we have

$$\epsilon^{\lambda_1 - 1} = -\lambda_2 \tag{59}$$

Next substituting Eqs. (57) and (59) into (47), we obtain

$$A = \int_0^\infty -x\lambda_2\epsilon^{\lambda_2 x} dx = -\left.\frac{\epsilon^{\lambda_2 x}}{\lambda_2}(\lambda_2 x - 1)\right|_0^\infty = -\frac{1}{\lambda_2} \tag{60}$$

Consequently

$$\lambda_2 = -\frac{1}{A} \tag{61}$$

and

$$\epsilon^{\lambda_1 - 1} = -\lambda_2 = \frac{1}{A} \tag{62}$$

Thus

$$p = \frac{1}{A}\epsilon^{-x/A} \quad \text{for } x \geq 0 \tag{63}$$

Finally

$$\begin{aligned} H &= -\int_0^\infty p \log p \, dx = -\int_0^\infty p\left[-\log A - \frac{x}{A}\right] dx \\ &= \log A \int_0^\infty p \, dx + \frac{1}{A}\int_0^\infty xp \, dx \\ &= \log A + 1 = \log(\epsilon A) \end{aligned} \tag{64}$$

PROBLEM

1. An ensemble of signals going through an amplifier is peak limited to the liniva

$$x = b$$

and is bottom limited to the liniva

$$x = a$$

In other words, all signal values in the ensemble must be between a and b. What is the maximum entropy per degree of freedom of the ensemble?

If the amplifier has a bandwidth W what is the entropy per unit of time of the ensemble?

4.4 Transformations of coordinates and Jacobians

(a) *Jacobians.* In the work which is to follow, we shall frequently have occasion to transform multiple integrals from one set of coordinates to another. We therefore wish to review for the reader some of the important features of these transformations. Suppose we have an integral

$$\iiint f(x,y,z)dx\,dy\,dz \tag{65}$$

and we want to transform it to a new set of coordinates u, v, w, where x, y, and z are each functions of u, v, and w. Then, we may write

$$f(x,y,z) = F(u,v,w) \tag{66}$$

At the same time, the differential of volume

$$dx\,dy\,dz \tag{67}$$

transforms into the new differential of volume

$$\left| J\left(\frac{x,y,z}{u,v,w}\right) \right| du\,dv\,dw \tag{68}$$

where

$$J\left(\frac{x,y,z}{u,v,w}\right) = \begin{vmatrix} \dfrac{\partial x}{\partial u} & \dfrac{\partial x}{\partial v} & \dfrac{\partial x}{\partial w} \\ \dfrac{\partial y}{\partial u} & \dfrac{\partial y}{\partial v} & \dfrac{\partial y}{\partial w} \\ \dfrac{\partial z}{\partial u} & \dfrac{\partial z}{\partial v} & \dfrac{\partial z}{\partial w} \end{vmatrix} \qquad (69)[1]$$

$J\left(\frac{x,y,z}{u,v,w}\right)$ is called the Jacobian of x, y, and z with respect to u, v, and w. The Jacobian is thus a scale factor of the transformation. The properties of Jacobians are defined and discussed in books on advanced calculus, and we will not go into these matters in detail

[1] The vertical bars in expression (68) are absolute-value signs, while in Eq. (69) they are the signs of a determinant.

here. We will, however, recall for reference the important formula

$$\left| J\left(\frac{x,y,z}{u,v,w}\right) \right| = \frac{1}{\left| J\left(\frac{u,v,w}{x,y,z}\right) \right|} \tag{70}$$

In accordance with Eqs. (66), (67), and (68), Eq. (65) transforms as

$$\iiint f(x,y,z)dx\,dy\,dz = \iiint F(u,v,w) \left| J\left(\frac{x,y,z}{u,v,w}\right) \right| du\,dv\,dw \tag{71}$$

While we have illustrated Jacobians in terms of three variables, all of the above discussion applies just as well to any number of variables.

(b) *Illustrative Example.* By way of illustration, let us find the Jacobian for the transformation from rectangular to polar coordinates. Here

$$x = r \cos \theta \tag{72}$$

$$y = r \sin \theta \tag{73}$$

Thus

$$\frac{\partial x}{\partial r} = \cos \theta \tag{74}$$

$$\frac{\partial y}{\partial r} = \sin \theta \tag{75}$$

$$\frac{\partial x}{\partial \theta} = -r \sin \theta \tag{76}$$

$$\frac{\partial y}{\partial \theta} = r \cos \theta \tag{77}$$

Consequently, the Jacobian of the transformation is

$$J\left(\frac{x,y}{r,\theta}\right) = \begin{vmatrix} \frac{\partial x}{\partial r} & \frac{\partial x}{\partial \theta} \\ \frac{\partial y}{\partial r} & \frac{\partial y}{\partial \theta} \end{vmatrix} = \begin{vmatrix} \cos \theta & -r \sin \theta \\ \sin \theta & r \cos \theta \end{vmatrix}$$

$$= r(\cos^2 \theta + \sin^2 \theta) = r \tag{78}$$

Therefore, if

$$f(x,y) = F(r,\theta) \tag{79}$$

$$\iint f(x,y)dx\,dy = \iint F(r,\theta)r\,dr\,d\theta \tag{80}$$

This result is, no doubt, well known to the reader.

In order to illustrate Eq. (70) let us now calculate $J\left(\frac{r,\theta}{x,y}\right)$. We start with

$$r = \sqrt{x^2 + y^2} \tag{81}$$

$$\theta = \tan^{-1}\left(\frac{y}{x}\right) \tag{82}$$

Thus

$$\frac{\partial r}{\partial x} = \frac{1}{2}\frac{2x}{\sqrt{(x^2 + y^2)}} = \frac{x}{\sqrt{(x^2 + y^2)}} \tag{83}$$

$$\frac{\partial r}{\partial y} = \frac{1}{2}\frac{2y}{\sqrt{(x^2 + y^2)}} = \frac{y}{\sqrt{(x^2 + y^2)}} \tag{84}$$

$$\frac{\partial \theta}{\partial x} = \frac{1}{1 + \frac{y^2}{x^2}}\left(\frac{-y}{x^2}\right) = \frac{-y}{x^2 + y^2} \tag{85}$$

and

$$\frac{\partial \theta}{\partial y} = \frac{1}{1 + \frac{y^2}{x^2}}\left(\frac{1}{x}\right) = \frac{x}{x^2 + y^2} \tag{86}$$

Therefore,

$$J\left(\frac{r,\theta}{x,y}\right) = \begin{vmatrix} \frac{x}{\sqrt{(x^2 + y^2)}} & \frac{y}{\sqrt{(x^2 + y^2)}} \\ \frac{-y}{x^2 + y^2} & \frac{x}{x^2 + y^2} \end{vmatrix} = \frac{x^2 + y^2}{(x^2 + y^2)^{3/2}}$$

$$= \frac{1}{\sqrt{(x^2 + y^2)}} = \frac{1}{r} \tag{87}$$

Consequently

$$\left| J\left(\frac{r,\theta}{x,y}\right) \right| = \frac{1}{\left| J\left(\frac{x,y}{r,\theta}\right) \right|} \tag{88}$$

in agreement with Eq. (70).

(c) *The Transformation of Probability Functions.* According to our definitions, a probability distribution function, say, $p(x_1,x_2, \ldots ,x_n)$, is always normalized so that

$$\int \cdots \int p(x_1,x_2, \ldots ,x_n)dx_1dx_2 \ldots dx_n = 1 \tag{89}$$

where the integration is taken over all possible values of $x_1,x_2, \ldots,$ and x_n. If now the coordinates are transformed from $x_1,x_2, \ldots ,x_n$ to $y_1,y_2, \ldots ,y_n$, let us find out what happens to the probability distribution function. Let

$$q(y_1,y_2 \ldots ,y_n)$$

be the distribution function in the new coordinates. Then according to the normalization requirement

$$\int \ldots \int q(y_1,y_2, \ldots ,y_n)dy_1dy_2 \ldots dy_n = 1 \tag{90}$$

As a consequence of Eqs. (68), (89), and (90)

$$\begin{aligned}\int \ldots \int p(x_1, \ldots ,x_n)dx_1 \ldots dx_n \\ = \int \ldots \int p(x_1, \ldots ,x_n) \left| J\left(\frac{x_1, \ldots ,x_n}{y_1, \ldots ,y_n}\right)\right| dy_1 \ldots dy_n \\ = \int \ldots \int q(y_1, \ldots ,y_n)dy_1 \ldots dy_n\end{aligned} \tag{91}$$

It follows that

$$q(y_1, \ldots ,y_n) = \left| J\left(\frac{x_1, \ldots ,x_n}{y_1, \ldots ,y_n}\right)\right| p(x_1, \ldots ,x_n) \tag{92}$$

Equation (92) gives the formula for transforming a probability function from one system of coordinates to another.

For example, suppose that

$$p(x) = \frac{1}{\sqrt{2\pi V^2}} \epsilon^{-x^2/2V^2} \tag{92.1}$$

is the probability distribution of the output signal (in volts) of an amplifier. If we wish to express the probability distribution in millivolts, we can then write

$$q(y) = \text{probability distribution in millivolts} = J\left(\frac{x}{y}\right) p(x) \tag{92.2}$$

In this case, y is the potential in millivolts while x is the potential in volts. Thus

$$y = 1000x \tag{92.3}$$

so that

$$J\left(\frac{x}{y}\right) = \frac{dx}{dy} = \frac{1}{1000} \tag{92.4}$$

Therefore

$$\begin{aligned} q(y) &= \frac{1}{1000\sqrt{2\pi V^2}} \epsilon^{-x^2/2V^2} = \frac{1}{1000\sqrt{2\pi V^2}} \epsilon^{-y^2/2(1000)^2V^2} \\ &= \frac{1}{\sqrt{2\pi (V')^2}} \epsilon^{-y^2/2(V')^2}\end{aligned} \tag{92.5}$$

where

$$V' = 1000V = \text{root-mean-square potential in millivolts} \tag{92.6}$$

4.5 Variation of entropy with the coordinate system

We have found in studying discrete information that entropy is a measure of the randomness of a chance variable. In the discrete case, the probability of an "event" has a definite numerical value, so that the entropy of a distribution has just one definite value. In the continuous case, on the other hand, entropy is measured in terms of a probability function. This function, as shown in Eq. (92), changes form if the coordinate system is changed. Consequently, we must not be surprised to find in the continuous case that entropy changes with the coordinate system. Another reason to expect this is the elimination of the last term in Eq. (3) in the definition of the entropy of a continuous distribution. Accordingly, let us now see how the entropy of a distribution changes, as we change from the coordinates $x_1, x_2, \ldots, x_n$ to new coordinates $y_1, y_2, \ldots, y_n$. We shall use the abbreviated notation x for $x_1, x_2, \ldots, x_n$ and y for $y_1, y_2, \ldots, y_n$. Now

$$H(x) = -\int p(x) \log p(x) dx \tag{93}$$

while

$$H(y) = -\int q(y) \log q(y) dy \tag{94}$$

If we substitute Eqs. (68), (70), and (92) into (94), we obtain

$$\begin{aligned} H(y) &= -\int J\left(\frac{x}{y}\right) p(x) \cdot \log \left[J\left(\frac{x}{y}\right) \cdot p(x)\right] J\left(\frac{y}{x}\right) dx \\ &= -\int J\left(\frac{x}{y}\right) p(x) \log \left[J\left(\frac{x}{y}\right)\right] J\left(\frac{y}{x}\right) dx \\ &\qquad - \int J\left(\frac{x}{y}\right) p(x) \log p(x) J\left(\frac{y}{x}\right) dx \\ &= -\int p(x) \log \left[J\left(\frac{x}{y}\right)\right] dx + H(x) \end{aligned} \qquad (95)^1$$

Thus the new entropy is the old entropy minus the expectation (*i.e.*, average value) of the logarithm of the above Jacobian.

In this connection we give the following quotation from Shannon:[2]

[1] When Jacobians appear in any of the equations for entropy in this chapter, the absolute value of the Jacobian should always be used.

[2] C. E. Shannon, "A Mathematical Theory of Communication," *Bell System Tech. Jour.*, vol. 27, October, 1948, p. 631.

In the continuous case the entropy can be considered a measure of randomness *relative to an assumed standard*, namely, the coordinate system chosen with each small volume element $dx_1 \ldots dx_n$ given equal weight. When we change the coordinate system the entropy in the new system measures the randomness when equal volume elements $dy_1 \ldots dy_n$ in the new system are given equal weight.

In spite of this dependence on the coordinate system the entropy concept is as important in the continuous case as the discrete case. This is due to the fact that the derived concepts of information rate and channel capacity depend on the *difference* of two entropies and this difference *does not* depend on the coordinate frame, each of the two terms being changed by the same amount.

The entropy of a continuous distribution can be negative. The scale of measurements sets an arbitrary zero corresponding to a uniform distribution over a unit volume. A distribution which is more confined than this has less entropy and will be negative. The rates and capacities will, however, always be non-negative.

4.6 The effect of a linear transformation on entropy

A case of particular interest in the transformation of coordinates is that of the linear transformations

$$\begin{aligned} y_1 &= a_{11}x_1 + a_{12}x_2 + \ldots + a_{1n}x_n \\ y_2 &= a_{21}x_1 + a_{22}x_2 + \ldots + a_{2n}x_n \\ &\vdots \\ y_n &= a_{n1}x_1 + a_{n2}x_2 + \ldots + a_{nn}x_n \end{aligned} \tag{96}$$

In this case the Jacobians $J\left(\frac{y}{x}\right)$ and $J\left(\frac{x}{y}\right)$ are readily calculated.

$$J\left(\frac{y}{x}\right) = \begin{vmatrix} \frac{\partial y_1}{\partial x_1} & \frac{\partial y_1}{\partial x_2} & \cdots & \frac{\partial y_1}{\partial x_n} \\ \frac{\partial y_2}{\partial x_1} & \frac{\partial y_2}{\partial x_2} & \cdots & \frac{\partial y_2}{\partial x_n} \\ \vdots & \vdots & & \vdots \\ \frac{\partial y_n}{\partial x_1} & \frac{\partial y_n}{\partial x_2} & \cdots & \frac{\partial y_n}{\partial x_n} \end{vmatrix} = \begin{vmatrix} a_{11} & a_{12} & \cdots & a_{1n} \\ a_{21} & a_{22} & \cdots & a_{2n} \\ \vdots & \vdots & & \vdots \\ a_{n1} & a_{n2} & \cdots & a_{nn} \end{vmatrix} = \frac{1}{J\left(\frac{x}{y}\right)} \tag{97}$$

Using the abbreviated notation $|a_{ij}|$ for the determinant of the coefficients, we have

$$J\left(\frac{x}{y}\right) = |a_{ij}|^{-1} \tag{98}$$

Therefore, from Eq. (95)

$$H(y) = H(x) - \int p(x) \log \left[J\left(\frac{x}{y}\right)\right] dx = H(x) + \log |a_{ij}| \tag{99}$$

Finally, we note that in the case of any measure preserving transformation, such as a rotation of coordinates or a shift of origin, $J = 1$ and $H(y) = H(x)$.

It is shown in Appendix VI that the transformation between the time and frequency sampling values can be made a linear measure preserving transformation. Under these circumstances it follows that the entropy of an ensemble has the same numerical value when expressed in terms of its frequency sampling values as when expressed in terms of its time sampling values.

4.7 Random noise and entropy power

For any given values of W and T, we now wish to show that random noise has the maximum entropy for a given average power. We have already shown in Sec. 4.3 that the distribution which has maximum entropy for a given average quadiva at a sampling point is a gaussian distribution.

In Appendix VIII it is shown that

$$H(x,y) \leq H(x) + H(y) \tag{100}$$

the equality sign applying only when x and y are independent. This derivation could be extended to show that

$$H(x,y,z) \leq H(x) + H(y) + H(z) \tag{100a}$$

and could then be extended to four or more variables. Consequently, *for given probability distributions at the individual sampling points, the entropy of an ensemble of signals each having a duration of n sampling points will be a maximum if there are no constraints between the signal values at different sampling points.*[1] Now if there are no

[1] This means that there is the equivalent of *no intersymbol influence.*

constraints between the signal values at the different sampling points in the time domain and if the probability distributions at all the sampling points are the same and are gaussian, then the signal has the characteristics of random noise. It follows that the ensemble of signals which has maximum entropy for a given average power and for a given W and T must have the characteristics of random noise.

Since random noise has no constraints between the signal values at the sampling points, it follows from Appendix VIII that the entropy of a random-noise signal having n sampling points is equal to the sum of the entropies at the individual sampling points. According to Secs. 3.8 and 4.3 the entropy at each sampling point is

$$h = \log \sqrt{2\pi\epsilon\sigma_I^2} \tag{101}$$

since the distribution is gaussian. We will use the notation

$$N = \sigma_I^2 \tag{101a}$$

for random noise where N is thus the average quadiva of the noise. Then Eq. (101) becomes

$$h = \log \sqrt{2\pi\epsilon N} \tag{102}$$

In the case of random noise, since the entropies at the sampling points are additive, we may therefore also write

$$\text{entropy/degree of freedom} = H_n = \log \sqrt{2\pi\epsilon N} \tag{103}$$

and

$$\text{total entropy} = H = 2TW \log \sqrt{2\pi\epsilon N} = TW \log 2\pi\epsilon N \tag{103a}$$

For the entropy per second, we have

$$H_t = 2WH_n = W \log 2\pi\epsilon N \tag{104}$$

With the aid of Eqs. (102) and (104) we will define what is called the *entropy power* of an ensemble. The entropy power ($\bar{N}$) of an ensemble is the average quadiva in a random-noise signal limited to the same time and bandwidth as the original ensemble and having the same entropy. Thus, if H_n is the entropy per degree of freedom of an ensemble, its entropy power $\bar{N}$ is

$$\bar{N} = \frac{1}{2\pi\epsilon}\,\epsilon^{2H_n} = \frac{1}{2\pi\epsilon}\,\epsilon^{H_t/W} \tag{105}$$

Since random noise has the maximum entropy of any ensemble for a given average power, the entropy power of any (signal or noise) ensemble is less than or equal to its actual average power.

4.8 Alternative calculation of the entropy of random noise

The entropy of random noise calculated in the foregoing section was based on our original Definition I of random noise which stated that random noise had no intersymbol influence between the sampling values in the time domain. We are about to show on the basis of entropy calculations that random noise also has no intersymbol influence between the frequency sampling values.

When there are no intersymbol influences, the entropies of different distributions are additive. We will now calculate the entropy of random noise by adding the entropies of the distributions of its frequency sampling values. If this gives the correct value for the total entropy, the calculation will have shown that random noise has, in fact, no intersymbol influence between its frequency sampling values. As a preliminary to our calculation we note that, according to Sec. 4.6, if the transformation between time and frequency coordinates is made measure preserving, then the correct value of entropy calculated from the frequency domain is numerically equal to the entropy calculated from the time domain.

In order to have a measure preserving transformation between frequency and time, we will, in accordance with Appendix VI, use

$$a' = 2\sqrt{\frac{W}{T}}\,a \tag{106}$$

and

$$b' = 2\sqrt{\frac{W}{T}}\,b \tag{107}$$

as our coordinates for the linivas in the frequency domain. According to Sec. 3.10 the distributions of a and b are

$$P(a) = \frac{1}{\rho\sqrt{\pi}}\,\epsilon^{-a^2/\rho^2} \tag{108}$$

and

$$P(b) = \frac{1}{\rho\sqrt{\pi}}\,\epsilon^{-b^2/\rho^2} \tag{109}$$

where

$$\rho^2 = \frac{T}{2W}\,\sigma_I^2 \tag{110}$$

Then according to Sec. 4.5, we have for the distributions in a' and b'

$$P(a') = \left| J\left(\frac{a}{a'}\right) \right| P(a)$$

$$= \frac{1}{2}\sqrt{\frac{T}{W}} \frac{1}{\rho\sqrt{\pi}} \epsilon^{-a^2/\rho^2}$$

$$= \frac{1}{\sigma_I\sqrt{2\pi}} \epsilon^{-a'^2/2\sigma_I^2} \tag{111}$$

and

$$P(b') = \frac{1}{\sigma_I\sqrt{2\pi}} \epsilon^{-b'^2/2\sigma_I^2} \tag{112}$$

The entropy of each distribution in a' and in b' is, thus, according to Sec. 4.3(b)

$$H_n(a') = \log\sqrt{2\pi\epsilon\sigma_I^2} = H_n(b') \tag{113}$$

This is precisely the same value as found in the preceding section for the entropy per degree of freedom in the time domain. In accordance with our foregoing argument, this equality of entropies establishes that there is no intersymbol influence between the sampling values in the frequency domain, since the number of degrees of freedom is the same in frequency as in time.

Exercise

Using the probability distributions of $|\Omega|$, ϕ, a and b of Sec. 3.10 show that

$$H(|\Omega|) + H(\phi) - \iint p(|\Omega|,\phi) \log \left| J\left(\frac{|\Omega|,\phi}{a,b}\right) \right| d|\Omega| d\phi = H(a) + H(b)$$

What is the significance of this result?

4.9 Entropy loss in linear filters

In Chap. II, we showed that the information in a signal could be expressed equally well in terms of the signal values at the sampling points in the time or the frequency domain. Then in Sec. 4.6, we showed that the value of the entropy was the same whether expressed in terms of time or frequency components. We will find that the expression of signals in the frequency domain greatly simplifies the calculation of the effect of filters on the transmission of information. As a background for later use in this connection we will now derive the following theorem:

THEOREM: *If an ensemble having an entropy H_n per degree of freedom in band W is passed through a linear filter with a transmission characteristic $Y(f)$, the output ensemble has an entropy H'_n per degree of freedom, where*

$$H'_n = H_n + \frac{1}{2W} \int_0^W \log |Y(f)|^2 df \tag{114}$$

Let

$$x_1, y_1, x_2, y_2, \ldots, x_{TW}, y_{TW}$$

be the values of an input signal at the sampling points in the frequency domain. The x's stand for in-phase or real components, while the y's are quadrature or imaginary components. Let the values of the output signal be

$$x'_1, y'_1, \ldots, x'_{TW}, y'_{TW}$$

Let us also write

$$Y(f) = G(f) + jS(f) \tag{115}$$

Then

$$\begin{aligned} x'_1 &= G(f_1)x_1 - S(f_1)y_1 \\ y'_1 &= S(f_1)x_1 + G(f_1)y_1 \\ &\cdots\cdots\cdots\cdots \\ x'_{TW} &= G(f_{TW})x_{TW} - S(f_{TW})y_{TW} \\ y'_{TW} &= S(f_{TW})x_{TW} + G(f_{TW})y_{TW} \end{aligned} \tag{116}$$[1]

The Eqs. (116) express a linear transformation from the unprimed to the primed coordinates. In accordance with Eq. (99) we may therefore write

$$H' = H + \log \begin{vmatrix} G(f_1) & -S(f_1) & 0 & 0 & \cdots \\ S(f_1) & G(f_1) & 0 & 0 & \cdots \\ 0 & 0 & G(f_2) & -S(f_2) & \cdots \\ 0 & 0 & S(f_2) & G(f_2) & \cdots \\ \cdot & \cdot & \cdot & \cdot & \cdots \\ \cdot & \cdot & \cdot & \cdot & \cdots \\ \cdot & \cdot & \cdot & \cdot & \cdots \end{vmatrix}$$

$$= H + \log \prod_{i=1}^{TW} |Y(f_i)|^2 = H + \sum_{i=1}^{TW} \log |Y(f_i)|^2 \tag{117}$$

If

$$TW \gg 1$$

[1] These follow from the equation

$$(x + jy)(G + jS) = (Gx - Sy) + j(Gy + Sx)$$

the values of f_i are closely and equally spaced throughout the band W. Then

$$\sum_{i=1}^{TW} \log |Y(f_i)|^2 = T \sum_{i=1}^{TW} \log |Y(f_i)|^2 \Delta f$$
$$= T \int_0^W \log |Y(f)|^2 df \quad \text{(approximately)} \tag{118}$$

where
$$\Delta f = \frac{1}{T} \tag{119}$$

according to Chap. II.

Thus
$$H' = H + T \int_0^W \log |Y(f)|^2 df \tag{120}$$[1]

and
$$H'_n = \frac{H'}{2TW} = H_n + \frac{1}{2W} \int_0^W \log |Y(f)|^2 df$$
$$= H_n + \frac{1}{W} \int_0^W \log |Y(f)| df \tag{121}$$

The term
$$\frac{1}{W} \int_0^W \log |Y(f)| df$$

which represents the difference in entropy is a property of the filter alone and does not depend upon the ensemble transmitted through it. Equation (121) is used in Sec. 5.3 in calculating a lower limit for the language transmission capacity of a system which has a peak power limitation.

Equation (121) may also be expressed in terms of the loss of entropy power. Thus if $\bar{N}$ is the entropy power of the ensemble entering the filter and $\bar{N}'$ is the entropy power of the output, then by Eqs. (105) and (121)

$$\bar{N}' = \frac{1}{2\pi\epsilon} \cdot \epsilon^{2H_n'} = \frac{1}{2\pi\epsilon} \epsilon^{2H_n} \cdot \epsilon^{\frac{1}{W}\int_0^W \log|Y(f)|^2 df} = \bar{N} \epsilon^{\frac{1}{W}\int_0^W \log|Y(f)|^2 df} \tag{122}$$

This loss in entropy power may conveniently be expressed in decibels.

The foregoing formulas are used in Sec. 5.3 to calculate the

[1] The fact that $|Y(f)|$ alone, without phase considerations, determines the entropy loss in a filter is tied up with the fact that the phase shift caused by the filter is the same for the real and imaginary components of the signal at any frequency.

entropy loss in a triangular filter and are used in Chap. VII in calculating the noise reduction in frequency modulation.

The theorem of this section should not lead the reader to the conclusion that two ensembles having the same average power, bandwidth, duration, and power spectrum will necessarily have the same entropy. If their probability distributions at the individual sample points are not the same, they will generally not have the same entropy. For example, white noise not having a gaussian distribution of values at the sample points will have less entropy than random noise of the same average power, bandwidth, duration, and power spectrum.

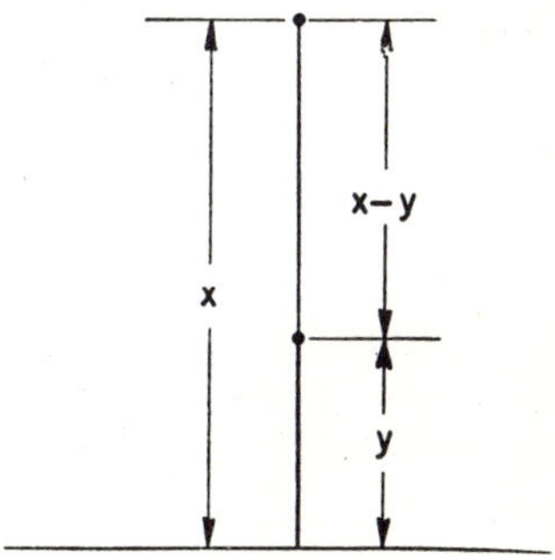

Fig. 4-1.

4.10 Entropy of the sum of two ensembles

If we have two independent ensembles of functions, we can form a new ensemble by superposition. Suppose the first ensemble has the probability distribution function $p(x_1, \ldots, x_n)$ and that of the second is $q(x_1, \ldots, x_n)$. Then, if the superposition is linear, the probability distribution function of the combined signals is (see Fig. 4-1)

$$r(x_1, \ldots, x_n) = \int \cdots \int p(y_1, \ldots, y_n) \cdot q(x_1 - y_1, \ldots, x_n - y_n) dy_1 \ldots dy_n \tag{123}$$

Let N_1 and N_2 be the average quadivas of the original ensembles, and let $\bar{N}_1$ and $\bar{N}_2$ be their entropy powers. Let N_3 be the average quadiva of the sum, and let $\bar{N}_3$ be its entropy power. Then if the original ensembles are *ergodic and independent,* we shall consider the following results to hold:

(I) $$N_3 = N_1 + N_2 \tag{124}$$

(II) $$\bar{N}_1 + \bar{N}_2 \leq \bar{N}_3 \leq N_1 + N_2 \tag{125}$$

Equation (124) has already been derived in Chap. III. It says that, if two incoherent ergodic ensembles are superimposed in a linear system, the average quadivas are additive. Inequality (125) says, in effect, that if two incoherent ergodic ensembles are superimposed the entropy per unit of average quadiva is increased or at least not decreased by the superposition. We do not have a proof of the inequality (125) but the following argument will show that it is probably true. Since random noise has the maximum possible average entropy per unit quadiva, we can expect that, when incoherent ergodic ensembles are superimposed, they tend in the direction toward becoming a random-noise ensemble. This means that (1) the intersymbol constraints become weaker and (2) the value distributions at the sample points tend toward normal distributions. These results are to be expected for, by hypothesis, the original ensembles are incoherent. The intersymbol constraints in the original ensembles therefore cannot match, so that we can expect them to become weaker in the superposition process. Furthermore the central-limit theorem would lead us to expect the liniva distribution to tend to become normal. On a few occasions, we shall find Eq. (125) of use, and we shall there assume its truth. The author has been unable to devise a proper derivation of the equation and has not found one in the literature. A proper derivation of this equation would be a valuable addition to the literature on information theory.

PROBLEMS

1. If a random-noise ensemble having the distribution

$$p(x) = \frac{1}{\sqrt{2\pi A^2}} \epsilon^{-x^2/2A^2}$$

at each sample point is superimposed on another random-noise ensemble having the distribution

$$q(x) = \frac{1}{\sqrt{2\pi B^2}} \epsilon^{-x^2/2B^2}$$

find the distribution of the resultant by using Eq. (123).

$$\textit{Ans.}\ r(x) = \frac{1}{\sqrt{2\pi(A^2 + B^2)}} \epsilon^{-x^2/2(A^2+B^2)}$$

2. Suppose we have an ensemble of functions with no intersymbol influence, having a probability distribution at each sample point

$$p(x) = \begin{cases} A \cos x & \left(-\frac{\pi}{2} < x < \frac{\pi}{2}\right) \\ 0 \quad \text{elsewhere} \end{cases}$$

Find the entropy per sample point.

3. Suppose we have an ensemble of functions with no intersymbol influence whose quadiva is peak limited to the value

$$x^2 = S$$

and whose average quadiva is

$$\widetilde{x^2} = K$$

Find the probability distribution $p(x)$, which gives maximum entropy subject to these limitations. Also find the value of this maximum entropy H.

4. Calculate the entropy power of white noise whose linivas are uniformly distributed from $(-A \leq x \leq +A)$ and whose probability distribution function is zero elsewhere.

5. Calculate the entropy of an ensemble of chains of pulses of the type shown in Fig. 4-A. These pulses are all of height a, and duration b, and occur at random at an average rate of c per sec. Let T be the duration of each chain.

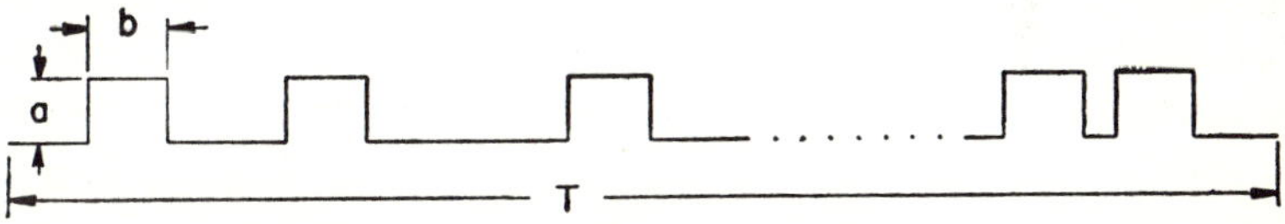

Fig. 4-A.

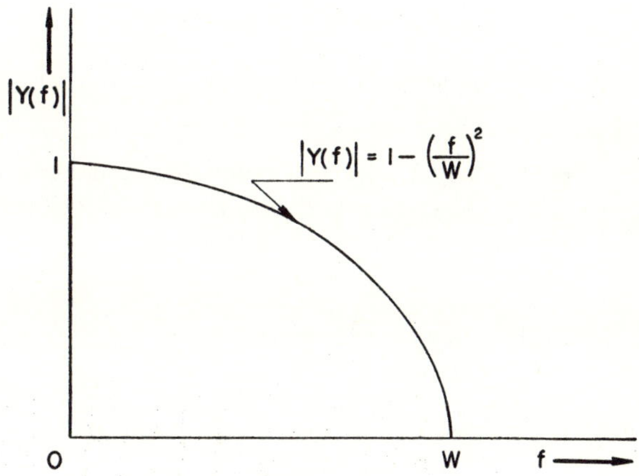

Fig. 4-B.

6. Find the loss in entropy caused by the filter shown in Fig. 4-B.

CHAPTER V

THE TRANSMISSION OF INFORMATION IN BAND-LIMITED SYSTEMS HAVING A CONTINUOUS RANGE OF VALUES[1]

5.1 The fundamental formulas

When a message i is transmitted and a message j is received, the methods developed in Chap. I give us for the amount of information received

$$\begin{aligned}&\text{quantity of language information}\\&\text{received in transmission}^{2}\ (i \to j)\\&= \log \left[\frac{\text{probability at the receiver of the transmission of } i \text{ after the message } j \text{ is received}}{\text{probability at the receiver of the transmission of } i \text{ before the message } j \text{ is received}}\right]\end{aligned} \tag{1}$$

Let us write x for the value of i and y for the value of j. We shall also write

$$p(x)dx = \left[\begin{array}{l}\text{probability that the transmitted signal has a}\\ \text{value between } x \text{ and } x + dx \text{ when nothing is}\\ \text{known about the received signal}\end{array}\right] \tag{2}$$

and

$$p_y(x)dx = \left[\begin{array}{l}\text{probability that the transmitted signal has a}\\ \text{value between } x \text{ and } x + dx \text{ when it is}\\ \text{known that the received signal has the}\\ \text{value } y\end{array}\right] \tag{3}$$

[1] This chapter is based largely upon material by C. E. Shannon in *Bell System Tech. Jour.*, vol. 27, October, 1948.

[2] The symbol $(i \to j)$ indicates that i is the transmitted message and j is the received message.

With this notation, we can rewrite Eq. (1) as

$$\begin{bmatrix}\text{quantity of language information} \\ \text{in transmission } (x \rightarrow y)\end{bmatrix} = \lim_{\Delta x \rightarrow 0} \log \left[\frac{p_y(x)\Delta x}{p(x)\Delta x}\right] = \log \left[\frac{p_y(x)}{p(x)}\right] \tag{4}$$

In case the transmitted signal gives the values $x_1, \ldots, x_n$ for the n sampling points which are received as $y_1, y_2, \ldots, y_n$, Eq. (4) may be rewritten as

$$\text{quantity of language information in transmission } (x_1, x_2, \ldots, x_n \rightarrow y_1, y_2, \ldots, y_n) = \log \left[\frac{p_{y_1, \ldots, y_n}(x_1, \ldots, x_n)}{p(x_1, \ldots, x_n)}\right] \tag{5}$$

In future analyses which apply equally well to Eqs. (4) or (5), we shall use the simpler form (4). The form (5) will generally be used only when the situation for several variables differs from that for one variable.

Before continuing with our analysis, we shall define a few more probabilities and point out some relations which exist between various probabilities. Thus

$$p(x,y)dx\,dy = \begin{bmatrix}\text{a priori probability that the transmitted signal will have a value between } x \text{ and } x + dx \\ \text{and that the received signal will have a value between } y \text{ and } y + dy\end{bmatrix} \tag{6}$$

$$p(y)dy = \begin{bmatrix}\text{probability that the received signal will have a value between } y \text{ and } y + dy \\ \text{when the transmitted signal is not known}\end{bmatrix} \tag{7}$$

$$p_x(y)dy = \begin{bmatrix}\text{probability that received signal will have a value between } y \text{ and } y + dy \\ \text{when it is known that } x \text{ is the value of the transmitted signal}\end{bmatrix} \tag{8}$$

The known laws of probability then give us the following relations:

$$\int_{-\infty}^{+\infty} p(x,y)dy = p(x) \tag{9}$$

$$\int_{-\infty}^{+\infty} p(x,y)dx = p(y) \tag{10}$$

$$p(x,y) = p_y(x)p(y) = p_x(y)p(x) \tag{11}$$

From Eq. (11) it follows that Eq. (4) may be rewritten as

$$\begin{bmatrix}\text{quantity of language information}\\ \text{in transmission } (x \to y)\end{bmatrix} = \log\left[\frac{p_x(y)}{p(y)}\right] \tag{12}$$

In any particular system, we then have for the average information per signal

$$\text{average information/signal} = \iint \log\left[\frac{p_y(x)}{p(x)}\right] p(x,y)dx\,dy \tag{13}$$

$$= \iint \log\left[\frac{p_x(y)}{p(y)}\right] p(x,y)dx\,dy \tag{14}$$

$$= \iint \log\,[p_x(y)]p(x,y)dx\,dy - \iint \log p(y)p(x,y)dx\,dy$$

$$= \iint \log\,[p_x(y)]p(x,y)dx\,dy - \int \log p(y)\left[\int p(x,y)dx\right]dy$$

$$= \iint \log\,[p_x(y)]p(x,y)dx\,dy - \int p(y)\log\,[p(y)]dy$$

$$= H(y) - H_x(y) \tag{15}$$

The quantity $H_x(y)$ is defined by the equation

$$H_x(y) = \text{ensemble average of } -\int p_x(y)\log p_x(y)dy \text{ over all possible values of } x$$

$$= -\int\left[\int p_x(y)\log p_x(y)dy\right]p(x)dx$$

$$= -\iint \log\,[p_x(y)]p(x,y)dx\,dy \tag{16}$$

The probability $p_x(y)$ describes the detailed noise properties of the system, while $H_x(y)$ describes the over-all effect of the noise. Simple manipulations of some of the foregoing equations will show that

$$H(y) - H_x(y) = H(x) - H_y(x) \tag{17}$$

In Chap. IV we found that the value of an entropy function such as $H(y)$ or $H_x(y)$ depends on the coordinate system used. We will now show, however, that the expressions $[H(y) - H_x(y)]$ and $[H(x) - H_y(x)]$ do not depend on the coordinate system, so that the average rate of transmission of information is independent of the coordinate system used.

Let $x_1,x_2, \ldots ,x_n,y_1,y_2, \ldots ,y_n$ be the original coordinates and let $x'_1, x'_2, \ldots, x'_n, y'_1, y'_2, \ldots, y'_n$ be the transformed coordinates. Let p's be used for probabilities in the original system of coordinates, and let q's be used for probabilities in the transformed system. Then

$$p(x,y) = \left| J\left(\frac{x',y'}{x,y}\right) \right| q(x',y') \tag{18}$$

$$p_x(y) = \left| J\left(\frac{y'}{y}\right) \right| q_{x'}(y') \tag{19}$$

$$p(y) = \left| J\left(\frac{y'}{y}\right) \right| q(y') \tag{20}$$

$$dx\,dy = \left| J\left(\frac{x,y}{x',y'}\right) \right| dx'dy' \tag{21}$$

and, of course,
$$\left| J\left(\frac{x,y}{x',y'}\right) \right| \cdot \left| J\left(\frac{x',y'}{x,y}\right) \right| = 1 \tag{22}$$

Therefore

$$\begin{aligned} H(y) - H_x(y) &= \iint \log\left[\frac{p_x(y)}{p(y)}\right] p(x,y)dx\,dy \\ &= \iint \log\left[\frac{q_{x'}(y')}{q(y')}\right] J\left(\frac{x',y'}{x,y}\right) q(x',y') J\left(\frac{x,y}{x',y'}\right) dx'dy' \\ &= \iint \log\left[\frac{q_{x'}(y')}{q(y')}\right] q(x',y')dx'dy' = H(y') - H_{x'}(y') \end{aligned} \tag{23}$$

The proof for $H(x) - H_y(x)$ is entirely analogous. We thus see that the average rate of transmission of information does not depend upon the coordinate system used for measurement.

It will be found convenient to introduce a normalization in the measurements of y and x. In order to determine the method of normalization, suppose we have an amplifier operating without noise. Then if x is a liniva of the input signal, the corresponding liniva of the output signal is Gx, where G is the gain of the amplifier. The normalization which we shall use is that the unit of measurement of the output signal shall be G times the unit of measurement of the input signal. In general, G need not be a pure numeric. The input signal may, for example, be sound pressure (measured in bars) at a microphone while the output signal is electric voltage. In this case, G is a conversion factor having the dimensions of volts per bar. In future analyses when it is important that the units of the received signal be normalized with respect to those of the transmitted signal, we will say that the received signal is normalized. Since the values of $[H(y) - H_x(y)]$ and of $[H(x) - H_y(x)]$ do not depend upon the coordinate system, normalization does not affect the calculated value of the rate of transmission of information.

According to Eqs. (16) and (11), we may write

$$\begin{aligned} H_x(y) &= -\int_{-\infty}^{\infty}\int_{-\infty}^{\infty} \log\,[p_x(y)]p(x,y)dx\,dy \\ &= -\int_{y=-\infty}^{+\infty}\int_{x=-\infty}^{+\infty} \log\,[p_x(y)]p_x(y)p(x)dx\,dy \end{aligned} \tag{24}$$

The $-\infty$ to $+\infty$ limits of integration have been implied all along, but we express them here in order to be precise.

If the noise described by $p_x(y)$ is independent of the transmitted signal and is combined with it in a linear superposition process, and if the received signal is normalized, then $p_x(y)$ is a function of $(y - x)$ only, and $p_x(y)$ does not depend on either x or y except in the combination $(y - x)$. The variable y in Eq. (24) for $H_x(y)$ may then be replaced by

$$z = y - x \tag{25}$$

Let us write

$$p_x(y) = f(z) = \text{a function of } z \text{ only} \tag{26}$$

We next introduce the notation

$$\int_{y=-\infty}^{+\infty} \log\,[p_x(y)]p_x(y)dy = \int_{z=-\infty}^{+\infty} \log\,[f(z)]f(z)dz = K \tag{27}$$

(where K is a constant independent of both x and y)

Equation (24) thus becomes

$$\begin{aligned} H_x(y) &= -\int\int \log\,[p_x(y)]p_x(y)p(x)dx\,dy = -K\int p(x)dx = -K \\ &= -\int \log\,[(p_x(y)]p_x(y)dz \end{aligned} \tag{28}$$

when the noise is independent of the transmitted signal and is combined with it in a linear superposition process and the received signal is normalized with respect to the transmitted signal. In this case $p_x(y)$ is the probability distribution that the noise has the value $(y - x)$, so that $H_x(y)$ is identical with the entropy of the noise alone. Thus, for this case, Eq. (15) becomes

$$\text{average information/signal} = H(y) - H_x(y)$$
$$= \text{entropy of received signal} - \text{entropy of noise} \quad (29)$$

If natural logarithms were used in calculating $H(y)$ and $H_x(y)$, then the answer must be multiplied by $\log_2 \epsilon (= 1.443)$ in order to get the average information in binits. Otherwise the unit of information may be called a *nepit* (contraction of neperian digit).

5.2 Language transmission capacities in systems with average power (quadiva) limitations

(a) *Random-noise Case.* If the noise present is random noise which is linearly superimposed on the signal, then it is a relatively simple matter to calculate $H_x(y)$. For a signal of duration T and bandwidth W, according to Eq. (103a) of Chap. IV, we have

$$H_x(y) = TW \log 2\pi\epsilon N \quad (30)$$

where N is the average quadiva (power) of the noise, and

$$\epsilon = 2.71828 \ldots \quad (31)$$

The average information per signal (of duration T) is thus from Eq. (29)

$$\text{average information/signal} = H(y) - TW \log 2\pi\epsilon N \quad (32)$$

We may also write

$$R_0 = \text{language transmission capacity of noisy channel}$$
$$= \lim_{T\to\infty} \frac{1}{T} [H(y_1, \ldots, y_{2TW}) - TW \log 2\pi\epsilon N] \quad (33)$$

It was shown in Chap. III that, if two incoherent signals are linearly superimposed, the quadratic contents are additive. The quadratic content of the y signal is therefore equal to that of the x signal plus that of the noise. If P is the average quadiva of the x distribution and N is the average quadiva of the noise, then the average quadiva of the y distribution is $P + N$, since all the ensembles involved are ergodic.

In Chap. IV we showed that, for a given average quadiva, $H(y_1 \ldots y_{2TW})$ will be a maximum if its probability distribution function is the same as that of random noise. Therefore R_0 will be a maximum if the y distribution is a random-noise distribution.

In Chap. III, we showed that if one random-noise distribution is linearly superimposed upon another, the resultant is also a random-noise distribution. It therefore follows that R_0 will be a maximum if the x distribution is a random-noise distribution.[1]

If the x distribution is a random-noise distribution of average quadiva P and the y distribution is a random-noise distribution of average quadiva $P + N$, then by Eq. (103a) of Chap. IV, we have

$$H(x) = TW \log 2\pi\epsilon P \tag{34}$$

and

$$H(y) = TW \log [2\pi\epsilon(P + N)] \tag{35}$$

In this case, the language transmission capacity is, according to Eqs. (33) and (35),

$$\begin{aligned} R_{\max} &= \lim_{T\to\infty} \frac{1}{T} [TW \log 2\pi\epsilon(P + N) - TW \log 2\pi\epsilon N] \\ &= W \log \left(\frac{P + N}{N}\right) \end{aligned} \tag{36}$$

This is the maximum rate at which language information can be sent through a channel having an average power limitation and disturbed by random noise. In Chap. VI we will show that by proper encoding it is possible to send information through the channel at this rate with a vanishingly small percentage of errors.

(b) *Arbitrary-noise Case.* Let us next consider the case when an arbitrary type of noise is present. We shall still assume that the system is linear.

Let P = average quadiva of the transmitted signal

N = average quadiva of the noise

$\bar{N}$ = entropy power of the noise

Thus, $P + N$ = average quadiva of the received signal[2]

assuming that the transmitted signal, noise, and received signal are all measured in normalized units. Equation (29) still applies in this case, so that the information per signal is the difference

[1] We have thus shown that a random-noise input is a sufficient condition for maximum R_0, but we have not shown that it is necessary. However, no input of the same quadratic content can yield a greater R_0 than a random-noise input.

[2] It is shown in Chap. III that, if any two incoherent signals are superimposed in a linear system, their quadratic contents are additive.

between the entropy of the received signal and the entropy of the noise.

The entropy per unit of time of the received signal cannot exceed $W \log [2\pi\epsilon(P + N)]$ which would be its value if its distribution characteristics were those of random noise. According to the definition of entropy power, the entropy per unit time of the noise is $W \log [2\pi\epsilon\bar{N}]$. Therefore, if R is the rate of transmission of language information, we know that

$$R \leq W[\log 2\pi\epsilon(P + N)] - W \log (2\pi\epsilon\bar{N}) = W \log \left[\left(\frac{P + N}{\bar{N}}\right)\right] \quad (37)$$

This determines a maximum possible value for R which can be attained only provided it is possible to make the received signal have the characteristics of random noise.

It is presumably always possible to make the transmitted signal have the properties of random noise. At least, the noise will not prevent it. If the transmitted signal is of a random-noise character, then by the theorem in Sec. 4.10, the entropy per unit of time of the received signal will be at least as great as $W \log [2\pi\epsilon(P + \bar{N})]$. We may, therefore, write

$$\begin{aligned} R_{\max} &\geq W \log [2\pi\epsilon(P + \bar{N})] - W \log [2\pi\epsilon\bar{N}] \\ &= W \log \left[\left(\frac{P + \bar{N}}{\bar{N}}\right)\right] \end{aligned} \quad (38)$$

Equations (37) and (38) give upper and lower bounds for the maximum rate of transmission of information in a noisy system. The actual value of $R_{\max}$ has not been worked out, except in the case of random noise and gaussian noise (see parts (a) and (c) of this section). If the noise is random, then $N = \bar{N}$ and the upper and lower bounds are the same. If the transmitted power is large in comparison with the noise, then even if the noise is not random, both Eqs. (37) and (38) approach the same value, so that we have

$$R_{\max} = W \log \left(\frac{P}{\bar{N}}\right) \quad \text{(approximately)} \quad (39)$$

Equation (39) indicates that the noise which will cause the greatest loss in language transmission capacity is the noise which has the greatest entropy power. For a given amount of power in a given bandwidth, no noise can be worse than random noise, since no other noise can have a greater entropy power.

Equations (36) and (39) are an improved version of what has been called the *Nyquist-Hartley law.*[1] This law states in effect that the time-bandwidth product (TW) determines the amount of information which can be transmitted. For fixed values of P and $\bar{N}$, this is certainly true. However, the above equations indicate that an increase in $\log (P/N)$ can compensate for a loss in (TW) when it is desired to transmit a certain amount of information. Nyquist and Hartley were probably aware of this, but they did not reduce it to a precise form as done by Shannon in the equations above. The practical applications of the formulas of this section will be discussed in Chap. VII.

(c) *Gaussian Noise.* It is also instructive to consider the case when the disturbing noise present is gaussian noise. As pointed out in Sec. 3.11, gaussian noise has a gaussian distribution of amplitudes at its frequency sampling points and has no intersymbol influence in the frequency domain. It differs from random noise in that its frequency spectrum need not be flat, and its spectrum does not have to start at zero frequency. If random noise is sent through a linear network, the output is gaussian noise.

Suppose that the desired signal power has a frequency spectral distribution function $s(f)$, so that

$$\int_{f_1}^{f_2} s(f)df = P \tag{40}$$

where P is the total signal power. Suppose that the noise power has a spectral distribution function $n(f)$, so that

$$\int_{f_1}^{f_2} n(f)df = N \tag{41}$$

where N is the total noise power. Let us now divide the frequency range into small bands, so that $n(f)$ is approximately constant in each. Let us get our information by examining the signal in the frequency domain, rather than the time domain. In the frequency

[1] H. Nyquist, "Certain Factors Affecting Telegraph Speed," *Bell System Tech. Jour.*, vol. 3, April, 1924, p. 324. H. Nyquist, "Certain Topics in Telegraph Transmission Theory," *A.I.E.E. Trans.*, vol. 47, April, 1928, p. 617. R. V. L. Hartley, "The Transmission of Information," *Bell System Tech. Jour.*, vol. 7, July, 1928, pp. 535–564.

domain, gaussian noise has a gaussian distribution of amplitudes and no intersymbol influence. Furthermore, the sampling theorem in the frequency domain is independent of whether or not the frequency range starts at $f = 0$. For maximum language transmission capacity, the desired signal should of course, be as unconstrained as possible. It will therefore also have no intersymbol influence in the frequency domain and will have a gaussian distribution of amplitudes. Consequently, the signal will have a gaussian distribution of amplitudes. Consequently, the signal will also have the properties of gaussian noise.

The entropy of the combined signal and noise in any narrow frequency band Δf will then be

$$T\,\Delta f \log 2\pi\epsilon[s(f)\Delta f + n(f)\Delta f]$$

while the entropy of the noise alone will be

$$T\,\Delta f \log 2\pi\epsilon[n(f)\Delta f]$$

The language transmission capacity in the band Δf will thus be

$$R_{\Delta f} = \Delta f \cdot \log \left[\frac{s(f) + n(f)}{n(f)} \right] \tag{42}$$

Since there is no intersymbol influence in the frequency domain in this case, the total language transmission capacity will be obtained by adding the values for the individual frequency bands. Thus

$$R = \int_{f_1}^{f_2} \log \left[\frac{s(f) + n(f)}{n(f)} \right] df \tag{43}$$

Let us next find what power spectrum of the desired signal will make R a maximum for a given value of total desired signal power P and a given gaussian noise $n(f)$. We thus wish to find the maximum value of the integral in Eq. (43) subject to the constraint of Eq. (40). According to Eq. (14) of Chap. IV, this requires that

$$\frac{\partial}{\partial s} \log \left(1 + \frac{s}{n}\right) + \lambda \frac{\partial s}{\partial s} = 0 \tag{44}$$

where we have written s for $s(f)$ and n for $n(f)$ in Eq. (44). From Eq. (44), we obtain

$$s(f) + n(f) = -\frac{1}{\lambda} \tag{45}$$

where λ is a constant and is thus independent of frequency. The value of λ is adjusted so that the total power in Eq. (40) is P. Equation (45) tells us that, if $n(f)$ is given, then for maximum language transmission capacity the power spectrum of the desired signal should be adjusted so that $s(f) + n(f)$ is independent of frequency. This condition is illustrated[1] in Fig. 5-1.

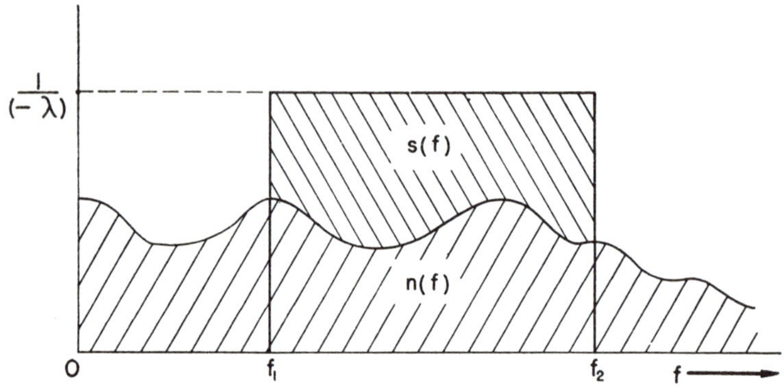

Fig. 5-1.

We will next show that the type of gaussian noise which is worst, in the sense that it makes the optimum value of Eq. (43) the smallest, is white gaussian noise. To show this, we have from Eqs. (43) and (45)

$$\begin{aligned} R_{\max} &= \int_{f_1}^{f_2} \log\left[\frac{s(f) + n(f)}{n(f)}\right] df \\ &= \int_{f_1}^{f_2} \log\,[s(f) + n(f)]df - \int_{f_1}^{f_2} \log\,[n(f)]df \\ &= [f_2 - f_1] \log\,(-1/\lambda) - \int_{f_1}^{f_2} \log\,[n(f)]df \end{aligned} \tag{46}$$

The first term on the right side of Eq. (46) is a positive constant, depending only on P, N, and $(f_2 - f_1)$ and is independent of the spectral distribution $n(f)$. The worst type of gaussian noise is therefore that which will make

$$\int_{f_1}^{f_2} \log n(f) df$$

[1] It is assumed here that the desired spectrum is restricted (by apparatus or legal restrictions) to the range from f_1 to f_2. Otherwise it should be allowed to spread over a wider frequency range.

a maximum, subject to the constraint Eq. (41). By Eq. (14) of Chap. IV, we then have

$$\frac{\partial}{\partial n}(\log n) + \lambda_1 \frac{\partial n}{\partial n} = 0 \tag{47}$$

or

$$n(f) = -\frac{1}{\lambda_1} = \text{constant} \tag{47.1}$$

Thus the worst type of disturbing gaussian noise is white gaussian noise.

5.3 The language transmission capacity of a system having a peak power limitation

(a) *Maximum Entropy of an Ensemble with Peak Power Limitation at the Sampling Points.* Let us first calculate the entropy of an ensemble whose peak power is limited to S (*i.e.*, the liniva ranges between $-\sqrt{S}$ and $+\sqrt{S}$) at all the sampling points. If this is the only constraint on the ensemble, then according to Sec. 4.3 for maximum entropy, the probability distribution is uniform between $-\sqrt{S}$ and $+\sqrt{S}$ at every sampling point and has the value $1/(2\sqrt{S})$ in this range. Consequently, the entropy at each sampling point is

$$H(x) = -\int p \log p \, dx = -\int_{-\sqrt{S}}^{+\sqrt{S}} \frac{1}{2\sqrt{S}} \log \frac{1}{2\sqrt{S}} \, dx$$
$$= -\frac{1}{2\sqrt{S}} \log\left(\frac{1}{2\sqrt{S}}\right) 2\sqrt{S} = \log 2\sqrt{S} = \frac{1}{2}\log 4S \tag{48}$$

For maximum entropy, the probabilities of the different sampling points are independent, and the entropies are therefore additive. In time T there are $2TW$ sampling points. Consequently, the maximum entropy per unit of time is

$$H = \frac{1}{T}(2TW)\frac{1}{2}\log 4S = W \log 4S \tag{49}$$

(b) *The Entropy of a Particular Ensemble which is Peak Limited Throughout Time T.* We will next calculate the entropy of an ensemble which is peak limited everywhere in time T, not only at the sampling points. We will first show that the ensemble found in part (a) will become such an ensemble after it is passed through a filter whose gain is unity at zero frequency and then declines linearly to a gain of zero at frequency W. In Appendix IX it is shown that if the signal

$$\frac{\sin 2\pi Wt}{2\pi Wt} \tag{50}$$

enters such a filter, the corresponding output will be

$$\frac{1}{2}\frac{\sin^2(\pi Wt)}{(\pi Wt)^2} \tag{51}$$

Now any of the signals considered in part (a) which were peak limited at the sampling points can, according to the sampling theorem, be expressed as

$$\sum_{n=0}^{2TW} A_n \frac{\sin(2\pi Wt - n\pi)}{(2\pi Wt - n\pi)} \tag{52}$$

where A_n is the liniva of the signal at the nth sampling point. A_n can thus not exceed $\sqrt{S}$ in magnitude. If this signal is passed through the filter, the output will be

$$\sum_{n=0}^{2TW} \frac{A_n}{2} \frac{\sin^2\left(\pi Wt - \frac{n\pi}{2}\right)}{\left(\pi Wt - \frac{n\pi}{2}\right)^2} \tag{53}$$

The terms in Eq. (53) do not change sign with time and always have the sign of A_n. The peak possible value of any output signal will therefore occur if all the A_n's have their maximum allowed positive value $+\sqrt{S}$. In this case, the input function must have been of constant amplitude, $+\sqrt{S}$, and since the filter has unit gain for zero frequency, the output is the same. Thus the output signal in this case has a peak quadiva S. Since, as just pointed out, this signal has the peak possible quadiva of any signal of the output ensemble, the output ensemble has a peak quadiva which is everywhere limited to S, not only at the sampling points.

We note in passing that the above input ensemble is not limited to $\pm\sqrt{S}$ everywhere even though it is so limited at the sampling points. This can easily be demonstrated with the aid of an example. Consider the signal

$$G(t) = \sum_{n=1}^{2TW} (-1)^n \sqrt{S} \frac{\sin(2\pi Wt - n\pi)}{(2\pi Wt - n\pi)} \tag{54}$$

This has the quadiva S at every sampling point. However, let us calculate its liniva at $t = 1/4W$ (which is not a sampling point).

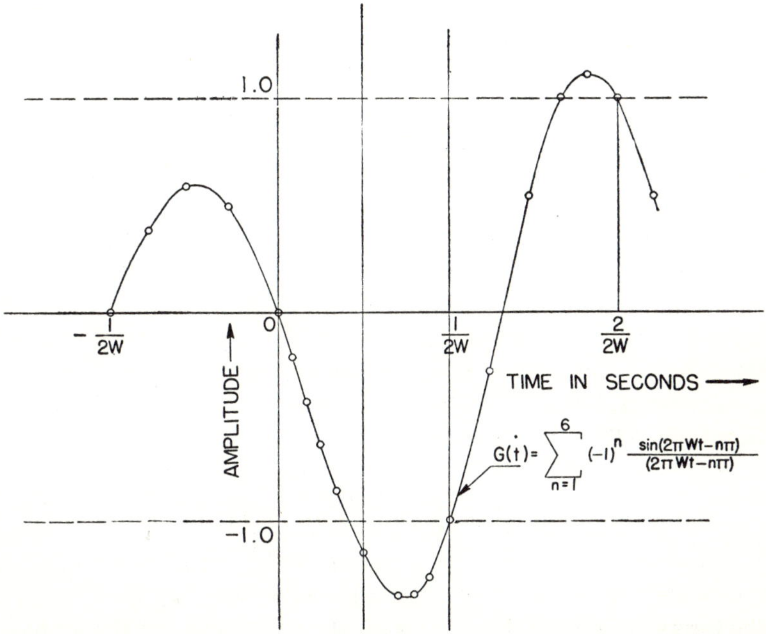

Fig. 5-2. A function which is peak limited to the quadiva 1.0 at the sample points only.

Thus

$$G\left(\frac{1}{4W}\right) = \sqrt{S} \sum_{n=1}^{2TW} (-1)^n \left[\frac{\sin\left(\frac{\pi}{2} - n\pi\right)}{\left(\frac{\pi}{2} - n\pi\right)}\right] \tag{55}$$

$$= \sqrt{S}\left\{-\frac{2}{\pi}\left[1 + \frac{1}{3} + \frac{1}{5} + \frac{1}{7} + \cdots + \frac{1}{4TW - 1}\right]\right\}$$

The series in the brackets on the right side of Eq. (55) does not approach a limit as the number of terms increases. Consequently, the signal quadiva is not peak limited to S in between the sampling points. A graph of Eq. (54) is shown in Fig. 5-2 for $2TW = 6$ and $S = 1$.

Let us now return to the output signal of the filter, which as we have shown is peak limited everywhere. According to Eq. (49) the entropy per unit time of the input is

$$H_i = W \log 4S \tag{56}$$

It then follows from Eq. (120) of Sec. 4.9 that the entropy per unit of time of the output is

$$\begin{aligned} H &= W \log 4S + \int_0^W \log \left(\frac{W-f}{W}\right)^2 df \\ &= W \log 4S - 2W = W \log \left(\frac{4S}{\epsilon^2}\right) \end{aligned} \tag{57}$$

Equation (57) is the entropy of a particular ensemble peak limited everywhere. This is in contrast with Eq. (49), which gives the maximum entropy for any possible ensemble which is peak limited at the sampling points. The maximum entropy for any possible ensemble which is peak limited everywhere to $\pm\sqrt{S}$ certainly cannot exceed that given by Eq. (49), so that its value must lie somewhere between those of Eqs. (49) and (57).

(*c*) *Language Transmission Capacities.* The rate of transmission of information is still given in the peak-limited case by Eq. (29), assuming that the system is linear and the units are normalized. If the only constraint on the transmitted signal is that it is peak limited to the quadiva S, then by Eq. (57) it is possible to get a value of $H(x)$ per unit time at least as great as $W \log (4S/\epsilon^2)$. This has, according to Eq. (105) of Chap. IV, an entropy power of

$$\frac{1}{2\pi\epsilon}\left(\frac{4S}{\epsilon^2}\right) = \frac{4S}{2\pi\epsilon^3} \tag{58}$$

If the noise has an entropy power $\bar{N}$, then by the theorem of Sec. 4.10 the entropy power of the received signal can be made as great as

$$\frac{4S}{2\pi\epsilon^3} + \bar{N} \tag{59}$$

whose corresponding entropy per unit time is

$$H(y) = W \log 2\pi\epsilon \left(\frac{4S}{2\pi\epsilon^3} + \bar{N}\right) \tag{60}$$

The rate of transmission of information can thus be made at least as great as

$$R = H(y) - H_x(y) = W \log 2\pi\epsilon\left(\frac{4S}{2\pi\epsilon^3} + \bar{N}\right) - W \log 2\pi\epsilon\bar{N}$$

$$= W \log\left(\frac{4S}{2\pi\epsilon^3\bar{N}} + 1\right) \tag{61}$$

The foregoing procedure illustrates the general methods which can be employed in various problems dealing with peak limiting. For a peak-limited ensemble, it is possible to get a value of entropy somewhere between the values given by Eqs. (49) and (57). With this as a starting point, and knowing the entropy power of the noise, one then can try to determine the range of possible values of the maximum rate R, making use of the theorem of Sec. 4.10 and the developments of the foregoing Chapter. The exact procedure to be followed will, of course, depend upon the conditions of the problem. It will usually not be possible to get an exact answer, but only to indicate a possible range of values.

Problem

1. Show by actual summation of Eqs. (52) and (53) of Sec. 5.3 that the output is of constant amplitude $\sqrt{S}$ when the input is.

CHAPTER VI

AN INTRODUCTION TO THE USE OF SIGNAL SPACE[1]

6.1 Some properties of signal space

We have shown that the $2TW$ numbers which are the values of a signal at the sampling points completely specify the signal. It is therefore possible to consider a space of $2TW$ dimensions in which every signal of duration T and maximum bandwidth W is represented by a different point. This is the point whose $2TW$ coordinates are the respective values of the signal at its sampling points. If we consider the coordinate axes of this space to be straight lines at right angles to each other, then the distance r from the origin to the point representing the signal has a simple interpretation as we shall now show. Suppose that the $2TW$ sampling points in question are sampling points in the time domain. Then

$$r^2 = \sum_{n=1}^{n=2TW} \left[G\left(\frac{n}{2W}\right)\right]^2 \tag{1}$$

It then follows from Eq. (52) of Chap. II that

$$r^2 = 2W \int_0^T [G(t)]^2\, dt = 2W \text{ times} \begin{bmatrix} \text{the quadratic} \\ \text{content of the signal} \end{bmatrix} \tag{2}$$

Next, instead of using $2TW$ sampling points in the time domain to represent a signal, let us use $2TW$ sampling points in the frequency domain. Let us use $\sqrt{\frac{2W}{T}}\, a_0$, $2\sqrt{\frac{W}{T}}\, a_m$, and $2\sqrt{\frac{W}{T}}\, b_m$ for our frequency sampling coordinates. Then according to Appendix VI, we need only use Eqs. (8), (9), and (10) of that appendix to trans-

[1] This chapter is based largely upon material in the article by C. E. Shannon, *Proc. I.R.E.*, vol. 37, January, 1949, pp. 10–21.

form from time sampling to frequency sampling coordinates. Each point in the same space of $2TW$ dimensions will then represent both a $G(t)$ and its corresponding $\Omega(\omega)$. The frequency sampling coordinate axes will also be at right angles to each other, and they will have the same origin as the time sampling coordinate axes. In other words, we may consider the frequency sampling coordinate axes as being obtained from the time sampling coordinate axes by a mere "rotation" in the $2TW$ dimensional space. These results are assured by Eqs. (12a) and (12b) of Appendix VI. Thus when the above frequency sampling coordinates are used, the time-sampling space is able to represent the frequency sampling space and vice versa just by a "rotation" of coordinates, with the points representing signals remaining fixed.

For a given signal let us use the notation

$$P = \overline{G\left[\left(\frac{n}{2W}\right)\right]^2} = \text{average value of } G\left[\left(\frac{n}{2W}\right)\right]^2 \qquad (3)^1$$

Then, from Eqs. (1) and (3) and (2)

$$P = \frac{1}{2TW}\sum_{n=1}^{2TW}\left[G\left(\frac{n}{2W}\right)\right]^2 = \frac{r^2}{2TW} \qquad (4)$$

Thus

$$r = \sqrt{2TWP} \qquad (5)$$

When considering the superposition of signals in linear systems, it is convenient to consider each signal as the vector extending from the origin of signal space to the point representing the signal as described above. The process of linear superposition is then merely vectorial addition, as illustrated in Fig. 6-1.

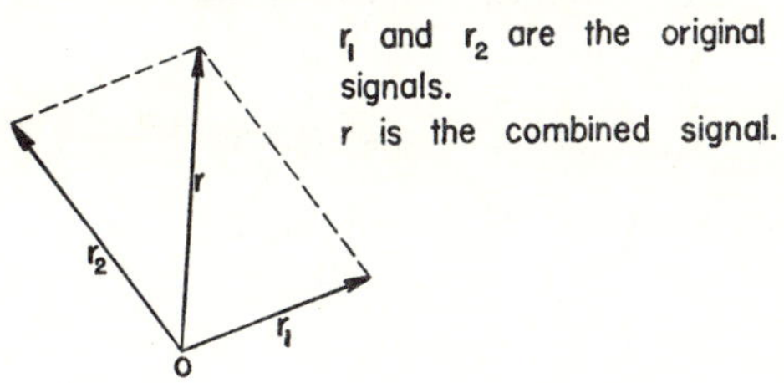

Fig. 6-1. Signal-space diagram of the superposition of signals in a linear system.

[1] P is thus the average quadiva and is proportional to the average power.

Next, consider an ensemble of signals. Since each of these signals can be represented by a point in signal space, the ensemble of signals is represented by a region of signal space. For example, if all the signals of the ensemble are of the same average power, then according to Eq. (5) they all lie on the surface of an n-dimensional sphere whose center is at the origin.

We note, in passing, that the vectors of incoherent signals are at right angles to each other in signal space since their quadratic contents are additive.

6.2 A language transmission capacity with almost no errors

In Sec. 5.2 we showed that a transmission system of bandwidth W and limited to average normalized transmitted power P and subjected to a linear superposition of random noise of average power N had a language transmission capacity

$$R_{\max} = W \log \left(\frac{P + N}{N} \right) \tag{6}$$

We will now show that by proper encoding it is possible to send information through the transmission system at this rate with a vanishingly small percentage of errors.

Let $M(T)$ be the number of distinguishable *available* signals from which an actual signal can be chosen to be sent through the transmission system in time T. In other words, meaning will be assigned to $M(T)$ only, of the different possible transmitted signals. Then according to Chap. I, if these $M(T)$ signals can be chosen so that there is a vanishingly small percentage of errors in transmission, the language transmission capacity $R_{\max}$ (measured in nepits) will be attained if, for large values of T,

$$M(T) = K\epsilon^{R_{\max} T} = K\epsilon^{TW \log\left(\frac{P+N}{N}\right)} \tag{7}$$

where K is any constant. Thus, we require

$$M(T) = K \left(\frac{P + N}{N} \right)^{TW} = K \left(\sqrt{\frac{P + N}{N}} \right)^{2TW} \tag{8}$$

Since the average transmitted power is limited to P, each transmitted signal of duration T may be represented as a point on the surface or within the hypervolume of an n-dimensional sphere of radius $\sqrt{2TWP}$. Since

$$n = 2TW \gg 1 \tag{9}$$

nearly all the volume of the n-dimensional sphere is very close to its surface.[1] Let us now choose the $M(T)$ required available transmitted signals at random from the possible transmitted signals, *i.e.*, at random within the volume of the n-dimensional sphere. Surprisingly enough we shall be able to show that with this random

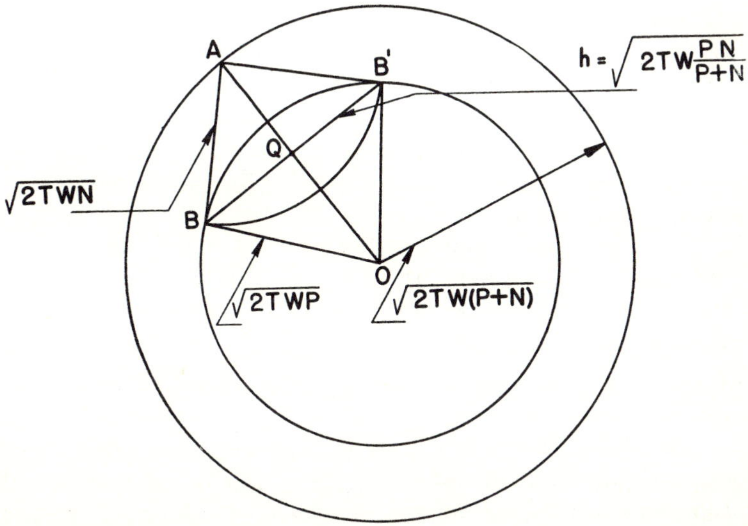

Fig. 6-2. Cross section of signal space showing transmitted and received signals with random noise and with an average power limitation on the transmitted signals.

choice, if $M(T)$ is no larger than required by Eq. (8), there will be a vanishingly small percentage of errors in transmission.

Let us consider the received signals arising from these $M(T)$ chosen transmitted signals. The former will each have an average power limited to $(P + N)$, so that by an argument similar to that used for the transmitted signals, almost all the received signals will be distributed over the surface of an n-dimensional sphere of radius $\sqrt{2TW(P + N)}$. In Fig. 6-2 we show a cross section of the n-dimensional sphere, with a transmitted signal B and its corresponding

[1] The reader can easily convince himself of this fact by noting the increase in the ratio of surface to volume for a given radius as n increases.

received signal A. The distance from B to A is $\sqrt{2TWN}$. The justification for the definite statements that B is on a sphere of radius $\sqrt{2TWP}$, A is on a sphere of radius $\sqrt{2TW(P + N)}$, and the distance from B to A is $\sqrt{2TWN}$ is that this situation is typical, and that all situations differing by an appreciable amount from this typical situation occur in a negligible percentage of the possible cases.[1]

We now wish to calculate the probability that some transmitted signal other than B may have given rise to the received signal A. If we can show that this is vanishingly small when there are $M(T)$ transmitted signals, then we will have shown that the language transmission capacity given by Eq. (6) can be achieved with a vanishingly small percentage of errors by the simple encoding choice of a random selection of the $M(T)$ signals. Now the received signal A arose from a transmitted signal located within the inner sphere of radius $\sqrt{2TWP}$, and located within a distance $\sqrt{2TWN}$ from A. The transmitted signal therefore was in the volume whose projection in Fig. 6-2 is the lens-shaped area. Now, no point in the lens shaped area is more than a distance

$$h = \overline{BQ} \tag{10}$$

from the point Q in Fig. 6-2. The volume in which any reasonably possible transmitted signal was located which gave rise to the received signal A is thus less than the volume of a sphere of radius h. By simple geometry,[2]

$$\sqrt{2TWP} \cdot \sqrt{2TWN} = h\sqrt{2TW(P + N)} \tag{11}$$

Therefore

$$h = \sqrt{2TW\frac{PN}{P + N}} \tag{12}$$

Since the transmitted signals were distributed at random, the probability of any particular transmitted signal other than B giving rise to A is less than the ratio of the volumes of the spheres of radius h and radius $\sqrt{2TWP}$, respectively. Now it is known that the

[1] This implies that T is large enough so that the signals are typical ergodic functions.

[2] The angle OBA must be a right angle since $\overline{OB}^2 + \overline{BA}^2 = \overline{OA}^2$ *i.e.*,

$$(\sqrt{2TWP})^2 + (\sqrt{2TWN})^2 = [\sqrt{2TW(P + N)}]^2$$

volume of an n-dimensional sphere of radius r is

$$V = \frac{\pi^{n/2}}{\Gamma\left(\frac{n}{2}+1\right)} r^n \tag{13}$$[1]

Thus the probability that any particular signal other than B gave rise to A is less than

$$\frac{V_h}{V_{OB}} = \frac{h^{2TW}}{(\sqrt{2TWP})^{2TW}} = \frac{\left(\sqrt{2TW\dfrac{PN}{P+N}}\right)^{2TW}}{(\sqrt{2TWP})^{2TW}} = \left(\frac{N}{P+N}\right)^{TW} \tag{14}$$

Altogether there are $[M(T) - 1]$ signals other than B. Therefore, the total probability that any signal other than B gave rise to A is less than

$$[M(T) - 1]\left(\frac{N}{P+N}\right)^{TW}$$

By Eq. (8)

$$\begin{aligned} [M(T) - 1]\left(\frac{N}{P+N}\right)^{TW} &= \left[K\left(\frac{P+N}{N}\right)^{TW} - 1\right]\left(\frac{N}{P+N}\right)^{TW} \\ &= K - \left(\frac{N}{P+N}\right)^{TW} \\ &= K \quad \text{(approximately)} \end{aligned} \tag{15}$$

if T is sufficiently large. The total probability that any signal other than B gave rise to A is thus less than K. Now K may be any constant and we will still have a language transmission capacity $R_{\max}$. Furthermore, the above analysis has shown that K is greater than the probability of a misinterpretation of a received signal. We can therefore choose K so that the percentage of errors in interpreting received signals will be smaller than any preassigned value, and we can still maintain a language transmission capacity $R_{\max}$, if T is sufficiently long. This is what we set out to prove.

The foregoing discussion is the analogue, for the continuous case, of the discussion in Sec. 1.10 of Chap. I. The attainment of a vanishingly small percentage of errors depends upon the use of

[1] D. M. Y. Sommerville *An Introduction to the Geometry of N-dimensions*, Dutton, New York (1929), p. 135.

very long (duration T) code groups, with their inherent time delays before any of the message is interpreted. The language transmission capacity is strictly limited to the value given by Eq. (6). Any attempt to increase $M(T)$ above that required by Eq. (6) leads to practically 100% errors, as was shown in the analogous case discussed in Chap. 1. It is nevertheless quite remarkable that, as originally shown by Shannon, one can show that, by proper encoding into long signals, one can attain the maximum possible language transmission capacity of a system while at the same time obtaining a vanishingly small percentage of errors.

CHAPTER VII

INFORMATION THEORY ASPECTS OF MODULATION AND NOISE REDUCTION

7.1 Introduction

Every communication engineer is familiar with the practice of modulation. In radio broadcasting, there must be some way of separating the various signals that are simultaneously present. This is accomplished by letting each desired signal modulate a carrier, and the separation of signals is then accomplished by the separation of carriers.

Besides making possible the separation of carriers, modulation has other useful characteristics in broadcasting. Thus the usual type of intelligence (audio, video, or other) transmitted in modern communication systems covers many octaves of frequency. Over such a wide range of frequency, propagation characteristics would vary greatly, so that the transmitting medium would introduce large amounts of uncontrollable frequency discrimination. This can largely be avoided by using a carrier, so that only the relatively narrow frequency range (considered in terms of octaves) of the modulated carrier is involved in transmission. Furthermore, if the energy is to be radiated into space, it is necessary to use a carrier because it is possible to obtain efficient radiation only at high frequencies. It is interesting to note that the human body has solved a corresponding problem in sound radiation in a similar manner. The movements of the muscles of the mouth and oral cavity occur at a rate below 10 per sec. However, the human mouth is too small an opening to be an efficient radiator at the wavelength corresponding to 10 cycles sec, that is, 110 ft. Consequently the vocal cords are used to generate a much higher carrier frequency, which can then be modulated by the other speech-regulating elements in the oral cavity.

In recent years a great deal of attention has been paid to the noise-reducing properties of various modulation systems. A certain amount of insight can be gained in studying these noise-reducing properties by using information theory. Accordingly, we shall deal with this application of information theory in the present chapter. We shall start out by calculating the entropies and information transmission rates of several cases in which the type of modulation is of special interest.

Once the concept of modulation has been recognized, its presence in many types of technical and natural phenomena is found to be widespread. Certain interesting and useful properties appear to be common to widely different types of modulation systems, and one system may give a revealing insight into another, as we will see both in this and in later chapters.

As a general definition of modulation we may use the following: *Modulation is the deviation of a signal from a prescribed standard pattern.* The pattern may be a time pattern, a space pattern, or a pattern in any set of coordinates whatever. Ordinarily, the prescribed standard (unmodulated) pattern of a system is considered to be known ahead of time and contains no information. The information in a modulated signal is in the modulation.

Modulation is often used as a means of adapting available apparatus to the transmission of desired information. From the point of view of information theory, a modulation system is a system of fixed constraints. In a modulated signal, the language constraints of the modulating signal are superimposed upon the fixed constraints of the modulation system.

7.2 Case I. The maximum entropy of an ensemble of signals limited to a given bandwidth

As mentioned in the preceding section, a modulation system is, among other things, a system of fixed constraints. These constraints will, of course, cause a loss of entropy, since they decrease the available choice of possible transmitted signals. Before studying particular modulation systems, we will first consider what is probably the least constrained of possible "practical" systems, namely, the case in which only the limitation in bandwidth and some limitation on transmitted power are specified as constraints. This can then serve as a standard, since no modulation system can have less constraints than the foregoing; therefore no modulation system can have greater entropy.

Let us first tabulate the constraints of the systems thus specified.

(1) Bandwidth limited to

$$\frac{\omega_2 - \omega_1}{2\pi} = W \quad \text{(not necessarily starting at zero)}$$

(2) Some limitation on total power, either peak or average power or both being limited.

(3) No other constraints except for the language constraints of the signal.

Let $\qquad H_I \equiv$ entropy of the ensemble

and $\qquad \bar{N}_I \equiv$ entropy power of the ensemble

In order to find the entropy H_I, we would like to add the entropies at the sample points. In Case I, since the passband does not start at zero frequency, the value of a function $G(t)$ of the ensemble is determined not only by the values of the function itself at its sample points but also by the values of its Hilbert transform at these points (see Sec. 2.3). Now in case we knew that the function $G(t)$ was peak limited,[1] we would still have no assurance that its Hilbert transform $G_1(t)$ was peak limited. However, we do know that the entropy of an ensemble of Hilbert transforms must be the same as the entropy of the ensemble of functions from which they were derived, since one ensemble determines the other completely. Furthermore we know that the average power of $G(t)$ is equal to that of $G_1(t)$, since they have the same power spectrum. While these items are of interest, they still do not enable us to calculate the total entropy of $G(t)$ in general.

If in addition to the limitation on bandwidth, there is only a limitation on the *average* power, then the maximum entropy can be calculated.[2] In this case, the ensemble of maximum entropy has the characteristics of a white gaussian noise ensemble. In fact

[1] This would place the maximum entropy per sample point of the $G(t)$ elements in the range

$$\log \sqrt{\frac{4S}{\epsilon^2}} \leq h_L \leq \log \sqrt{4S}$$

where S is the peak power.

[2] In an ergodic ensemble, the average power of a typical function is determined by the symbol (quadiva) probabilities and must therefore be the same for all typical functions of the ensemble.

both $G_1(t)$ and $G(t)$ are white gaussian noise ensembles. The entropies of $G(t)$ and of $G_1(t)$ at the sample points are equal. They are also additive, since the sampling values of $G(t)$ and $G_1(t)$ are independent of one another. In fact, since this ensemble has the properties of a white gaussian noise ensemble, its entropy is[1]

$$H_I = TW \log 2\pi\epsilon P \tag{1}$$

where P is its average power, and

$$\bar{N}_I = P \tag{2}$$

is its entropy power.

If white gaussian noise of average power N is superimposed on this ensemble, the maximum language transmission capacity will be

$$R_{\max} = W \log \frac{P + N}{N} \tag{3}$$

the derivation being the same as that of Eq. (36) of Chap. V.

Actually, in practice, the power limitation on a modulated signal is more likely to be a peak power limitation than an average power limitation. In this case, there is no assurance that the entropy of the $G_1(t)$ sampling values is equal to that of the $G(t)$ sampling values. Furthermore, these entropies may not be additive, since there may be constraints between the sampling values. The situation is therefore not amenable to accurate calculation in this case. In order to get some idea of orders of magnitude in this case, let us assume that the entropies of the $G_1(t)$ values are equal to those of the $G(t)$ values and that these entropies are additive. This would put the entropy in the range

$$TW \log \frac{4S}{\epsilon^2} \leq H_L \leq TW \log 4S \tag{4}$$

and the entropy power would be in the range

$$\frac{1}{2\pi\epsilon}\left(\frac{4S}{\epsilon^2}\right) \leq \bar{N}_L \leq \frac{1}{2\pi\epsilon}\,(4S) \tag{5}$$

according to Sec. 5.3.

[1] In writing Eq. (1), we assume that the entropy of a white gaussian noise ensemble is the same as that of a random-noise ensemble for the same duration, bandwidth, and average power. Although this fact is almost obvious, the proof is tedious and will not be given here.

If white gaussian noise of average power N is superimposed on this ensemble, then for the entropy power $\bar{N}'$ of the combination, we have

$$\bar{N}' \geq \bar{N}_L + N \tag{6}$$

Thus for the entropy H' of the combination

$$H' \geq TW \log \left\{ 2\pi\epsilon \left[\frac{1}{2\pi\epsilon} \left(\frac{4S}{\epsilon^2} \right) + N \right] \right\} = TW \log \left[\frac{4S}{\epsilon^2} + 2\pi\epsilon N \right] \tag{7}$$

The entropy of the noise alone is

$$H_N = TW \log 2\pi\epsilon N \tag{8}$$

Thus the language transmission capacity is approximately

$$R = \frac{H' - H_N}{T} \geq W \log \left[\frac{4S}{2\pi\epsilon^3 N} + 1 \right] \tag{9}$$

7.3 Case II. Amplitude modulation (double sideband)

We wish next to consider the very practical case of double-sideband amplitude modulation with a peak power limitation. Here we have the following constraints:

1. Bandwidth limited (by law or due to apparatus limitation) to

$$\frac{\omega_2 - \omega_1}{2\pi} = 2W_a$$

2. Peak quadiva (peak power) limited to A^2.
3. Characteristics of amplitude modulation, *i.e.*, no anti-symmetrical sidebands.

A general amplitude-modulated signal of this type may be written

$$\begin{aligned} G(t) &= A[1 + g(t)] \cos 2\pi Ft \\ &= A \cos 2\pi Ft + Ag(t) \cos 2\pi Ft \end{aligned} \tag{10}$$

where $F = (\omega_2 + \omega_1)/4\pi$.

The degrees of freedom of $G(t)$ lie entirely in $g(t)$, since, for a given transmitter, A and F are presumably fixed.[1] The entropy of $G(t)$ is then the same as the entropy of $g(t)$ provided that they are both expressed in the same system of coordinates. Another way of realizing this is to note that the probability of obtaining $G(t)$ for the signal is the same as the probability of obtaining $g(t)$ for the mod-

[1] The reader can show that if $p_A = 1$ and $p_F = 1$, the entropy corresponding to these two "degrees of freedom" vanishes.

ulation. $g(t)$ is a peak-limited signal. Its extreme negative value is (-1), and its maximum positive value is limited by the apparatus to the value $\left(\frac{\sqrt{S}}{A} - 1\right)$, where S is the peak power of $G(t)$.

The signal $g(t)$ has a possible frequency spectrum starting at 0 and extending to the maximum possible modulation frequency for the system, namely, $W_a = (\omega_2 - \omega_1)/4\pi$. It then has $2TW_a$ sample points, and the ensemble of signals $g(t)$ have a maximum entropy in the range

$$TW_a \log \frac{4S'}{\epsilon^2} \leq H_{\text{AM}} \leq TW_a \log 4S' \tag{11}$$

where $S' = \frac{S}{4}$ and the coordinates used are those of $Ag(t)$ rather than $g(t)$.

The entropy power of the ensemble is then in the range

$$\frac{1}{2\pi\epsilon}\left(\frac{S}{\epsilon^2}\right) \leq \bar{N}_{\text{AM}} \leq \frac{1}{2\pi\epsilon}(S) \tag{12}$$

Let us next consider that white gaussian noise due to thermal noise, etc., may be superimposed upon the amplitude-modulated signal. Let the average noise quadiva be

$$N = 2KW = 2K\left(\frac{\omega_2 - \omega_1}{2\pi}\right) \tag{13}$$

If $n(t)$ is the noise signal, then

$$\overline{n(t)^2} = N \tag{14}$$

In this case, it is convenient to separate the noise into symmetrical and antisymmetrical sidebands of the carrier, and combine into an in-phase and quadrature term.[1] Thus, we obtain for the noise signal

$$\text{noise signal} = n_I(t) \cos 2\pi Ft + n_Q(t) \sin 2\pi Ft \tag{15}$$

where $n_I(t)$ and $n_Q(t)$ are stochastic functions of time having no frequency components above

$$W_a = \frac{W}{2} \tag{16}$$

and

$$\overline{n_I^2} = \overline{n_Q^2} = N \tag{17}$$

[1] *F.A.,M., and N.*, pp. 167–175.

Both $n_I(t)$ and $n_Q(t)$ have a normal distribution of linivas, since they arise from the linear superposition of a large number of random-frequency components to which the central-limit theorem can be applied.[1] In fact, both $n_I(t)$ and $n_Q(t)$ are random noise.

When the noise is superimposed on the signal, we have for the combined signal

$$G(t) + n(t) = A \cos 2\pi Ft + Ag(t) \cos 2\pi Ft + n_I(t) \cos 2\pi Ft + n_Q(t) \sin 2\pi Ft \qquad (18)$$

The envelope of this signal is

$$\sqrt{[A + Ag(t) + n_I(t)]^2 + [n_Q(t)]^2} \qquad (19)$$

For large values of signal-to-noise ratio, this envelope reduces to

$$A + Ag(t) + n_I(t) \quad \text{(approximately)} \qquad (20)$$

In the ensemble of envelope signals we then have the random noise $n_I(t)$ superimposed upon the desired signal $Ag(t)$. If $\bar{N}'$ is the entropy power of the envelope, we have

$$\overline{N'} \geq \bar{N}_{\text{AM}} + N \qquad (21)$$

since

$$\overline{n_I^2} = N = 2K\left(\frac{\omega_2 - \omega_1}{2\pi}\right) \qquad (22)$$

Thus the entropy of the envelope signal is

$$H' \geq TW_a \log\left[\left(\frac{S}{\epsilon^2}\right) + 2\pi\epsilon\, N\right] \qquad (23)$$

The language transmission capacity of the envelope signals is then

$$R_{\text{AM (max)}} = \frac{H' - H_N}{T} \geq W_a \log\left[\frac{\frac{S}{\epsilon^2} + 2\pi\epsilon N}{2\pi\epsilon N}\right] = W_a \log\left[\frac{S}{2\pi\epsilon^3 N} + 1\right]$$

$$= W_a \log \frac{S}{2\pi\epsilon^3 N} \quad \text{(approximately)} \qquad (24)$$

[1] It follows that the envelope of pure white gaussian noise, namely, $\sqrt{n_I^2 + n_Q^2}$ has the same probability distribution as the magnitude of the radius vector of a two-dimensional normal distribution. The detected envelope will therefore not have a normal distribution.

We emphasize that this formula holds only for large signal-to-noise ratios. If the signal-to-noise ratio is not large, the superposition of desired signal and noise is not linear and the whole situation is more complicated.

We note that for similar transmitted-power conditions, the foregoing rate of information transmission is a little less than half of that of Case I [see Eq. (9)], since

$$W_a = \frac{W}{2} \tag{25}$$

The original noise N in both cases has the same numerical value

$$N = 2K\left(\frac{\omega_2 - \omega_1}{2\pi}\right) = 2KW = 4KW_a \tag{26}$$

PROBLEMS

1. By direct application of the use of symmetrical and antisymmetrical sidebands, show that a white gaussian noise signal may be expressed as

$$n_I(t) \cos \frac{\omega_2 + \omega_1}{2} t + n_Q(t) \sin \frac{\omega_2 + \omega_1}{2} t$$

where $n_I(t)$ and $n_Q(t)$ are random noise of bandwidth $\left(\frac{\omega_2 - \omega_1}{4\pi}\right)$ and

$$\overline{n_I^2} = \overline{n_Q^2} = N = 2K\left(\frac{\omega_2 - \omega_1}{2\pi}\right)$$

2. Find the entropy of the detected envelope of pure white gaussian noise of magnitude

$$N = 2K\left(\frac{\omega_2 - \omega_1}{2\pi}\right)$$

7.4 Case III. Amplitude modulation (single sideband transmission)

This is similar to Case I if $\omega_2 - \omega_1$ is also the peak-modulation frequency of interest. However, in the present case, a large carrier is locally added at the receiver and the information is obtained by detecting the envelope of the carrier plus signal plus noise. If an ideal linear detector is used, a particular frequency component of the signal

$$B_s \cos [2\pi(F + \alpha)t + \psi_s] \tag{27}$$

superimposed on the large carrier

$$A \cos 2\pi Ft \tag{28}$$

would give rise to an envelope amplitude

$$A\left[1 + \frac{B_s}{A}\cos\left(2\pi\alpha t + \psi_s\right)\right] \tag{29}$$

approximately, provided that

$$A \gg B_s$$

If a noise component of this frequency

$$B_n \cos\left[2\pi(F + \alpha)t + \psi_n\right] \tag{30}$$

were also present, the two together would give rise to an envelope amplitude

$$A\left[1 + \frac{B_s}{A}\cos\left(2\pi\alpha t + \psi_s\right) + \frac{B_n}{A}\cos\left(2\pi\alpha t + \psi_n\right)\right] \tag{31}$$

In other words the effect of addition of the local carrier plus detection is merely one of shifting frequencies, but it does not change the relative values of the different frequencies components of either the signal itself or the noise itself or the relations between the two. Since the entropy can be calculated from the properties of the frequency components, the entropies of the signal and noise and the information transmission rate are all unaffected by the addition of the local carrier combined with detection.

The formulas of Case I therefore apply to the case of single sideband transmission provided that W is taken as the maximum modulation frequency.

Even though the frequency band of the detected envelope starts at zero, so that we do not need to know the characteristics of the Hilbert transform of the detected envelope in order to calculate the entropy of the envelope ensemble, the limitations (such as peak-value limitations) on the quadiva of the envelope are not the same as those on the original high-frequency signal. We therefore know as much, but no more, about the entropy in Case III as in Case I.

Exercise

Show that the entropy in the single-sideband transmission of narrow-band angle modulation is identical with that of single-sideband transmission of amplitude modulation.

7.5 Case IV. Frequency modulation

As in Case II, we will limit our analytical discussion to the case in which the superimposed noise is small in comparison with the signal.

A pure frequency-modulated signal may be expressed as

$$A \sin \theta = A \sin [2\pi F t + 2\pi \int G(t) dt] \tag{32}$$

where A is a constant, called the signal amplitude, and F is a constant, called the carrier frequency. The *instantaneous frequency* of the above signal is, by definition,

$$\text{instantaneous frequency} = \frac{1}{2\pi}\frac{d\theta}{dt} = F + G(t) \tag{33}$$

Since A and F are the same constant for an ensemble of frequency-modulated signals, the information is carried by $G(t)$.

Let us now assume that, during transmission, white gaussian noise is linearly superimposed upon the frequency-modulated signal. We will assume that the quadratic content of the noise is small in comparison with that of the signal. Let us first find the result of the superposition of any particular small frequency component

$$B \sin [2\pi(F + \mu)t] \tag{34}$$

of the noise on the signal. The combined signal may be written

$$\begin{aligned} A \sin [2\pi F t &+ 2\pi \int G(t) dt] + B \sin [2\pi F t + 2\pi \mu t] \\ &= \{A \cos [2\pi \int G(t) dt] + B \cos 2\pi \mu t\} \sin 2\pi F t \\ &\quad + \{A \sin [2\pi \int G(t) dt] + B \sin 2\pi \mu t\} \cos 2\pi F t \\ &= \sqrt{\{A \cos [2\pi \int G(t) dt] + B \cos 2\pi \mu t\}^2 + \{A \sin [2\pi \int G(t) dt] + B \sin 2\pi \mu t\}^2} \\ &\quad \sin \left\{2\pi F t + \tan^{-1} \frac{A \sin [2\pi \int G(t) dt] + B \sin 2\pi \mu t}{A \cos 2\pi \int G(t) dt + B \cos 2\pi \mu t}\right\} \end{aligned} \tag{35}$$

Accordingly, the instantaneous frequency of the combined signal is

$$\frac{1}{2\pi}\frac{d}{dt}\left\{2\pi F t + \tan^{-1} \frac{A \sin [2\pi \int G(t) dt] + B \sin 2\pi \mu t}{A \cos [2\pi \int G(t) dt] + B \cos 2\pi \mu t}\right\} \tag{36}$$

Since $B/A \ll 1$, we can expand

$$\tan^{-1} \frac{A \sin [2\pi \int G(t)dt] + B \sin 2\pi\mu t}{A \cos [2\pi \int G(t)dt] + B \cos 2\pi\mu t}$$

in a Taylor series and neglect higher power terms of (B/A). We thus obtain[1]

$$\tan^{-1} \frac{A \sin [2\pi \int G(t)dt] + B \sin 2\pi\mu t}{A \cos [2\pi \int G(t)dt] + B \cos 2\pi\mu t}$$
$$= 2\pi \int G(t)dt + \frac{B}{A} \sin \left\{2\pi[\mu t - \int G(t)dt]\right\} \quad \text{approximately} \tag{37}$$

Using this value, the instantaneous frequency of the combined signal is

$$f = F + G(t) + \frac{B}{A} [\mu - G(t)] \cos \{2\pi[\mu t - \int G(t)dt]\} \tag{38}$$

[1] If $x = B/A$,

$$\frac{\sin 2\pi \int G(t)dt + \frac{B}{A} \sin 2\pi\mu t}{\cos 2\pi \int G(t)dt + \frac{B}{A} \cos 2\pi\mu t} \text{ is of the form } \frac{a + bx}{c + dx}$$

For small x

$$\frac{a + bx}{c + dx} = \frac{a}{c} + \left[\frac{bc - ad}{c^2}\right] x + \cdots$$

This, in turn, is of the form $\alpha + \beta x$.
For small x

$$\tan^{-1} (\alpha + \beta x) = \tan^{-1} \alpha + x \frac{\beta}{1 + \alpha^2} + \cdots$$

Using, the foregoing values of α, β, a, b, c, d, and x, we have for small x

$$\tan^{-1} (\alpha + \beta x) = \tan^{-1} \frac{a}{c} + \frac{bc - ad}{c^2} \frac{1}{1 + \left(\frac{a}{c}\right)^2} x$$
$$= \tan^{-1} \frac{a}{c} + \frac{(bc - ad)}{c^2 + a^2} x$$
$$= 2\pi \int G(t)dt + \frac{B}{A} \left[\sin 2\pi\mu t \cos 2\pi \int G(t)dt - \cos 2\pi\mu t \sin 2\pi \int G(t)dt\right]$$
$$= 2\pi \int G(t)dt + \frac{B}{A} \sin \left\{2\pi[\mu t - \int G(t)dt]\right\}$$

Equation (38) shows that the frequency modulation produced by

$$B \sin [2\pi(F + \mu)t] \tag{39}$$

has a magnitude of

$$\frac{B}{A} [\mu - G(t)] \tag{40}$$

and that it is linearly superimposed on the signal.

The instantaneous frequency of the added modulation is obtained by taking $\frac{1}{2\pi} \frac{d}{dt}$ of the argument of the cosine in Eq. (38). Thus

$$\begin{aligned} \text{instantaneous added modulation frequency} \\ &= \frac{1}{2\pi} \frac{d}{dt} \{2\pi[\mu t - \textstyle\int G(t)dt]\} \\ &= \mu - G(t) \qquad (41) \end{aligned}$$

The entropy of the frequency-modulated signal before superposition of the noise is that of $G(t)$. The peak value of $G(t)$ is ΔF, the maximum permitted deviation frequency, and its bandwidth is W_m, the maximum useful modulation frequency. The entropy of the ensemble of signals $G(t)$, is that of a peak-limited ensemble, and is therefore in the range

$$TW_m \log [4(\Delta F)^2] \geq H_{FM} \geq TW_m \log \left[\frac{4(\Delta F)^2}{\epsilon^2}\right] \tag{42}$$

This entropy is measured using frequency-deviation coordinates. The constraints of the FM system have ruled out all frequency-deviation components in $G(t)$ which are of higher frequency than W_m.

Let us next find the power and the entropy of the noise. Assume that the original noise ensemble superimposed on the system was white gaussian noise whose frequency components were uniformly distributed from $F - \Delta F$ to $F + \Delta F$. According to Eqs. (34) to (41) these give rise to FM components of frequencies ranging from 0 to $2\Delta F$. However, those components of modulation frequencies higher than W_m will be removed (by the audio amplifier) after detection. They will therefore not appear in the final output.

In order to be able to calculate the magnitude of the noise quantitatively, we will calculate it for periods of time when $G(t)$ is zero. If that is not done, it is difficult to carry through the analysis and to interpret any results which may be obtained. The periods of time when $G(t)$ is zero are times when the signal modulation is zero. This would, for example, correspond to a pause between sentences, in an audio program.

The original noise signal, due to white gaussian noise, has an average quadiva

$$N = 2K\left(\frac{\omega_2 - \omega_1}{2\pi}\right) = 4K\,\Delta F \tag{43}$$

since, in this case

$$\left(\frac{\omega_2}{2\pi}\right) = F + \Delta F \tag{44}$$

and

$$\left(\frac{\omega_1}{2\pi}\right) = F - \Delta F \tag{45}$$

The average over the noise ensemble of the spectrum amplitude $\overline{|\Omega(\omega)|}$ is flat with frequency. The frequency components of this

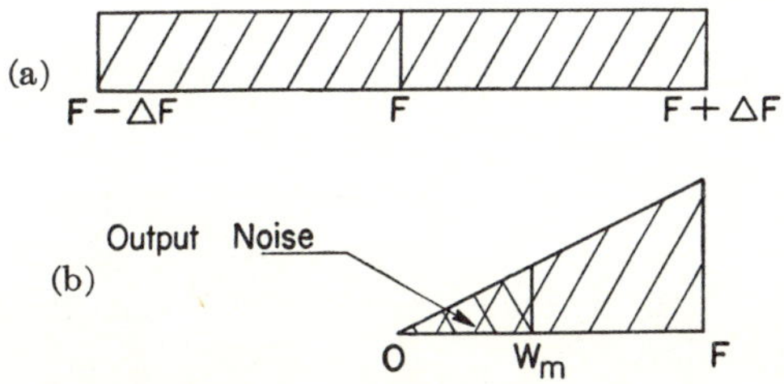

Fig. 7-1. (a) Frequency distribution of the spectrum amplitude of the superimposed white gaussian noise. (b) Frequency modulation spectrum due to the distribution in (a).

noise ensemble will give rise to frequency modulation of the desired carrier, as already stated. Frequency components of the original noise signal equidistant from but on opposite sides of the original carrier will give rise to the same noise-modulation frequency. The mean-square values of these noise-modulation components are additive, since their sources are incoherent. According to Formula (40), with $G(t) = 0$, the actual frequency modulation of the carrier will have a triangular spectrum as shown in Fig. 7-1(b) rather than a flat spectrum. Of this triangular spectrum, only those frequency components in the cross-hatched area in Fig. 7-1(b) will have frequencies low enough to pass through the audio (or video) amplifier.

Let us write $2K\,df$ for the original white gaussian noise average quadiva in the narrow spectrum range df, since

$$\int_{F-\Delta F}^{F+\Delta F} 2K\,df = 4K\,\Delta F \tag{46}$$

This narrow band noise is equivalent to a frequency component

$$\sqrt{4K\,df}\,\sin\,(2\pi ft + \theta_f) \tag{47}$$

and gives rises to frequency modulation of the desired carrier of magnitude

$$\frac{\sqrt{4K\,df}}{A}\,|f - F| \tag{48}$$

according to Formula (40). The FM average quadiva (output power) arising from (48) is therefore

$$\frac{1}{2}\,\frac{(f - F)^2}{A^2}\,4K\,df \tag{49}$$

The output noise average quadiva due to the entire noise spectrum is therefore

$$\int_{F-W_m}^{F+W_m} \frac{(f - F)^2}{2A^2}\,4K\,df = \frac{4}{3}\,\frac{K}{A^2}\,W_m^3 \tag{50}$$

On the same basis, the average output signal quadiva of a 100% modulated desired signal is

$$\tfrac{1}{2}(\Delta F)^2 \tag{51}$$

so that the output signal-to-noise power ratio is approximately

$$\frac{3A^2(\Delta F)^2}{8K(W_m)^3} \tag{52}$$

for a 100% modulated signal if we assume that the presence of the signal modulation does not affect the average noise. In the case of double-sideband amplitude modulation, the corresponding output signal-to-noise power ratio for a 100% modulated signal in the same strength noise field is

$$\frac{(A')^2}{8KW_m} \tag{53}$$

according to Eqs. (13), (17), and (20). A' in Eq. (53) stands for the peak value of the unmodulated AM carrier. For 100% modulation and the same average transmitted power in the AM and FM cases, we have

$$(A')^2 = \tfrac{2}{3}A^2 \tag{54}$$

since 100% amplitude modulation increases the transmitted power by 50%. From Eqs. (52), (53), and (54) we therefore obtain

$$\frac{3}{8}\frac{A^2(\Delta F)^2}{K(W_m)^3}\bigg/\frac{(A')^2}{8KW_m} = \frac{1}{2}\left(3\,\frac{\Delta F}{W_m}\right)^2 \tag{55}$$

as the noise-improvement factor of frequency modulation over double-sideband amplitude modulation (on a power basis).

Let us now calculate the entropy of the output noise in the FM case. According to the preceding discussion, the noise output is the same as we would get by sending random noise of average quadiva

$$\frac{4KW_m^3}{A^2} \tag{56}$$

through a triangular filter. According to Secs. 4.7 and 4.9, the entropy of the output noise is therefore

$$\begin{aligned} H_{\text{noise}} &= TW_m \log 2\pi\epsilon \left[\frac{4K(W_m)^3}{A^2}\right] + 2T\int_0^{W_m} \log\left(\frac{f}{W_m}\right) df \\ &= TW_m \log 2\pi\epsilon \left[\frac{4K(W_m)^3}{A^2}\right] - 2TW_n. \\ &= TW_m \log \frac{2\pi}{\epsilon}\left[\frac{4K(W_m)^3}{A^2}\right] \end{aligned} \tag{57}$$

The rate of transmission of language information is equal to the entropy of the signal-plus-noise minus the entropy of the noise. Since we are assuming that the signal-to-noise ratio is large, the entropy of the signal plus noise is approximately that of the signal alone. Thus,

$$\text{rate of transmission of language information} = \frac{H_{\text{FM}} - H_{\text{noise}}}{T}$$

Now from Eqs. (42) and (57), we have

$$W_m \log\left[\frac{4(\Delta F)^2}{\epsilon^2}\right] - W_m \log\frac{2\pi}{\epsilon}\left[\frac{4K(W_m)^3}{A^2}\right]$$

$$= W_m \log\left[\frac{2}{\pi\epsilon}\cdot\frac{\Delta F^3}{W_m^3}\,\frac{A^2}{4K\,\Delta F}\right] \leq \frac{H_{\text{FM}} - H_{\text{noise}}}{T} \tag{58a}$$

$$\leq W_m \log\left[\frac{2\epsilon}{\pi}\cdot\frac{A^2}{4K\,\Delta F}\,\frac{\Delta F^3}{W_m^3}\right] \tag{58b}$$

This gives an idea of the rate of transmission of language information in frequency modulation. It should be noted that $A^2/8K\ \Delta F$ is the signal-to-noise power ratio of the original received signal before detection.

Comparing Eq. (55) with Eqs. (24) and (58a), we see the quantitative effect of noise reduction on the rate of transmission of language information.

7.6 Case V. Pulse code modulation[1]

In pulse code modulation (PCM) the message signal is sampled at equidistant points as shown in Fig. 7-2(a), and the signal values (linivas) at these points are replaced with the nearest integral (or other of a set of chosen discreet) values, as shown in Fig. 7-2(b). The integral values are then coded into pulse groups. For example in Fig. 7-2(c), the pulse groups have four possible pulse positions. Each pulse group indicates the signal value at a corresponding sample point according to the code in Fig. 7-3. At each of the possible pulse positions, a pulse group either has or has not a pulse. The presence of a pulse is indicated by a 1, its absence by a zero in the equivalent code. These pulse groups can thus express all integral values of signal level from zero to 15. Each pulse group may be preceded by a synchronizing pulse of characteristic shape (for example, longer duration) showing when the pulse group starts. At the decoder, the pulse groups are reconverted into the original signal. There is, however, some discrepancy between the original and the decoded signal, since the latter is limited to only 16 possible levels. This discrepancy may be considered a type of ergodic noise[2] and is called quantizing noise. It is predictable on the average, but not in detail, if only the decoded (output) signal is known. Pulse code modulation is ordinarily used in a specialized type of service in which quantizing noise turns out to be the principal type of

[1] For information regarding pulse code modulation, see: W. M. Goodall, "Telephony by Pulse Code Modulation," *Bell System Tech. Jour.*, vol. 26, July, 1947, pp. 395–409; L. A. Meacham and E. Peterson, "An Experimental Multichannel Pulse Code Modulation System of Toll Quality," *Bell System Tech. Jour.*, vol. 27, January, 1948, pp. 1–43; and B. M. Oliver, J. R. Pierce, and C. E. Shannon, "The Philosophy of PCM," *Proc. I.R.E.*, vol. 36, November, 1948, pp. 1324–1331.

[2] A reasonable argument could also be given for calling the discrepancy either cross talk or distortion (see Chap. III, p. 124).

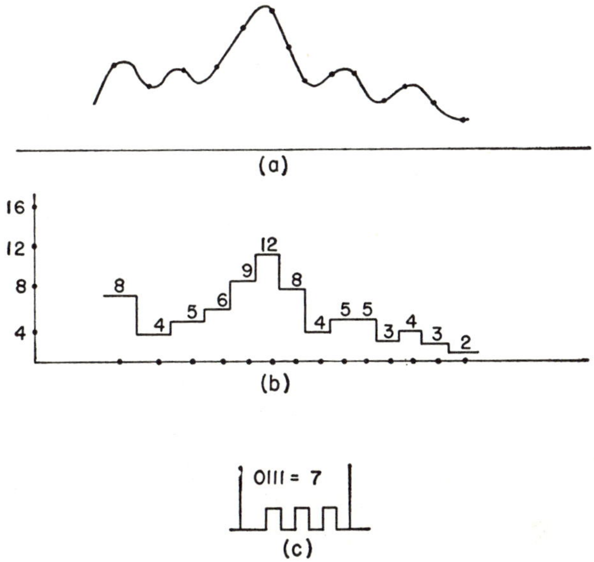

Fig. 7-2. Aspects of pulse code modulation: (a) original signal showing sample points; (b) quantized form of (a); (c) a pulse group representing the level 7.

Pulse Group	Code	Amplitude Represented	Pulse Group	Code	Amplitude Represented
	0000	0		1000	8
	0001	1		1001	9
	0010	2		1010	10
	0011	3		1011	11
	0100	4		1100	12
	0101	5		1101	13
	0110	6		1110	14
	0111	7		1111	15

Fig. 7-3. Binary code for four-pulse groups in PCM.

noise present.[1] We are principally interested, in this section, in showing how increasing the number of pulses in a pulse group decreases the effect of quantizing noise. We will therefore assume in the following analysis that quantizing noise is the only discrepancy between the input and output signals in a pulse code modulation system.

In transforming a signal to quantized form, as for example in going from Fig. 7-2(a) to Fig. 7-2(b), the maximum error which can be made at any sample point is $\pm\ h/2$, where h is the magnitude of one quantizing step. All values of error up to this maximum value are equally likely. The mean-square value of the noise is therefore

$$\begin{bmatrix}\text{mean-square} \\ \text{quantizing noise}\end{bmatrix} = \frac{1}{h}\int_{-h/2}^{h/2} x^2 dx = \frac{h^2}{12} \tag{59}$$

Thus

$$\text{root-mean-square noise} = \frac{h}{2\sqrt{3}} \tag{60}$$

The decoded signal on the other hand can have any integral signal value between 0 and $(2^n - 1)h$, where n is the number of possible pulse positions. ($n = 4$ in Figs. 7-2 and 7-3). Assuming that all allowed signal values are equally likely, which is the maximum entropy condition, the average signal power will be approximately $[(2^n - 1)h]^2/3$ and the signal-to-noise power ratio will be approximately

$$\frac{\text{average signal power}}{\text{average noise power}} = [2(2^n - 1)]^2 \tag{61}$$

The maximum rate of transmission of information, on the other hand, is simply

$$\log\,(2^n) = n \log 2 \tag{62}$$

binits per pulse group, since there are 2^n different possible signals which may be represented by a pulse group, and the decoded signals are essentially all correct to the limit of accuracy of the quantizing process.

The foregoing system of pulse code modulation is the one most widely used in practice. A more general system is, however, theoretically possible. This system would allow a larger number of

[1] See Oliver, Pierce, and Shannon, *Proc. I.R.E.*, vol. 36, November, 1948, pp. 1324–1331.

pulse levels, say L, rather than just 0 and 1 as shown in Fig. 7-3. Such a system would allow L^n quantizing levels instead of just 2^n as in Fig. 7-3. In this way the maximum language transmission capacity could be pushed closer to the limit imposed by the superimposed random noise. It is shown by Oliver, Pierce, and Shannon[1] that the maximum number of pulse levels which can be used so that the pulse error rate will still be negligible, as required by pulse code modulation is approximately

$$L = \sqrt{0.1\left(\frac{P}{N}\right) + 1} \tag{63}$$

provided that P/N is reasonably large in comparison with unity. In Eq. (63),

$$P = \text{average signal power} \tag{64}$$

$$N = \text{average power of superimposed (white gaussian) noise} \tag{65}$$

In computing the average in Eq. (64) it is assumed that all of the L possible levels are equally probable.

In this generalized system of pulse code modulation, the maximum language transmission capacity is

$$\log L^n = n \log L = \frac{n}{2} \log \left(1 + 0.1\,\frac{P}{N}\right) \tag{66}$$

binits per pulse group.

Now suppose that F_0 is the bandwidth of the original signal [as in Fig. 7-2(a)]. Then according to the sampling theorem in the time domain (see Sec. 2.1), the required number of sampling points is

$$\begin{bmatrix}\text{number of sampling} \\ \text{points per unit of time}\end{bmatrix} = 2F_0 = \begin{bmatrix}\text{number of pulse} \\ \text{groups per unit of time}\end{bmatrix} \tag{67}$$

Since there are n pulse positions in a code group, we then have

$$\begin{bmatrix}\text{number of pulse positions} \\ \text{per unit of time}\end{bmatrix} = 2F_0 n \tag{68}$$

The minimum bandwidth of the pulse-code-modulation signal is then approximately

$$W = F_0 n \tag{69}$$

[1] *Proc. I.R.E.*, vol. 36, November, 1948, pp. 1324–1331.

since the bandwidth required for satisfactory representation of a number of off-on pulses of equal duration is approximately equal to $\frac{1}{2}$ the number of pulses per second. Assuming that the power of the white gaussian noise is proportional to the bandwidth used, as is usually the case, we also have

$$N = 2KW \tag{70}$$

where K is the noise strength constant introduced in Sec. 7.3. Substituting Eqs. (69) and (67) into Eq. (66), we have

$$\begin{aligned}\begin{bmatrix}\text{maximum language} \\ \text{transmission capacity}\end{bmatrix} &= \frac{W}{2F_0} \log\left(1 + 0.1\,\frac{P}{N}\right) \text{ per pulse group} \\ &= W \log\left(1 + 0.1\,\frac{P}{N}\right) \text{ per unit of time}\end{aligned} \tag{71}$$

Finally, substituting Eq. (70) into (71), we obtain

$$\begin{aligned}&\text{maximum language transmission capacity} \\ &\qquad = R_0 = W \log_2\left(1 + 0.1\,\frac{P}{2KW}\right) \text{ binits/unit of time} \\ &= 2W \log_2 \ (\textit{maximum number of pulse levels}) \text{ binits/unit of time}\end{aligned} \tag{72}$$

We note in passing that the rate R_0 in pulse code modulation is approximately proportional[1] to the available bandwidth W. The other system which we have analyzed and which uses increased available bandwdith to increase R_0 is frequency modulation. The available bandwidth in frequency modulation was $2\Delta F$ in Eq. (58). In the case of frequency modulation [see Eq. (58)], the rate increased only as the logarithm of the available bandwidth. The reason for this difference is that pulse code modulation can assign a different meaning to every different permutation of the components of a pulse group and thus gets a maximum available choice. A detailed examination of the operation of frequency modulation, on the other hand, would show that the latter uses the number $2\Delta F/\mu$ of sidebands present for a given modulation frequency for increasing available choice, but does not make effective use of the different possible permutations of these sidebands.

[1] $\log_2 (1 + 0.1P/2KW)$ varies much more slowly with W than does W for relatively large values of $P/2KW$.

PROBLEM

1. Discuss the reduction of noise modulation due to superimposed white gaussian noise by pulse code modulation and derive a formula to show how it varies with bandwidth.

7.7 Noise reduction in modulation systems—derivation of the limit formula[1]

As has been stated earlier, there are various reasons for using the different types of modulation systems. However, the reason which is of most interest from the standpoint of information theory is noise reduction. Many types of modulation have the characteristic that they can use increased available bandwidth for noise-reduction purposes. Wideband frequency modulation is typical of these modulation systems, and as shown by Eq. (55), the wide bandwidth in frequency modulation can be used for noise-reduction purposes.

Various types of pulse-modulation systems also have noise-improvement factors which increase with the bandwidth utilized. It therefore becomes a matter of interest to know whether it is possible to set a limit to the rapidity with which the signal-to-noise ratio can increase with the utilized bandwidth so as to have some idea of how a given system compares with the best possible system. We will now use information theory for such an investigation.

In the first place, we may point out that signal-to-noise ratio loses its significance if we are not dealing with linear system. Thus if the signal in Fig. 7-4(a) is sent through an amplifier with a square-law characteristic, the signal-to-noise ratio is immediately improved, as shown in Fig. 7-4(b). This process, however, distorts the signal, and the noise improvement would be lost if we sent the signal through an inverse network which restored the undistorted signal.

As an alternative to the signal-to-noise ratio, we will now try to show that the amount of information in a signal is a true measure of signal quality and of the extent to which it resembles or matches the desired signal. Consider the transmission system in Fig. 7-5. Here a phonograph record at the transmitter calls for a given wave-shape of sound pressure vs. time in the room where a broadcast receiver is located. By means of the phonograph turntable and pickup arm, the information on the phonograph record is trans-

[1] A paper based on the remainder of this chapter was published by the author in *Proceedings Second London Symposium on Applications of Communication Theory*, Butterworth Scientific Publication, London, 1953.

formed into an electrical voltage vs. time signal in the audio amplifier. This in turn, after several intermediate transformations, is transformed into a modulated RF electromagnetic wave in the space between the transmitter and the receiver. After further transformations the signal finally appears as sound pressure vs. time in the room where the loudspeaker is located.

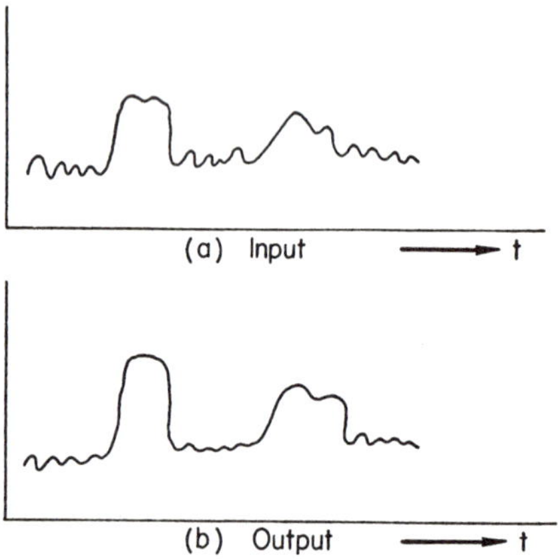

Fig. 7-4. Improvement of signal-to-noise ratio by a square law amplifier, without gain in information.

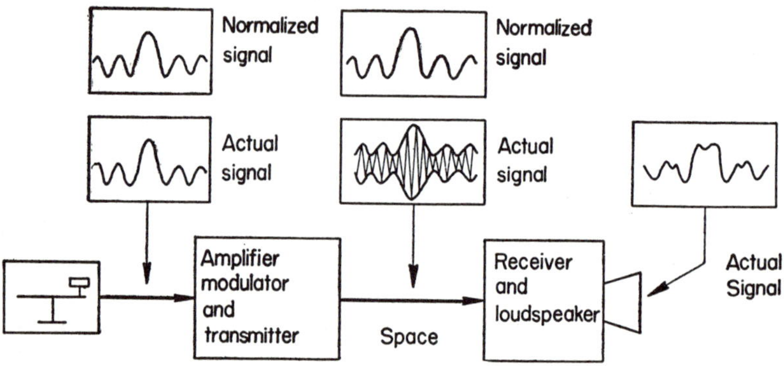

Fig. 7-5. An information transmission system, showing the ultimate output called for by the signal at various points as a normalized signal.

The actual signal called for by the phonograph record is recorded in code on the record. This code includes a convention regarding turntable speed, and includes conventions regarding the translation of mechanical displacement on the record into electrical voltage in the amplifier and, in turn, into sound pressure at the destination. In a similar manner there is a code by means of which the signal anywhere in the transmission system can be translated into an equivalent sound signal at the output. In other words a signal anywhere in the transmission system calls for a particular output signal which we may say is its coded equivalent. In studying the transmission of information through the system we can therefore normalize all the signals by expressing them as the output signal which they call for. In this way we can readily show where the signal is corrupted by distortion or cross talk or noise.

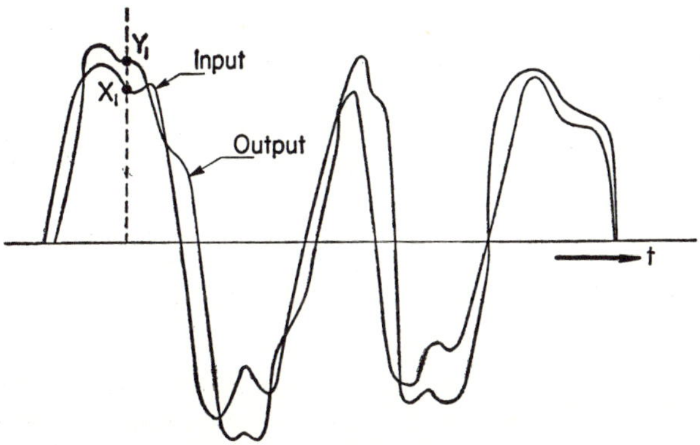

Fig. 7-6.

The fundamental measure of the freedom from noise in the output signal is determined by how closely the output signal matches the signal called for by the input to the over-all transmission system. The discrepancy between the output signal and the normalized input signal is the noise. In a linear system, this discrepancy (noise) is just the algebraic difference between the output and the input. However, the present discussion is not limited to linear systems. In Fig. 7-6 is shown an output signal and the normalized form of the input signal which gave rise to it. If x is the coordinate of the signal value at a sample point, then the information at the sample point having the particular normalized input value $x = x_1$ and the output value $x = y_1$ is

$$\log \left[\frac{p(x_1)_{y_1} dx}{p(x_1) dx} \right] = \log \left[\frac{p(x_1)_{y_1}}{p(x_1)} \right] \tag{73}$$

where

$$p(x_1)\,dx = \left[\begin{array}{l}\text{probability that the sample point has an input value be-}\\ \text{tween } x_1 \text{ and } x_1 + dx \text{ before the output value } y_1 \text{ is known}\end{array}\right] \tag{74}$$

$$p(x_1)_{y_1}\,dx = \left[\begin{array}{l}\text{probability that the sample point has an input value be-}\\ \text{tween } x_1 \text{ and } x_1 + dx \text{ after the output value } y_1 \text{ is known}\end{array}\right] \tag{75}$$

The ratio of probabilities $\left[\dfrac{p(x)_y}{p(x)}\right]$ is a measure of the increase in knowledge of the desired signal on the basis of receiving the actual signal. In order to know how much noise is present, one must know these probabilities.

In any system having linear superposition of signal and noise, if the noise is random noise, we showed in Chap. V that for the probability distribution having maximum language transmission capacity

$$\overline{\log \frac{p(x)_y}{p(x)}} = \log \sqrt{1 + \frac{P}{N}} = \overline{\log \frac{p(y)_x}{p(y)}} \tag{76}$$

where P is the average signal power and N is the average noise power and the bar over a quantity stands for the ensemble average. The ratio of probabilities is thus a measure of the P/N ratio in this particular case. As already pointed out, the (P/N) ratio can be improved by artificial means, such as sending the signal through certain non-linear systems, but nothing like this can be done to the probability ratio. The latter is a true (absolute) measure of the signal quality, in all cases, at least as far as its information is concerned.[1] This information can be decreased but never increased in later parts of the transmission system. The average value of $\log\left[\dfrac{p(x)_y}{p(x)}\right]$ is proportional to the maximum language transmission

[1] If there is intersymbol influence, then x and y can stand for complete input and output signals, respectively, instead of values at a particular sample point. It is then necessary to divide

$$\overline{\log\left[\frac{p(x)_y}{p(x)}\right]}$$

by the number of sample points.

capacity of a system. Thus, the formulas for the language transmission capacity of a system using a specified type of modulation are a more significant measure of the freedom from noise than the formulas for signal-to-noise ratio.

The conventional formula for the noise (voltage) improvement of a modulation system is the square root of the ratio of the average transmitted power required with double-sideband amplitude modulation divided by the average transmitted power required with the modulation system in question in order to obtain the same signal-to-noise ratio in the output.

In accordance with our foregoing discussion, however, we will expect a more absolute measure for the noise (voltage) improvement of a modulation system to be given by *the square root of the ratio of the average transmitted power required with double-sideband amplitude modulation divided by the average transmitted power required with the modulation system in question in order to obtain the same language transmission capacity.*

With this new point of view, let us now calculate the noise improvement of frequency modulation. We recall that the formula for the maximum rate of information transmission in double-sideband amplitude modulation is (for large signal-to-noise ratios) approximately [see Eq. (24)]

$$(R_{\max})_{\text{AM}} \geq W_m \log \frac{S}{2\pi\epsilon^3 N} = W_m \log \frac{\frac{4}{3}P_{\text{AM}}}{\pi\epsilon^3(4KW_m)} \tag{77}$$

where W_m is the maximum modulation frequency and P_{AM} is the average signal power for a 100% modulated signal. The corresponding formula for frequency modulation is [Eq. (58a)]

$$(R_{\max})_{\text{FM}} \geq W_m \log \left[\frac{2}{\pi\epsilon}\left(\frac{\Delta F}{W_m}\right)^3 \frac{A^2}{4K\,\Delta F}\right]$$

$$= W_m \log \left[\frac{2}{\pi\epsilon}\left(\frac{\Delta F}{W_m}\right)^3 \frac{P_{\text{FM}}}{2K\,\Delta F}\right] \tag{78}$$

While Formulas (77) and (78) are inequalities rather than equations, the quantities on the right do correspond to each other and they show how the information rate varies with the controlling factors. The ratio $(P_{\text{AM}}/P_{\text{FM}})$ for equal values of $(R_{\max})_{\text{AM}}$ and $(R_{\max})_{\text{FM}}$ will give the noise-improvement factor on a power basis and its square root will give the noise-improvement factor on a voltage basis. We should get a good expression for this factor by equating the right sides of Formulas (77) and (78). Thus

$$W_m \log \left[\frac{2}{\pi\epsilon}\left(\frac{\Delta F}{W_m}\right)^3 \frac{P_{\text{FM}}}{2K\,\Delta F}\right] = W_m \log \frac{\frac{4}{3}P_{\text{AM}}}{\pi\epsilon^3(4KW_m)} \tag{79}$$

Thus

$$\frac{P_{\text{AM}}}{P_{\text{FM}}} = 3\epsilon^2\left(\frac{\Delta F}{W_m}\right)^2 \tag{80}$$

so that

$$\sqrt{\frac{P_{\text{AM}}}{P_{\text{FM}}}} = \sqrt{3}\epsilon\,\frac{\Delta F}{W_m} \tag{81}$$

The quantity $(\Delta F/W_m)$ on the right of Eq. (81) is called the deviation ratio of the wideband FM system. While certain approximations were made in the derivation of Eq. (81), these approximations would not affect the vital result that the noise *liniva* improvement factor is proportional to the deviation ratio. We note that this same result was also obtained in Eq. (55) which was derived using the conventional point of view.

In the use of extra bandwidth for noise-improvement purposes, one question to which information theory might be hoped to give an answer is the question of what is the maximum noise improvement theoretically possible with any type of modulation. In other words, is there a maximum noise improvement which cannot be exceeded regardless of the type of modulation. If we limit the discussion to cases of linear superposition of the signal with random (or white gaussian) noise, the answer is "yes." This improvement depends on the relative value of the quantity of information per unit time which can be transmitted in the modulation system as compared with the quantity of information per unit time which can be transmitted in a double-sideband AM system when both have the same normalized input signal, *i.e.*, when both are attempting to transmit the same message. Thus, if W_s is the maximum bandwidth used by the modulation system, then the maximum information rate in the ideal modulation system is

$$W_s \log\left(\frac{P + N_s}{N_s}\right) = W_s \log\left(1 + \frac{P}{N_s}\right)$$

$$= W_s \log\left(\frac{P}{N_s}\right) \quad \text{approximately} \tag{82}$$

for large signal-to-noise ratios, where

$$P = \text{average signal power} \tag{83}$$

and

$$N_s = 2KW_s \tag{84}$$

is the average noise power superimposed on the signal. For a given input signal power and given available receiver quality and carrier frequency, it may be assumed that the N_s is proportional to W_s. This accounts for the form of Eq. (84). Suppose that the desired message (*i.e.*, desired output signal) has a maximum bandwidth W_m. Now in view of Eqs. (11), (77) and (82) a message ensemble of bandwidth W_m transmitted by double sideband amplitude modulation and having a signal-to-noise ratio P'/N' can be coded into one-to-one correspondence with the above ideal system provided that

$$W_s \log \left(\frac{P}{2KW_s}\right) \geq W_m \log \frac{\frac{4}{3}P'}{\pi\epsilon 4KW_m} \tag{85}$$

or

$$\left(\frac{P}{2KW_s}\right)^{W_s} \geq \left(\frac{\frac{4}{3}P'}{\pi\epsilon 4KW_m}\right)^{W_m} \tag{86}$$

Equation (86) indicates that the noise-improvement factor $\sqrt{(P'/P)}$ in an ideal system can be expected to improve exponentially[1] with the bandwidth W_s. No modulation system can do better than this ideal system. Frequency modulation, for example, has a noise-improvement factor which as shown in Eq. (81) increases only linearly with bandwidth.

In the early receiver stages where signal and noise are superimposed, this superposition is essentially linear in all systems of modulation. Therefore Eq. (86) is essentially a true description of the upper limit for the noise improvement against random noise which is possible with the use of wide bandwidths. This limit formula is an important result of information theory.

7.8 Introductory considerations concerning information loss and reliability

In this section we wish to consider some ways in which information can be lost and also some ways in which the reliability of infor-

[1] The presence of W_s and of W_m in the denominators of the opposite sides of this inequality keeps the variation from being exactly exponential, but it is still approximately exponential.

mation can be improved. This will, among other things, help us to study the mechanisms for noise reduction. In Fig. 7-7 is a diagram of a general system for the transmission of information. The output of the source is a sequence of messages, drawn from the message alphabet α. In particular cases, these messages may be individual symbols, or they may be continuous signals of specified waveshape and duration, or they may be messages of other types. During the process of transmission of information, these messages go through a

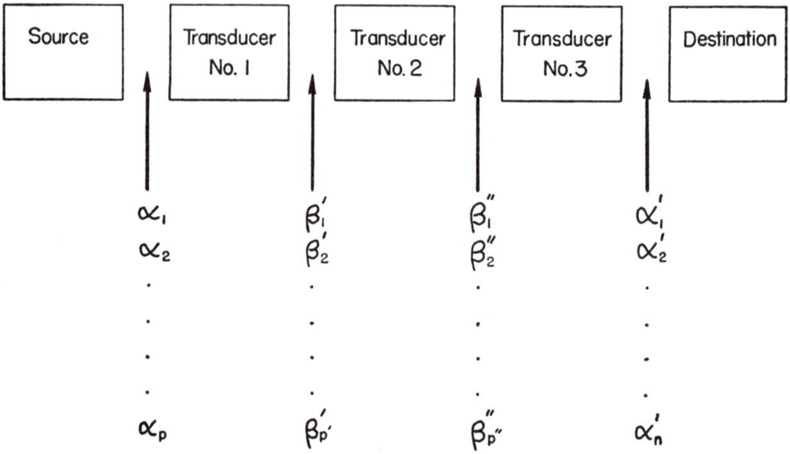

Fig. 7-7. A general transmission system, showing the alphabets at various locations.

series of transducers before reaching the final destination. These transducers in technical cases may, for example, be amplifiers, modulators, detectors, loudspeakers, or transmission paths. Furthermore, a group of transducers in cascade may also be considered as a single transducer. Thus a whole radio receiver may be considered a single transducer, or the individual stages of the receiver may be considered as separate transducers.

At a transducer, an input message is changed into an output message. Thus, at transducer No. 1 (Fig. 7-7) an input message α is changed into a message selected from the β' alphabet. *If the particular β' message to which an α message gives rise is controlled by definite probabilities other than zero and unity, then we say that there is ergodic noise in the transducer.* If the particular β' message to which an α message gives rise is not certain but also is *not* controlled by definite probabilities, then we say that there is *non-ergodic* noise in the transducer. If every particular α message is certain to give rise to a respective particular β' message, we say that the transducer is noiseless. If every different α message will give rise to a different particular β' message, then not only is the transducer noiseless but it also loses no information. If several different α messages will give rise to the same β' message, then information will be lost in the transducer. There is one general conclusion which we may draw immediately from Fig. 7-7. This is that, in order that there shall be no loss of information, it is necessary that the number of messages in the α alphabet shall be less than (or equal to) that in the β' alphabet, which in turn must be less than (or equal to) that in the β'' alphabet, and so on.

Let $p(\alpha)$ be the probability that the symbol α is transmitted by the source. Then the amount of information which is in the message as it leaves the source is

$$\log \frac{1}{p(\alpha)} = -\log p(\alpha) \tag{87}$$

If the input message α causes the output message β', then the information in the output message is

$$\log \left[\frac{p(\alpha)_{\beta'}}{p(\alpha)}\right] \tag{88}$$

where $p(\alpha)_{\beta'}$ is the probability that message α was transmitted by the source on the basis of β' being the output of the transducer. If

$$p(\alpha)_{\beta'} = 1 \tag{89}$$

then we know for sure that α was transmitted when β' is the output of the transducer, and therefore the transducer must be noiseless.

Suppose, however, that for a given input α to the transducer the output β' is certain, but that several different inputs, say α_1, α_2, and α_3, all have the same output. In this case

$$p(\alpha)_{\beta'} < 1 \tag{90}$$

even though there is no noise in the transducer. It is clear that the output of the transducer will have lost some of the information that was in the input in this case. This may be described as a loss of information due to *loss of accuracy*. We shall use the term *alphabet deficit* (or *ambiguity*) to describe the foregoing phenomenon. In the event that certain input messages α can give rise to no output β', this may be considered as a special case of the foregoing in which zero output is one of the β' messages.

It would appear that noise and alphabet deficit are the only possible causes of loss of information in a transducer. We note, in passing, that information cannot be created in a transducer, but it can be lost.

Let us define the reliability of a transmission system as

reliability = average probability that an input message will be correctly interpreted in the output (91)

It is well known that the reliability of a transmission system can be increased by the use of redundancy. The most elementary and probably the most common expression of this redundancy is just to repeat the message. For a given transmission channel, such repetition will of course reduce the effective rate of sending messages.

Let us now examine in more detail the use of redundancy for the increase of reliability. Consider a transducer whose alphabet of messages at the input is B, C, D, E, and F and at the output is A, B, C, D, E, F, and G (see Fig. 7-8). Suppose that there is a 0.8 probability that any given input message will give rise to the same output message and a 0.1 probability that any given message will give rise to each of the output messages adjacent to it in the alphabet. The reliability of the transmission system in this case is

$$\text{reliability} = 0.8 \tag{92}$$

If we now decide to send only groups of the same message repeated three times (Case I of Table I), and if we interpret each output group as that message which is present the largest number of times in the group, then the average reliability will certainly be increased to about 0.9. The rate of sending messages will, however, be reduced by a factor of 3.

We next note that we have an alternative method (Case II of Table I) of improving reliability, namely, by loss of apparent accuracy, *i.e.*, reduction of alphabet size. Thus if we use only the symbols, B and E as input symbols, the interpretation of the output signals is certain and the reliability has been increased to 1.0. In fact, not only is the reliability greater, but the possible information per input message is also greater in the latter case. This is clear because there are eight different possible input messages (in groups of three) in the second case and only five in the first as shown in Table I. In order to achieve this better method of improving reliability, we should note that we made use of the known characteristics

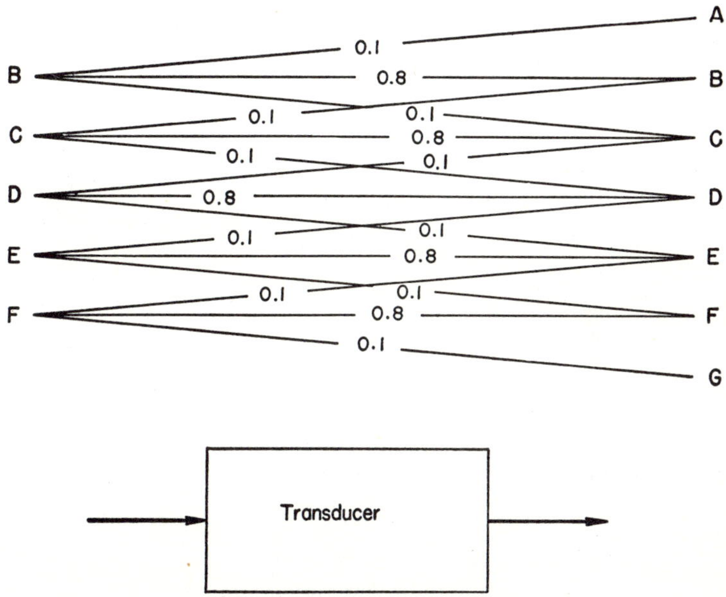

Fig. 7-8. A noisy transducer, showing input and output alphabets, as well as transition probabilities.

of the noise, namely, the relative probabilities of the transformations between letters.

TABLE I

POSSIBLE INPUT MESSAGES IN VARIOUS METHODS OF IMPROVING RELIABILITY

Case I Reliability about 0.9	Case II Reliability 1.0	
BBB	BBB	BEE
CCC	BBE	EBE
DDD	BEB	EEB
EEE	EBB	EEE
FFF		
Maximum language transmission capacity $< \frac{1}{3} \log_2 5$ binit/symbol	Maximum language transmission capacity = 1 binit/symbol	

Both of the methods of using redundancy shown in Table I give less language transmission capacity than could be obtained by using all possible permutations of the letters B, C, D, E, and F. Furthermore, we know by the Theorem on page 58 of Chap. I that by sufficiently involved encoding we can attain a reliability as close to 1.0 as we please, while retaining the maximum language transmission capacity. We will find the foregoing considerations of value when discussing the noise reduction of various modulation systems.

7.9 Noise reduction by modulation systems—Generalized selectivity[1]

It is well known that there exists a certain correspondence between signals and transmission networks, such that many networks have a particularly large ratio of output over input signal magnitude for special-signal waveshapes. Thus a high-Q tuned LRC circuit will transmit sinusoidal signals of the frequency $f = \dfrac{1}{2\pi\sqrt{LC}}$ with less attenuation than any other frequency. As another example, an FM detector will give more output for signals of the waveshape

$$A \cos \left(2\pi \mathrm{ft} + \frac{D}{\mu} \sin 2\pi\mu t\right)$$

than for any other signal waveshape. However, whereas in the case of the tuned circuit the output signal is of the same waveform as the input, in the case of the FM detector this is not true. In the case of the latter, the output is just the frequency modulation of the input signal. As a general description of the degree to which a given transmission system will give a greater response to certain signals than to others, we will use the term *generalized selectivity*. This is an outgrowth of the conventional concept of the ordinary *selectivity* of a tuned circuit.

Every tuned radio receiver is an example of the use of ordinary selectivity to reduce the relative amount of undesired signals and noise superimposed on a desired signal. We will shortly show that

[1] Most of the ideas presented here were originally presented by Goldman, "Some Fundamental Considerations Concerning Noise Reduction and Range in Radar and Communication," *Proc. I.R.E.*, vol. 36, May, 1948, pp. 584–594. This paper appeared before Shannon's publications on the statistical theory of information. Its terminology and, in particular, its use of the word "information" are now obsolete.

noise-reducing modulation systems are examples of the use of *generalized selectivity* for the same purpose.

It is instructive to consider the graphical representations of selectivity. In Fig. 7-9 is a curve showing the selectivity curve of an ordinary tuned LRC circuit. This type of selectivity has certain special simplifying properties. In the first place the output-signal magnitude is proportional to the input-signal magnitude. In the second place, the frequency components of the output signal have the same waveshape as the corresponding components of the input signal. These two special characteristics of ordinary selectivity are

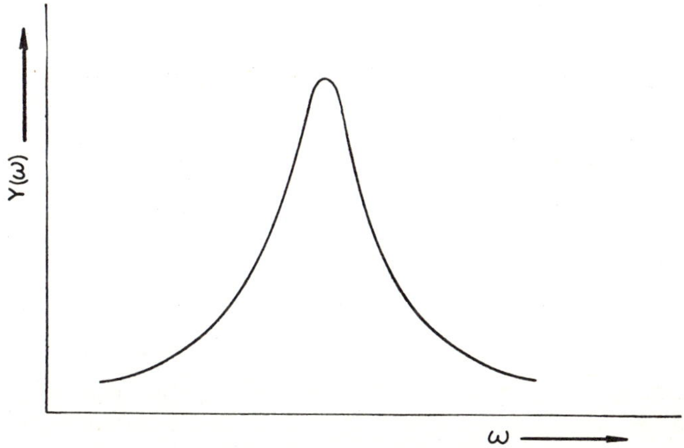

Fig. 7-9. Selectivity curve of a tuned LRC circuit.

not usually characteristic of generalized selectivity. The property which is common to both ordinary and generalized selectivity is that both give a larger ratio of preferred to unpreferred signal in the output than in the input.

The space in which generalized selectivity may be shown pictorially is signal space of $2TW$ dimensions. Figures 7-10 and 7-11 show diagrams of signal space, Fig. 7-11 illustrating generalized selectivity. In Fig. 7-10, the preferred signal is depicted as the direction of the vector s. This characterizes the situation, when the preferred signal is chosen by waveshape alone and not by signal magnitude. All points lying in the direction of this vector specify signals of the same waveshape, and differ from each other only by a constant multiplier. The preferred signals of most

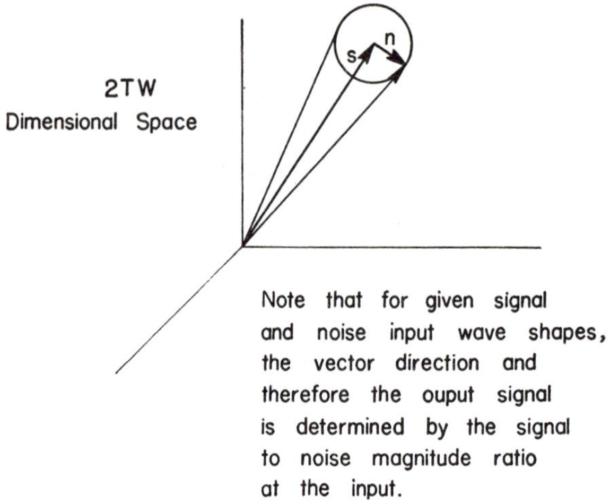

Fig. 7-10. The superposition of desired signal and noise in signal space.

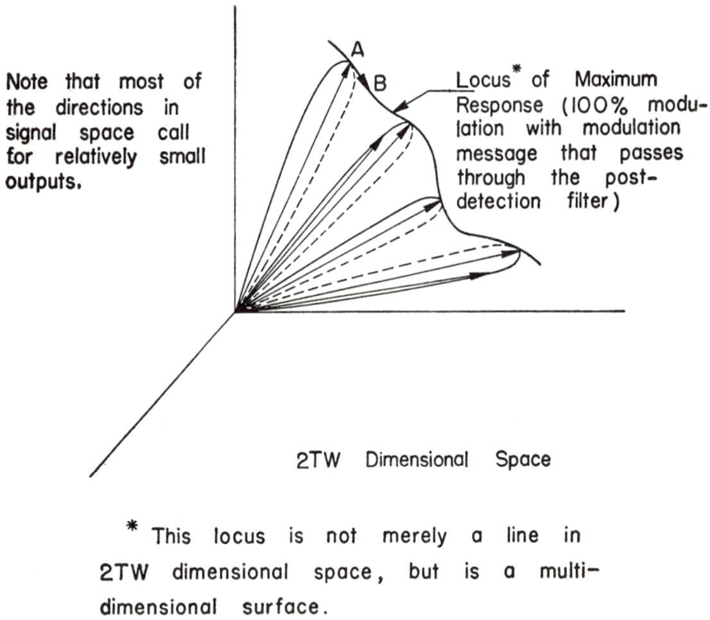

Fig. 7-11. Diagram in signal space showing the generalized selectivity of a transmission system.

technical modulation systems can be specified in this way.[1] In most cases (such as frequency modulation, for example) the magnitude of the output signal does not depend on the magnitude of the input signal. The output-signal waveshape and usually its magnitude depend only on the *direction* of the input-signal vector in signal space.

In the early stages of a receiver where noise is usually superimposed on the signal, the superposition process is ordinarily linear. Accordingly the noise vector may be added to the signal vector in signal space by the usual method of vector addition. If the noise is random noise, we have shown in Chap. VI that the locus of the reasonably probable noise vectors lie on the surface of a $2TW$ dimensional sphere centered about the end point of the signal vector. As long as the noise is small in comparison with the signal, the direction of the combined signal-plus-noise vector is not very different from that of the signal vector alone.

In the case of the tuned circuit whose selectivity characteristic is shown in Fig. 7-9, there is only one signal waveshape of maximum response (tuned frequency), and it is a property of the circuit alone. In a case of generalized selectivity as shown in Fig. 7-11 on the other hand, there is a whole continuum of waveshapes of maximum response. For example, in wideband frequency modulation, practically any signal having the peak available deviation frequency and a low modulation frequency is a waveshape of maximum response. If some other signal of smaller magnitude such as the noise in Fig. 7-10 is linearly added to such a signal of maximum response, the resultant change in output signal is very small—much smaller than the ratio of the two vectors. The selectivity curve for generalized selectivity would look something like the directional pattern of an antenna drawn in a space of many dimensions (see Fig. 7-11 for an example). However, the selectivity is usually not of a linear type. In other words, the output due to the noise in Fig. 7-11, shows interaction effects between the signal and noise.

The numerical value of the possible generalized selectivity is measured as the output (S/N) ratio divided by the input (S/N) ratio. As thus defined the generalized selectivity usually increases with the available bandwidth. The reason for this may be described

[1] In an even more general type of selectivity, the preferred signal would be specified by a point in signal space, and not by all points in a given direction.

by saying that, as the bandwidth increases, there are more degrees of freedom available in which to find differences between the signal and the noise in order to distinguish them for the purpose of generalized selectivity.

The ultimate measure of the noise improvement of a modulation system is determined by the amount of information that finally gets through the transmission system, despite the noise. The maximum increase in information made possible by using extra bandwidth is limited by Eq. (86). The extent to which a given modulation system

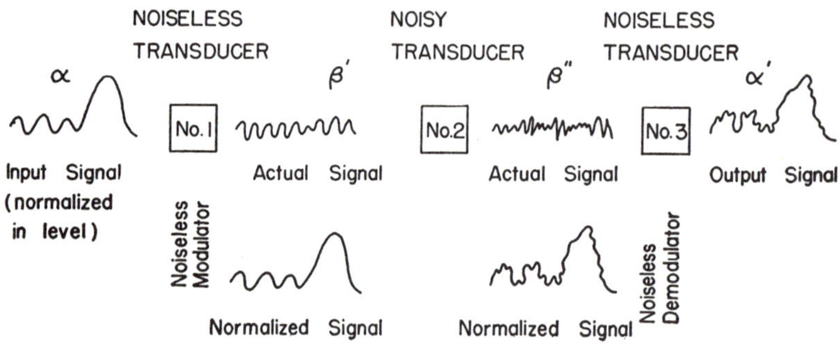

Fig. 7-12. The transmission system whose generalized selectivity is depicted in Fig. 7-13.

makes use of this extra information determines how great its generalized selectivity will be. To get an insight into the mechanism of generalized selectivity, consider Fig. 7-12. The input to the transmission system consists of a noiseless signal or one of very low noise level. To keep the analysis manageable, we will assume that the input signal has a very large, but finite, alphabet of possible signal levels. The maximum information per message of duration T in an ensemble of input messages is then

$$2TW_m \log L_m$$

where L_m is the number of discrete signal levels available and W_m is the bandwidth of the input system.

Since each different input-signal message is certain to give rise to a different output message from the noiseless transducer No. 1, the information of the output signal of the transducer No. 1 must also be $2TW_m \log L_m$. We therefore must have

$$2TW_m \log L_m \leq 2TW \log L$$

where W and L are the respective bandwidth and number of discrete signal levels in the modulated signal. Actually $2TW \log L$ is usually much greater than $2TW_m \log L_m$, but most of the possible messages in the modulated-signal alphabet do not correspond to any possible input signal. These unused available messages of the modulated-signal alphabet have zero probability, and they therefore do not affect the information transmission.

Next the signal goes through the noisy transducer No. 2. The output alphabet β'' is ordinarily the same as the input alphabet β'. However, because of the noise, more of the available output alphabet is used. If the particular α message which gave rise to each used β'' symbol is certain, then no information has been lost. Ordinarily, however, there is some multiple spread (*i.e.*, duplication, triplication, etc.) so that some information is lost.

Finally the signal goes through transducer No. 3. This transducer is noiseless, but information can be lost here by alphabet deficit. In fact such a loss commonly occurs. Many possible β'' messages are usually transformed into the same α' message. In fact this type of information loss is an important factor in keeping noise-reducing modulation systems from reaching the limit in effectiveness given in Eq. (86). The most usual fault of modulation systems in this respect is that they do not cause an adequate difference in the α' alphabet between groups of β' messages whose sample point values are, respectively, permutations of each other. In any case, however, if the information coming out of transducer No. 3 is large, this means that the system can effectively still separate a large number of messages of the original α alphabet. Accordingly the effective noise present in the output signals must then be small.

In Fig. 7-13 is a graphical representation of noise reduction by the use of extra bandwidth.[1] A group of messages of bandwidth W_m and the same waveshape but different signal levels are shown in Fig. 7-13(a). The locus of the representations of these messages

[1] While we have always referred to noise reduction by the use of extra bandwidth, a similar noise reduction can be obtained by using extra time to transmit a signal. The important requirement for noise reduction to be possible is that the number of sample points must be increased. The trouble with using extra time is that it makes the reception lag behind the transmission by an ever-increasing amount. Cases of this type are practical when there is time to spare to repeat the information. Radar and television use noise reduction of this type.

in α signal space (of $2TW_m$ dimensions) is shown in Fig. 7-13(b). The corresponding locus of the representations of these messages in β' signal space (of $2TW$ dimensions) for modulated signals of the same average transmitted power is shown in Fig. 7-13(c) where W is the bandwidth of the modulated signals. Since the modulated signals representing different degrees of modulation do not usually have the same waveshape, this locus is not a straight line. The

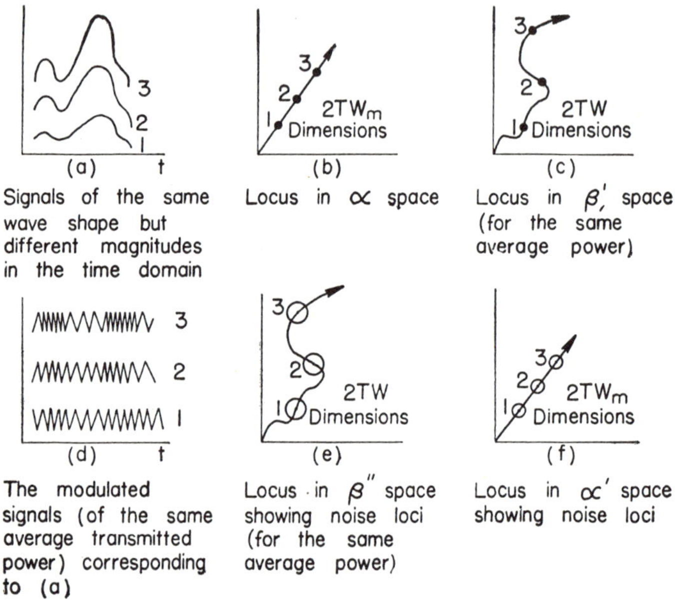

Fig. 7-13. Geometrical illustration in signal space of the use of extra bandwidth for noise reduction.

modulated signals themselves are shown in Fig. 7-13(d). In Fig. 7-13(e) the loci of uncertainty due to noise are shown in β'' signal space for three particular degrees of modulation. Finally the transformation of the signal locus as well as the noise loci into α' signal space is shown in Fig. 7-13(f). The particular thing to note is that the elongation of the locus in modulated signal (β'') space decreases the size which the noise loci will have in α' space and thus increases the signal-to-noise ratio in the output. Since the locus can conceivably be tremendously elongated by winding back and forth in the higher dimensional $2TW$ space, the possibility for improvement of the signal-to-noise ratio using wide frequency bands is pictorially evident.

Next let us consider the mechanism of the well-known "clean-up" effect in the receivers of many noise-reducing modulation systems. When such a receiver has a white gaussian or other noise input, there is noise in the output. However, if a large unmodulated signal of the type to which the system is most sensitive is added to the noise input, the noise in the output is greatly reduced. This removal of the noise by an unmodulated signal is described as the clean-up effect. An example of this clean-up effect is the extreme quiet in the output of an FM sound receiver when a strong unmodulated carrier is tuned in. The explanation of the clean-up effect is easily understood by reference to Fig. 7-10. The desired unmodulated signal in this case largely determines the direction in signal space of the $S + N$ vector. Since the desired signal in this case is unmodulated, it calls for zero output.

When one input signal is much larger than any other signal superimposed upon it and therefore effectively determines the direction of the resultant, it supplies a kind of "standard of coherence." The effect of any superimposed small component then depends on its relation to this "standard of coherence," *i.e.*, on the relation of its direction in signal space to that of the standard of coherence. For a relatively small number of directions[1] of the former, there will be an appreciable effect of the small superimposed component on the final output. For most directions, the effect will not be appreciable. Since the direction in signal space of the standard of coherence is determined by the detailed order and relation to each other of all the sample point values in the controlling signal, it is usually necessary for a superimposed signal to be coherent with the controlling signal in order that the former shall have an appreciable effect on the output. This is the reason for the name "standard of coherence."[2]

[1] These will be the directions such as AB in Fig. 7-11 which vary a given direction (of the standard of coherence) of large output toward other directions of large output (but where the modulation is different).

[2] Generalized selectivity requires many degrees of freedom to determine the standard of coherence, and these are given by a signal. On the other hand, ordinary selectivity requires only a few degrees of freedom to determine a standard of coherence, and these can be given by a simple circuit.

EXERCISE

(Pre-detection vs. post-detection radar noise at ultimate range)

In a linear detector for amplitude modulation, when the noise is large in comparison with the signal, show that the random RF phase relations between signal and noise will cause the noise to tend to remove the signal. In particular if the input to the detector is

$$S \cos wt + N \cos (wt + \phi)$$

where w is the carrier frequency and ϕ is a random RF phase angle, show that the average value of the envelope of signal plus noise will be

$$N + \left(\frac{S}{4N}\right) S$$

approximately, when $N \gg S$.

7.10 Improvement thresholds[1]

All known noise-reducing *modulation* systems will operate effectively only above a certain threshold value of the input (S/N) ratio.[2] A diagram illustrating the improvement threshold in the case of pulse frequency modulation is shown in Fig. 7-14. The improvement threshold occurs when the signal pulses are twice as high as the peak noise fluctuations that are likely to occur during the reception of a message. At this level it is possible to use top and bottom limiters to remove all noise except that which occurs during the time of rise and fall of the pulses. If the signal falls below the improvement threshold level, the reception of noise between pulses causes a sudden great increase in the noise output, usually so great that it will blanket the desired signal.

In the case of frequency modulation, when the signal exceeds the noise entering the frequency detector, the frequency modulation of the signal by the noise is relatively small. If, however, the noise exceeds the signal, then it is the noise which supplies the standard of coherence, *i.e.*, controls the RF phase of the signal-plus-noise combination. The various sidebands of the desired signal

[1] A discussion of thresholds from a different point of view is given by C. E. Shannon, *Proc. I.R.E.*, vol. 37, January, 1949, p. 15.

[2] The only noise-improvement methods which do not have a threshold are those in which the noise improvement depends upon an absolute measurement of frequency or time. Frequency selectivity and autocorrelation are examples of methods of this type. These are not noise-improving modulation systems.

are then no longer coherent in phase with the effective carrier and can no longer operate effectively in unison to give large amounts of frequency modulation. The transition level between larger signal and larger noise is the approximate level of the improvement threshold in frequency modulation.

In the case of double-sideband amplitude modulation, there is also an improvement threshold. Above the threshold, the upper and lower sets of sidebands of the desired signal, because of their phase

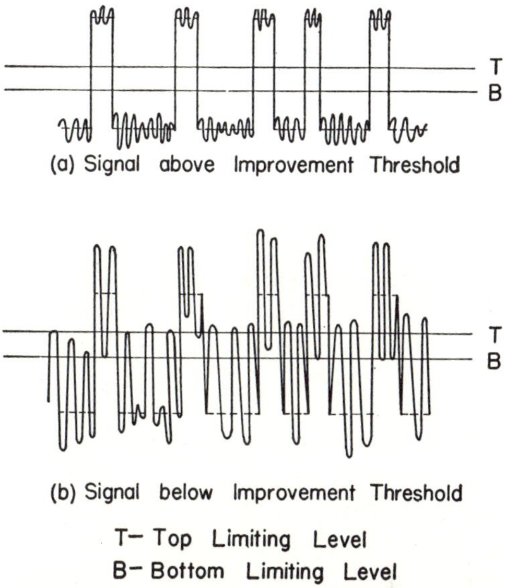

Fig. 7-14. Noise improvement threshold in pulse frequency modulation.

relation with respect to each other and with respect to the effective carrier, have double efficiency in causing amplitude modulation. Below this threshold, the noise is large enough so that the carrier is modulated more than 100% most of the time by the noise components. This overmodulation eliminates most of the desired signal or transforms it into distortion.

The foregoing examples illustrate the phenomenon of improvement thresholds. When the input signal-to-noise ratio falls below the improvement threshold, the desired signal no longer controls the standard of coherence, discussed in Sec. 7.9, and therefore cannot benefit from the generalized selectivity of the modulation system.

In fact, when the desired signal loses the standard of coherence, the generalized selectivity usually acts against it and tends to remove it from the output.

The improvement threshold in most modulation systems occurs around the level when the signal power is equal to the power of the noise superimposed upon it. Independent of any threshold, frequency selectivity and sometimes time gating can be used to reduce the superimposed noise. These will decrease the noise by some numerical factor, but they will still leave an average value of residual noise. When the signal level falls below this residual noise level, it ordinarily loses control of the standard of coherence. The reason for this is illustrated in the signal space diagrams shown in Fig. 7-15. We see that, when the signal power exceeds the noise power by a considerable amount, the signal

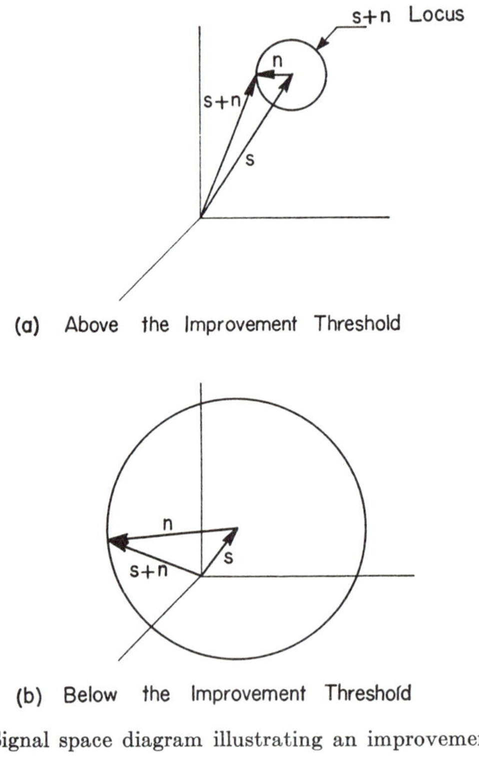

Fig. 7-15. Signal space diagram illustrating an improvement threshold.

effectively controls the direction in signal space of the $S + N$ vector and therefore controls the waveshape of the signal output. When, however, the noise power exceeds the signal power by a considerable amount, the situation is reversed, and the noise largely determines the waveshape of the signal output.

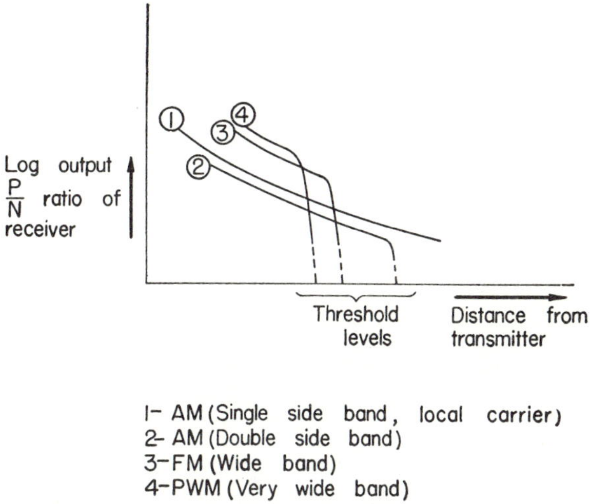

Fig. 7-16. Threshold versus range for a given average power in various types of modulation. (The threshold distance limits the maximum useful range.)

The maximum operating range of any communication system is determined by the location at which the signal falls below the improvement threshold. When this occurs, even in double-sideband amplitude modulation, there is a relatively sudden large rise of the noise[1] level which effectively blankets the signal. The presence of the threshold is illustrated in Fig. 7-16. Below the improvement

[1] This added noise is due to an irreversible process, and it cannot be removed by frequency selectivity. A realization of this fact is important in the design of communication systems. For example, in the design of an amplitude-modulated communication or radar receiver, if the signal-to-noise ratio is much higher than the improvement threshold, the bandwidth of the receiver prior to the second detector can be increased manyfold in order not to lose the signal in case there is frequency drift of either the transmitter or the receiver oscillator. There will be no loss in signal-to-noise ratio in the final output, despite the increased predetection noise caused by the increased bandwidth, so long as the bandwidth is again narrowed to its optimum value by the audio (or video) amplifier. The situation, however, is quite different in case the signal-to-noise

threshold, a major part of what comes into the receiver as desired signal is transformed by the detector into distortion or noise because of the loss of the coherence standard. This is a distinguishing characteristic of the improvement threshold.

In certain types of communication systems, there are artificial means of extending the range before the improvement threshold is reached. The most obvious practical method of doing this is to protect the standard of coherence separately, by making it a narrow band signal so that it can be separated from the extra noise due to wide bandwidth by frequency selectivity, since frequency selection by a tuned circuit has no threshold.[1] The amount of noise reduction possible is, of course, still limited by the fact that one cannot manufacture information.[2]

7.11 Perception selectivity[3]

In the practical reception of signals, whether audio, television, or radar, the final human observer usually adds a large and important amount of effective noise reduction. A human observer will weigh intelligible speech much higher than "gibberish" as being parts of the signal, and the human observer will even fill in the gaps where noise makes the signal unintelligible. As soon as the signal can be recognized as belonging to a customary type of communication signal, such as speech or music, the human perception mechanism gives greatly preferred weighting, on the average, to true signals as compared with noise and greatly decreases the previous probability that the signal is a noise fluctuation. This human "perception selectivity" is thus another example of generalized selectivity.

---

ratio is near the improvement threshold. In that case, widening of the predetection bandwidth to the extent that the signal-to-noise ratio falls into the range of the improvement threshold will cause an irreversible rise in noise which cannot be erased by narrowing the bandwidth of the audio (or video) amplifier. For this reason, if maximum range is desired, the predetection bandwidth should not be increased beyond its normal value of twice the modulation frequency range except in so far as is absolutely necessary because of frequency drift.

[1] A reliable clock, if it could be used, would also have no threshold in giving a time standard.

[2] An important practical example of special protection of a standard of coherence is the application of automatic frequency control to the horizontal synchronizing signal of a television receiver.

[3] This section is taken from Goldman, *Proc. I.R.E.*, vol. 36, May, 1948, p. 593.

Perception selectivity, like other noise-reducing systems, has a threshold below which it will not operate. The threshold at which perception selectivity begins may be roughly described as the S/N level at which the signal can be recognized as belonging to the transmission "language." At this level the human perception mechanism recognizes parts of the mixed signal and noise as coming from a common origin; *i.e.*, it recognizes them as parts of a coherent signal. This level may be described as the intelligibility threshold, and is the noise-improvement threshold of a human being as a signal detector. It is probable that perception selectivity begins at a lower S/N ratio than any non-living type of noise-improvement system so far devised.

Perception selectivity for speech does not require any extra bandwidth in addition to that normally used, because the normal audio range of, say, 10 kc is many times more than adequate to carry the actual amount of information in a speech signal. The actual message information in speech is generated at a rate of about 5 to 25 significant time intervals per second. The audio bandwidth is thus of the order of one thousand times as much as necessary for the purposes of carrying the pure "message information" in speech. The excess bandwidth is thus available for the generation of characteristic human-speech waveshapes which can be modulated by the "message information." The human detector then gives preferred weighting to these characteristic human-speech waveshapes.

It is found, in the study of non-human noise-improvement systems such as frequency modulation, pulse phase modulation, and many others that coherent repetition of the information in several different frequency channels is a common characteristic of the signals used. At least part of the noise reduction of audio perception selectivity is obviously due to the use of signals with this same characteristic.

Perception selectivity is also used in radar. For example, in the use of a radar "A" presentation, the fact that a slightly greater signal amplitude is apparently characteristic of a particular location along the horizontal axis in many repeated intervals is interpreted by the human observer as indicating an echo. It is interesting to note that repetition of the information in the time domain, rather than in the frequency domain, is here used for noise reduction.

Perception selectivity is an important part of almost every communication system in which a human observer is the final receiver.

Problems

1. For a present-day television system, give reasonable values for the following items in each box of the block diagram in Fig. 7-A.

(1) Code capacity (in binits per second)
= maximum rate possible if language constraints cause no loss in rate.

(2) Language transmission capacity (in binits per second)
= actual average rate during television transmissions.

Assume typical values for picture characteristics, signal strengths, etc., at various points. State what values you are using and show how you

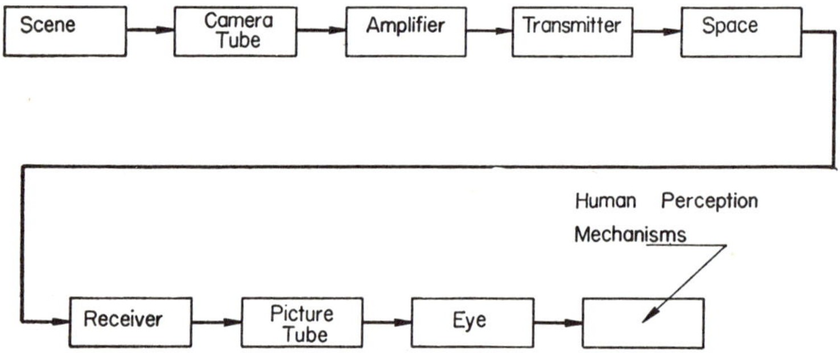

Fig. 7-A.

calculate the code capacity and language transmission capacity in each case. What conclusions can you draw on the basis of information theory regarding the possibility of reducing the bandwidth required in television without a reduction in picture quality?

2. A system of pulse frequency modulation uses 100,000 pulses/sec when unmodulated. This rate can be changed to 50,000 or 150,000 pulses/sec during maximum modulation. The pulses are 1 microsecond long and the RF carrier is amplitude modulated with the pulses. The modulation of the pulse train is an audio signal of 5 kc maximum bandwidth. Assume that the pulses reach the grid of the first RF tube at a level of 250 microvolts peak to peak. Assume that the receiver has a noise level, normalized to the grid of the first RF tube, of 2 microvolts squared per kilocycle.

(a) What is the approximate maximum entropy of the ensemble of pulse frequency modulated signals.

(b) What is the approximate maximum rate of transmission of information of the system.

CHAPTER VIII

LINEAR CORRELATION, FILTERING, AND PREDICTION

8.1 The unit impulse

An important function in our analysis in this chapter will be the unit impulse. The unit impulse $\delta(t)$ is a function whose value is zero except in an arbitrarily small interval around $t = 0$, where it becomes infinite in such a way that

$$\int_{-a}^{+b} \delta(t)dt = 1 \tag{1}$$

In Eq. (1), a and b are any finite positive quantities.

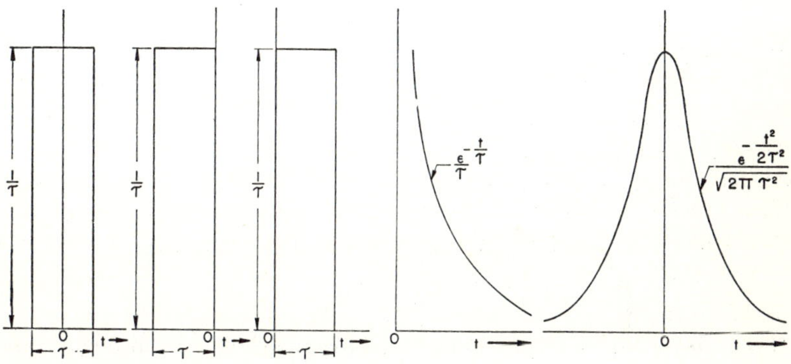

Fig. 8-1. Functions which become unit impulses as the parameter τ approaches zero.

True unit impulses cannot be generated experimentally, but they are nevertheless very useful in signal and circuit analysis. In Fig. 8-1 are various pre-limit forms of unit impulses. All these functions would become unit impulses as the parameter τ approaches zero. The net result of using any of these forms is always the same in practice when $\tau \to 0$, so that the particular form should be chosen which is most easily manipulated in the problem at hand. The function $\delta(t - t_0)$ represents a unit impulse at $t = t_0$. If $P(t)$ is an impulse function in general (*i.e.*, not necessarily a unit impulse), then

$$\int_{t_1}^{t_2} P(t)dt$$

is called the *strength* or *magnitude* of the impulse, where the impulse

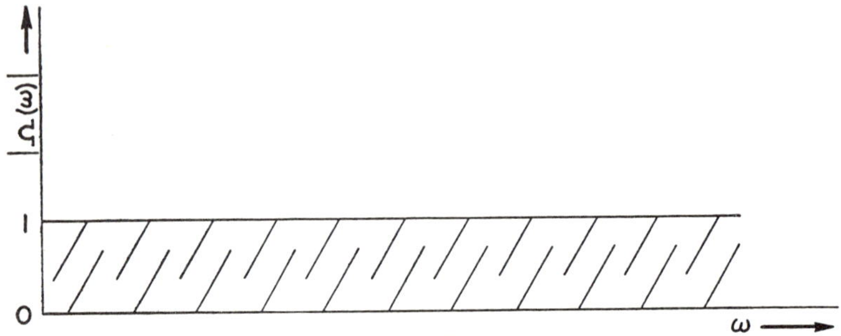

Fig. 8-2. Amplitude spectrum of a unit impulse.

lies entirely in the small interval $t_2 - t_1$. Thus a unit impulse is an impulse of unit strength.

The spectrum of the unit impulse is obtained with the aid of the Fourier integral. This gives

$$\begin{aligned} \Omega(\omega) &= \int_{-\infty}^{+\infty} \delta(t)\epsilon^{-j\omega t}\,dt \\ &= \int_{-\infty}^{+\infty} \delta(t)\cos\omega t\,dt + j\int_{-\infty}^{+\infty} \delta(t)\sin\omega t\,dt \\ &= \cos(0) + j\sin(0) = 1 \end{aligned} \tag{2}$$ [1]

Thus the spectrum $\Omega(\omega)$ of the unit impulse has unit amplitude and zero phase for all frequencies. The spectrum is plotted in Fig. 8-2.

The energy of a unit impulse in any frequency band $\omega_1/2\pi$

[1] In general

$$\begin{aligned} \int_{-\infty}^{+\infty} g(t)\delta(t - C)dt &= \int_{C-a}^{C+a} g(t)\delta(t - C)dt \\ &= g(C)\int_{C-a}^{C+a} \delta(t - C)dt = g(C) \end{aligned}$$

where a is an arbitrarily small quantity.

$< \omega/2\pi < \omega_2/2\pi$ is proportional to

$$\int_{\omega_1}^{\omega_2} |\Omega(\omega)|^2 d\omega = \int_{\omega_1}^{\omega_2} d\omega = \omega_2 - \omega_1 \tag{3}$$

and is thus proportional to the bandwidth. The total energy for all frequencies below any frequency $\omega_0/2\pi$ is accordingly proportional to ω_0, while the total energy above this frequency is infinite. The energy of the frequency components in a unit impulse is therefore concentrated at the extremely high frequencies. It is worthy of note that, although

$$\int_{-\infty}^{+\infty} \delta(t)\,dt$$

is finite, the energy that it represents is proportional to

$$\int_{-\infty}^{+\infty} [\delta(t)]^2 dt$$

which is infinite.

If we express the unit impulse as a Fourier integral, with the aid of Eq. (2), we obtain

$$\begin{aligned} \delta(t) &= \frac{1}{2\pi} \int_{-\infty}^{+\infty} \Omega(\omega) \epsilon^{j\omega t} d\omega \\ &= \frac{1}{2\pi} \int_{-\infty}^{+\infty} \epsilon^{j\omega t} d\omega = \frac{1}{\pi} \int_0^{\infty} \cos \omega t \, d\omega \end{aligned} \tag{4}$$

Equation (4) shows that all the frequency components of the unit impulse are in phase at $t = 0$, which is the reason for the great height of the impulse.

When a unit impulse is sent through a low-pass filter of bandwidth W, the output is, according to Eq. (4),

$$G(t) = \frac{1}{\pi} \int_0^{2\pi W} \cos \omega t \, d\omega = \frac{\sin 2\pi W t}{\pi t} = 2W Sa(2\pi W t) \tag{5}$$

Thus the output is the sampling function multiplied by twice the bandwidth. This is a most interesting result, in the light of the sampling theorem in the time domain.

With this preliminary knowledge of the unit impulse, we proceed now to outline the properties of linear filters.

8.2 The properties of physically realizable linear filters[1]

It is assumed that the reader has a general knowledge of linear networks.[2] The present section is in the nature of a review, and its purpose is to present an over-all picture of those characteristics of linear filters which will be useful to us.

The steady-state transmission properties of a linear filter are expressed by means of its *complex transfer function* $Y(\omega)$. We may write

$$Y(\omega) = G(\omega) + jS(\omega) = \epsilon^{a(\omega)} \cdot \epsilon^{j\phi(\omega)} = |Y(\omega)|\epsilon^{j\phi(\omega)} \tag{6}$$

where

$$|Y| = \sqrt{G^2 + S^2} = \epsilon^a \tag{7}$$

and

$$\phi = \tan^{-1}\left(\frac{S}{G}\right) \tag{8}$$

If a sine wave of frequency $(\omega_1/2\pi)$ and amplitude E is the input of a linear filter, the output is also a sine wave of the same frequency, but its amplitude will be $|Y(\omega_1)| \cdot E$ and its phase will be advanced by the angle $\phi(\omega_1)$. With the aid of Fourier integral analysis it is

[1] Linearity is not involved in analyzing a single function by such methods as Fourier analysis. Linearity only becomes involved when a response function is found for a given input, and one attempts to superimpose the responses of the components.

Generally speaking, the superposition of two different signals, such as desired signal and noise, implies a response function and therefore involves linearity. However, the analysis of a single signal into coherent and incoherent parts does not necessarily involve a response and therefore does not necessarily involve linearity.

[2] The properties of linear networks are discussed in H. W. Bode, *Network Analysis and Feedback Amplifier Design*, Van Nostrand, New York (1945); W. R. LePage and S. Seely, *General Network Analysis*, McGraw-Hill, New York (1952); and in Goldman, *Transformation Calculus and Electrical Transients*, Prentice-Hall, New York (1949). Chaps. IV and VII.

The discussion given here is heavily indebted to H. W. Bode and C. E. Shannon, "A Simplified Derivation of Linear Least Square Smoothing and Prediction Theory," *Proc. I.R.E.*, vol. 38, April, 1950, p. 417–425.

There is a valuable report by W. H. Huggins, "Network Approximation in the Time Domain," Report No. E5048A of the Air Force Cambridge Research Laboratories. This report discusses certain intimate relations between probability theory and circuit theory and gives a fresh point of view in network analysis. As a forerunner of things to come in network theory, Huggins' paper is of considerable interest. It also contains some discussion of the questions dealt with in the present chapter.

shown that a knowledge of $Y(\omega)$ for all frequencies determines the transient as well as the steady-state properties of linear filters.

Instead of using $Y(\omega)$ to express the properties of a filter, we can also use its inverse Fourier transform $B(t)$, where

$$B(t) = \frac{1}{2\pi} \int_{-\infty}^{+\infty} Y(\omega) \epsilon^{j\omega t} d\omega \tag{9}$$

$B(t)$ is called the *impulsive response* of the filter, because it is the response of the filter to the unit impulse $\delta(t)$, since

$$\delta(t) = \frac{1}{2\pi} \int_{-\infty}^{+\infty} \epsilon^{j\omega t} d\omega \tag{10}$$

We can then also, of course, express $Y(\omega)$ in terms of $B(t)$ using the direct Fourier transformation

$$Y(\omega) = \int_{-\infty}^{+\infty} B(t) \epsilon^{-j\omega t} dt \qquad (11)^{1}$$

Although $Y(\omega)$ and $B(t)$ are mathematically equivalent to one another, from an experimental point of view the direct measure-

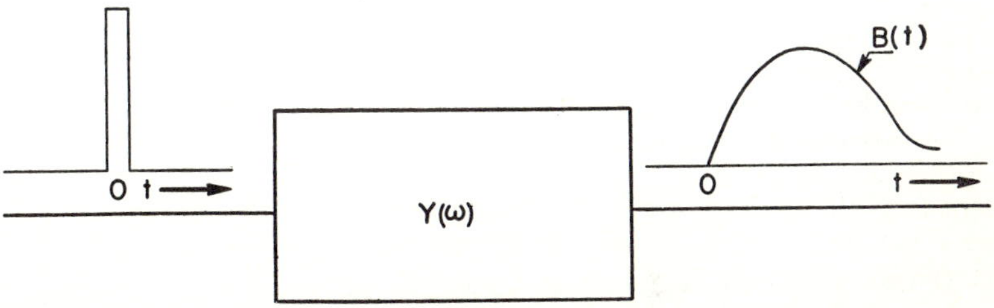

Fig. 8-3. Diagrammatic representation of the response of a linear filter to a unit impulse.

ment of $Y(\omega)$ represents a steady-state test while the measurement of $B(t)$ represents a transient test.

The impulsive response $B(t)$ is the output obtained from the filter if a unit impulse at time $t = 0$ is the input (see Fig. 8-3). Since the properties of the filter are independent of time, if a unit impulse at time $t = t_1$ is the input of the filter, the output will be

[1] Equation (11) tells us that $Y(\omega)$ is the spectrum of $B(t)$. This is, of course, to be expected since $B(t)$ is the output when the input is a unit impulse, for the latter has a constant spectrum.

$B(t - t_1)$. Since the filter is linear, the principle of superposition may be applied to it. By this we mean that the combined output of the filter due to a number of different inputs will be equal to the sum of the outputs obtained from the individual inputs. Thus, if the function $f(t)$ in Fig. 8-4 is the input to the filter, $f(t)$ may be considered to be the superposition of the thin vertical slices of height $f(t)$ and width dt at the various times t. Each of these vertical slices may be considered an impulse of strength $f(t)dt$ at the time t.

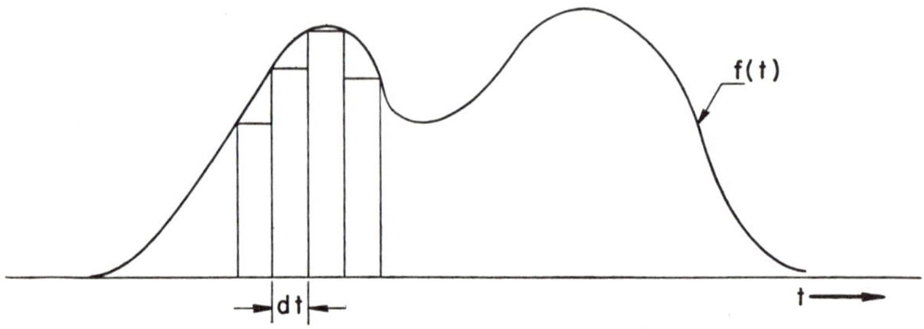

Fig. 8-4. Representation of a function as a succession of impulses.

The response of the filter at a subsequent time t' to this particular vertical slice will therefore be $f(t)dt\, B(t' - t)$. Then by the principle of superposition, the total response $g(t')$ at time t' to the entire input $f(t)$ will be given by the addition of the responses to all the slices occurring prior to t'. Thus

$$g(t') = \int_{-\infty}^{t'} f(t)B(t' - t)dt \qquad (12)^1$$

If we change the variable of integration in Eq. (12) from t to

$$\tau = t' - t \qquad (13)$$

Eq. (12) becomes

$$g(t') = \int_0^{\infty} f(t' - \tau)B(\tau)d\tau \qquad (14)$$

This expression for $g(t')$ has a great deal of physical significance. In Eq. (14), τ stands for the age of the slice $f(t' - \tau)d\tau$ at the time t'. The factor $B(\tau)$ is a weighting function by which the voltage inputs of age τ must be multiplied to determine their contributions to

[1] For all physically realizable networks, $B(t' - t) = 0$ when $t > t'$, so that Eq. (12) may also be written

$$g(t') = \int_{-\infty}^{+\infty} f(t)B(t' - t)dt$$

the present output. Equation (14) is called a superposition integral. It will be particularly valuable for us because it interprets the impulsive response as a weighting function for past inputs. In this way in information theory we can use the impulsive response to express that part of the intersymbol influence in the output which is caused by the filter.

Certain important characteristics of $B(t)$ and $Y(\omega)$ are found by making use of the fact that the networks which they represent must be *physically realizable.*[1] As shown in books on network theory, the requirement of physical realizability puts two important restrictions on the possible form of $B(t)$. In the first place $B(t)$ must be zero for $t < 0$. In other words, the network cannot respond to an impulse before the impulse arrives. In the second place, $B(t)$ must approach zero (with reasonable rapidity) as t approaches $+\infty$. This means that the effect of an impulse will eventually die out. Translated into the language of information theory, this ensures that the intersymbol influence caused by a realizable filter will be of finite range. Therefore if a signal which is a typical function of an ergodic ensemble is the input to a linear physically realizable filter, the output will also be a typical function of an ergodic ensemble.

The above two restrictions on $B(t)$ are sufficient as well as necessary conditions for physical realizability. Any impulsive response $B(t)$ which satisfies these conditions can be approximated as closely as desired with a passive lumped element network, together with an ideal amplifier.

The requirement of physical realizability also puts restrictions on the possible form of $Y(\omega)$. The principal restriction[2] is the stability requirement,[3] namely, that $Y(\omega)$, considered as a function of the

[1] A physically realizable network is one that can be synthesized from ideal passive circuit elements. In addition, the network may include an ideal amplifier whose only function is to change the absolute value of its gain. This definition ensures that a physically realizable network will be stable.

[2] In Appendix X it is shown that this restriction is equivalent to the restrictions on $B(t)$ mentioned in the foregoing paragraphs.

[3] In the more usual way of stating this restriction, use is made of the function

$$W(j\omega) = Y(\omega)$$

thereby changing the formal variable from ω to $j\omega$. The restriction then states that $W(s)$ considered as a function of the complex variable s shall be analytic in the half plane Re $(s) > 0$.

complex variable ω, must be an analytic function in the half plane defined by Im $(\omega) < 0$. Instead of stating the restriction in terms of the location of the singularities of $Y(\omega)$ in the complex ω plane, we can use an alternative and essentially equivalent criterion which involves only real values of ω. This is the Paley-Wiener criterion which states that if

$$\int_0^\infty \frac{\log |Y(\omega)|}{1 + \omega^2} d\omega \tag{15}$$

is a finite number,[1] then $|Y(\omega)|$ is the amplitude function of a physically realizable filter.[2] If $Y(\omega)$ is the transfer function of a physically realizable filter, then for a given gain function $a(\omega)$, where

$$a(\omega) = \log |Y(\omega)| \tag{16}$$

there is a minimum possible phase function $\phi(\omega)$, and if $Y(\omega)$ has this minimum phase function, the filter is said to be of minimum-phase type. For a network of minimum-phase type, $Y(\omega)$, considered as a function of the complex variable ω, has no zeros in the half plane defined by Im $(\omega) < 0$. We already know from the foregoing paragraph that the condition of physical realizability requires that it shall have no poles in this half plane. For a network

[1] We note that a low-pass filter with sharp cutoff is not physically realizable. This follows from the fact that, for the low-pass filter, $|Y(\omega)| = 0$ and log $|Y(\omega)| = -\infty$ for frequencies above cutoff. Therefore

$$\int_0^\infty \frac{\log |Y(\omega)|}{1 + \omega^2} d\omega$$

is not finite.

The properties of $B(t)$ also show that the low-pass filter with sharp cutoff is not physically realizable. For the low-pass filter, $B(t)$ is the sampling function

$$2W \left(\frac{\sin 2\pi Wt}{2\pi Wt}\right)$$

Since this does not vanish for $t < 0$, the filter is not physically realizable.

Although this filter is not physically realizable, it is useful in mathematical analysis. Also it can be approximated by some physically realizable filters, which, however, involve time delays.

[2] Appendix (A) by Henry Wallman of Vol. 18 Radiation Laboratories Series on "Vacuum Tube Amplifiers" gives an excellent discussion of the practical aspects of physical realizability from the standpoint of the Paley-Wiener criterion. Derivation of the Paley-Wiener criterion will be found in the book by R. E. A. C. Paley and N. Wiener, *Fourier Transforms in the Complex Domain*, American Math. Soc. Colloquium Publication, vol. 19, 1934.

of minimum-phase type, Bode has shown that the phase $\phi(\omega_0)$, at the frequency $(\omega_0/2\pi)$, may be calculated from the gain characteristic by the equation

$$\phi(\omega_0) = \frac{2\omega_0}{\pi} \int_0^\infty \frac{a(\omega) - a(\omega_0)}{\omega^2 - \omega_0^2}\, d\omega \tag{17}$$ [1]

In some of the later analyses of this chapter, it will be required to find (in theory at least) a filter of minimum-phase type, for which $|Y(\omega)|^2$ is given as a function of ω. The Paley-Wiener criterion shows that if $|Y(\omega)|^2$ is given and

$$\int_0^\infty \frac{\log |Y(\omega)|^2}{1 + \omega^2}\, d\omega = 2 \int_0^\infty \frac{\log |Y(\omega)|}{1 + \omega^2}\, d\omega \tag{18}$$

is a finite number, then $|Y(\omega)|^2$ is the square of the magnitude of a physically realizable transfer function $Y(\omega)$. Thus

$$|Y(\omega)|^2 = Y(\omega) \cdot Y^*(\omega) \tag{19}$$

where $Y^*(\omega)$ is the complex conjugate of $Y(\omega)$. In other words, if the expression $|Y(\omega)|^2$ makes the integrals in Eq. (18) finite quantities, then $|Y(\omega)|^2$ can be factored in the form of Eq. (19). It will then be possible, as shown in an example below, to choose $Y(\omega)$ in such a way that it has no singularities in the half plane $\mathrm{Im}(\omega) < 0$.

In case $|Y(\omega)|^2$ can, in addition, be approximated sufficiently closely as a rational fraction, *i.e.*, if

$$|Y(\omega)|^2 = \frac{Q_1(\omega)}{Q_2(\omega)} \tag{20}$$

where both $Q_1(\omega)$ and $Q_2(\omega)$ are polynomials with real coefficients, then it is a relatively easy matter to express $Y(\omega)$ in a minimum-phase-shift form. Thus we can write (see also part (b) of exercise below)

$$|Y(\omega)|^2 = \frac{Q_1(\omega)}{Q_2(\omega)} = k^2 \frac{(\omega - \alpha_1)(\omega - \alpha_1^*)(\omega - \alpha_2)(\omega - \alpha_2^*) \cdots}{(\omega - \beta_1)(\omega - \beta_1^*)(\omega - \beta_2)(\omega - \beta_2^*) \cdots} \tag{21}$$

We note that both the zeros and poles of $|Y(\omega)|^2$ considered as a function of the complex variable ω occur in conjugate pairs. Let us use the terms $\alpha_1, \alpha_2, \ldots, \beta_1, \beta_2, \ldots$ for those zeros and poles

[1] See Goldman, *Transformation Calculus and Electrical Transients*, pp. 242 *et seq.* for the derivation of this equation and alternative forms.

whose imaginary parts are positive. Then

$$Y(\omega) = k \frac{(\omega - \alpha_1)(\omega - \alpha_2) \ . \ . \ .}{(\omega - \beta_1)(\omega - \beta_2) \ . \ . \ .} \tag{22}$$

will be the complex transfer function of a physically realizable filter of minimum-phase type with the prescribed $|Y(\omega)|^2$. Equation (22) is thus the desired solution of the problem.

A minimum-phase network has the useful property that its inverse, with the transfer function

$$Y^{-1}(\omega) = \frac{1}{k} \frac{(\omega - \beta_1)(\omega - \beta_2) \ . \ . \ .}{(\omega - \alpha_1)(\omega - \alpha_2) \ . \ . \ .} \tag{23}$$

is also physically realizable.[1] If a signal $f(t)$ is the input of the filter $Y(\omega)$, we can recover $f(t)$ in its original form by passing the output of $Y(\omega)$ through the inverse filter $Y^{-1}(\omega)$. Furthermore, the recovery is accomplished *with no over-all time delay.* There is, however, no physically realizable exact inverse for a non-minimum-phase network. The best that can be done in this case is to provide a network which will recover the original signal, but *with a time delay.*

By way of illustration,[2] Fig. 8-5 shows a filter of non-minimum-phase-shift type and also a filter of minimum-phase-shift type, both having transfer functions with the same absolute values. The inverse of the minimum-phase-shift filter is also shown.

The foregoing survey of the properties of physically realizable linear filters will serve as a basis for the analysis of the use of these filters in later sections of the present chapter.

EXERCISE

(a) The quantity $|Y(\omega)|^2$ is a real quantity. If it can be expressed as the quotient of two polynomials, which must therefore also be real, show that $|Y(\omega)|^2$ may be expressed as

$$|Y(\omega)|^2 = k^2 \frac{(\omega - \alpha_1)(\omega - \alpha_1^*)(\omega - \alpha_2)(\omega - \alpha_2^*) \ . \ . \ .}{(\omega - \beta_1)(\omega - \beta_1^*)(\omega - \beta_2)(\omega - \beta_2^*) \ . \ . \ .}$$

(b) Show that for any lumped constant network, if α_1 has both a real

[1] In this and the preceding paragraph we have ignored certain special cases in which the analysis would run into difficulty. In all these cases, the present analysis can be made to apply by making certain pragmatic approximations.

[2] This example was suggested by Dr. W. R. LePage.

Fig. 8-5. (a) Filter of minimum phase shift type.

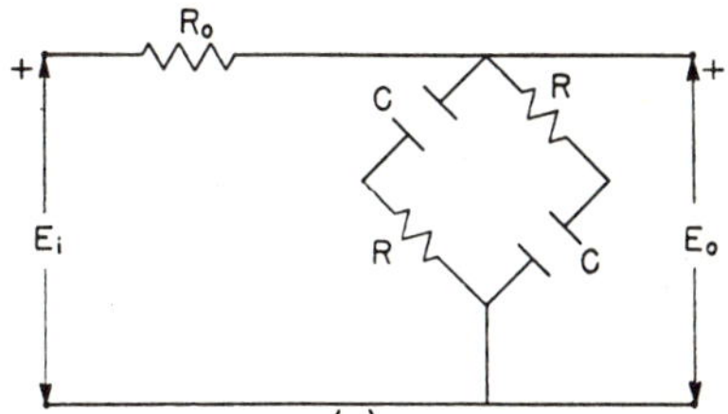

(a)

$$\text{Transfer function} = \frac{E_0}{E_i} = \frac{R}{2R_0+R} \cdot \frac{\left(\omega - \dfrac{j}{RC}\right)}{\omega - \dfrac{j}{(2R_0+R)C}},$$

(b) Filter of non-minimum phase shift type.

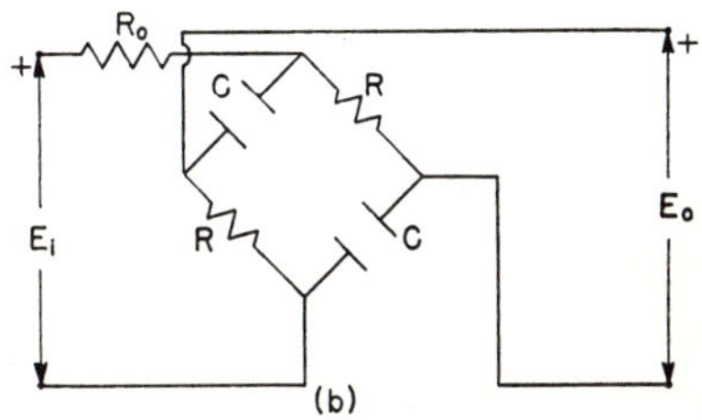

(b)

$$\text{Transfer function} = \frac{E_0}{E_i} = \frac{R}{2R_0+R} \cdot \frac{\left(\omega + \dfrac{j}{RC}\right)}{\omega - \dfrac{j}{(2R_0+R)C}}$$

(c) Filter of minimum phase shift type. If this is followed by an ideal amplifier of gain $(2R_0 + R)(r + r_1)/Rr_1$, the combination is the inverse of (a), provided $c_1 = 2R_0Cc/(RC - rc)$ and $r_1 = (2R_0 + R)(RC - rc)/2R_0c$.

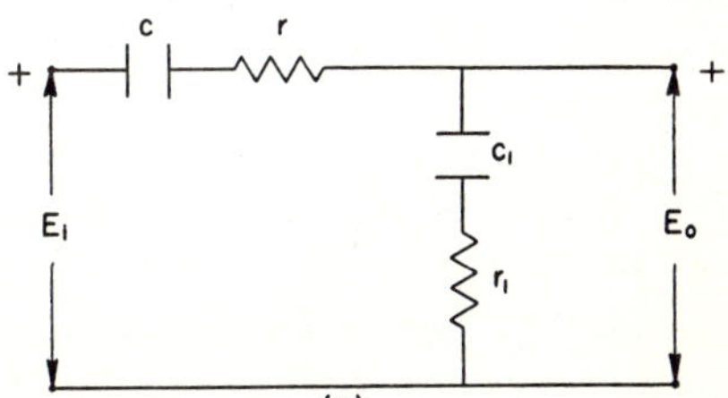

(c)

$$\text{Transfer function} = \frac{E_0}{E_i} = \frac{r_1}{r+r_1} \cdot \frac{\omega - \dfrac{j}{r_1 c_1}}{\omega - j\dfrac{c+c_1}{c\,c_1(r+r_1)}}$$

$$= \frac{r_1}{r+r_1} \cdot \frac{\omega - \dfrac{j}{(2R_0+R)C}}{\omega - \dfrac{j}{RC}}$$

and an imaginary part, then

$$(\omega - \alpha_1)(\omega + \alpha_1)(\omega - \alpha_1^*)(\omega + \alpha_1^*)$$

is a factor of $|Y(\omega)|^2$, and an analogous result holds for β_1.

8.3 The Fourier transform criterion of Van Vleck and Middleton and North

Suppose we have a signal function $s(t)$ which is superimposed on a noise function $n(t)$ to give a combined signal $s(t) + n(t)$. This is a very common occurrence in technical work, and a closely related analogue occurs in the transmission of information under practically all circumstances. A number of investigations have been made to determine whether filters can be designed for use with the above combined signal which will make it easier to distinguish the signal

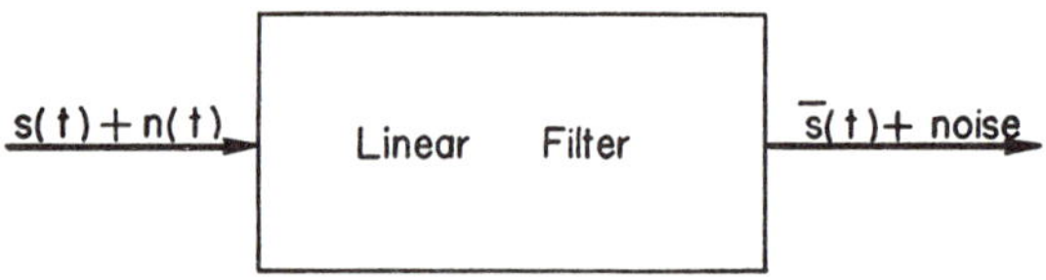

Fig. 8-6.

from the noise. In the present section we shall determine the characteristics of the *linear* filter which will give the greatest ratio of peak signal to root-mean-square noise in the output when the input signal is $s(t) + n(t)$, if $n(t)$ is random (or at least white gaussian) noise. Such a filter will be particularly valuable in fields like radar where the recognition of the signal peaks is highly important.

The signal may be expressed in terms of its frequency components by the Fourier integral

$$s(t) = \frac{1}{2\pi} \int_{-\infty}^{+\infty} S(\omega) \epsilon^{j\omega t} \, d\omega = \frac{1}{2\pi} \int_{-\infty}^{+\infty} |S(\omega)| \epsilon^{j\phi_s(\omega)} \epsilon^{j\omega t} d\omega \quad (24)$$

The exact waveshape of the noise, of course, cannot be predicted ahead of time, but we do know that its spectrum is uniformly distributed through the frequency range.

Suppose that the combined signal plus noise is sent (see Fig. 8-6) through a linear filter having the transfer function

$$Y(\omega) = |Y(\omega)| \epsilon^{j\phi_Y(\omega)} \quad (25)$$

The output of the filter due to the signal $s(t)$ alone will then be

$$\bar{s}(t) = \frac{1}{2\pi} \int_{-\infty}^{+\infty} |Y(\omega)||S(\omega)|\epsilon^{j[\omega t+\phi_s(\omega)+\phi_Y(\omega)]}\, d\omega \tag{26}$$

In particular, at the time t_0, the output (signal liniva) due to $s(t)$ will be

$$\bar{s}(t_0) = \frac{1}{2\pi} \int_{-\infty}^{+\infty} |Y(\omega)||S(\omega)|\epsilon^{j[\omega t_0+\phi_s(\omega)+\phi_Y(\omega)]}\, d\omega \tag{27}$$

On the other hand, the output noise power, and therefore the mean-square value of the noise, will be proportional to

$$\int_{-\infty}^{+\infty} |Y(\omega)|^2 d\omega$$

and may therefore be expressed

$$\text{mean-square-noise output} = k^2 \int_{-\infty}^{+\infty} |Y(\omega)|^2 d\omega \tag{28}$$

where k^2 is a positive constant of proportionality. The ratio of signal to root-mean-square noise at the time t_0 will therefore be

$$\frac{\text{signal liniva}}{\text{root-mean-square noise}} = \frac{\left| \dfrac{1}{2\pi} \displaystyle\int_{-\infty}^{+\infty} |Y(\omega)||S(\omega)|\epsilon^{j[\omega t_0+\phi_s(\omega)+\phi_Y(\omega)]}\, d\omega \right|}{\left[k^2 \displaystyle\int_{-\infty}^{+\infty} |Y(\omega)|^2 d\omega \right]^{\frac{1}{2}}} \tag{29}$$

Our problem is to find the form of $Y(\omega)$ so that the above ratio will be a maximum. In order to do this we will make use of the *Schwarz inequality.*[1] According to one formulation of this inequality, if $F(\omega)$ and $G(\omega)$ are either real functions of ω or are Fourier transforms of real functions of t, then

$$\left| \int_{-\infty}^{+\infty} F(\omega) \cdot G(\omega) d\omega \right|^2 \leq \int_{-\infty}^{+\infty} |F(\omega)|^2 d\omega \int_{-\infty}^{+\infty} |G(\omega)|^2 d\omega \tag{30}$$

Therefore, if we set

$$F(\omega) = Y(\omega) = |Y(\omega)|\epsilon^{j\phi_Y(\omega)} \tag{31}$$

and

$$G(\omega) = |S(\omega)|\epsilon^{j[\omega t_0+\phi_s(\omega)]} \tag{32}$$

it follows that

[1] Margenau and Murphy, *The Mathematics of Physics and Chemistry*, Van Nostrand, New York, (1943), pp. 130–131.

$$\left| \int_{-\infty}^{+\infty} |Y(\omega)||S(\omega)|\epsilon^{j[\omega t_0+\phi_s(\omega)+\phi_Y(\omega)]}d\omega \right|^2 \leq \int_{-\infty}^{+\infty} |Y(\omega)|^2 d\omega \cdot \int_{-\infty}^{+\infty} |S(\omega)|^2 d\omega \quad (33)$$

Thus

$$\frac{\left| \dfrac{1}{2\pi} \displaystyle\int_{-\infty}^{+\infty} |Y(\omega)||S(\omega)|\epsilon^{j[\omega t_0+\phi_s(\omega)+\phi_Y(\omega)]}d\omega \right|}{\left[k^2 \displaystyle\int_{-\infty}^{+\infty} |Y(\omega)|^2 d\omega \right]^{\frac{1}{2}}} \leq \left[\frac{1}{4\pi^2 k^2} \int_{-\infty}^{+\infty} |S(\omega)|^2 d\omega \right]^{\frac{1}{2}} \quad (34)$$

The right side of the inequality (34) depends only on the input signal and does not change with $Y(\omega)$. The ratio Eq. (29) which is also the left side of Eq. (34) will therefore have its maximum value when the inequality (34) becomes an equality. This occurs when

$$F(\omega) \cdot G(\omega) = |F(\omega)|^2 = |G(\omega)|^2 \quad (35)$$

that is, when

$$|Y(\omega)||S(\omega)|\epsilon^{j[\omega t_0+\phi_s(\omega)+\phi_Y(\omega)]} = |Y(\omega)|^2 = |S(\omega)|^2 \quad (36)$$

This occurs when

$$Y(\omega) = |Y(\omega)|\epsilon^{j\phi_Y(\omega)} = |S(\omega)|\epsilon^{-j\omega t_0-\phi_s(\omega)} = S^*(\omega)\epsilon^{-j\omega t_0} \quad (37)^1$$

Equation (37) is therefore the required form of $Y(\omega)$ to make the output ratio of signal to root-mean-square noise a maximum at the time t_0. Since the value of this maximum ratio, which is the right side of Eq. (34), does not depend on the value of t_0, we have the surprising result that we get the same ratio for all values of time for which we can design a $Y(\omega)$ to satisfy Eq. (37). The right side of Eq. (34) is therefore the maximum ratio of *peak* signal to root-mean-square noise obtainable with a linear filter, since no greater value is possible. We note, also, that for a given input noise level k^2 the ratio depends only on the "energy" of the input signal,

$$\frac{1}{2\pi} \int_{-\infty}^{+\infty} |S(\omega)|^2 d\omega$$

[1] The more general condition $F(\omega)G(\omega) = |F(\omega)||G(\omega)|$ where $|F(\omega)| = k|G(\omega)|$ and k is a constant independent of ω also makes Eq. (30) an equality. Therefore the equality sign in Eq. (37) can be replaced by a proportionality sign.

Equation (37) tells us that the filter which will give the maximum value of peak signal to root-mean-square noise in its output must have a transfer function $Y(\omega)$ which is the complex conjugate of the Fourier transform of the desired input signal multiplied by a phase factor $\epsilon^{-j\omega t_0}$. Equation (37) is therefore called the *Fourier transform criterion.* This result was apparently discovered independently by D. O. North, W. W. Hanson, N. Wiener and by J. H. Van Vleck and D. Middleton.[1]

When Eq. (37) is satisfied, the output signal according to Eq. (26) will be

$$\bar{s}(t) = \frac{1}{2\pi}\int_{-\infty}^{+\infty} |S(\omega)|^2 \epsilon^{j\omega(t-t_0)}d\omega \tag{38}$$

In the next section we shall define and discuss the correlation functions. We mention here however,[2] that the finite-auto-correlation function of the input signal $s(t)$ displaced by the time t_0 is precisely the output signal $\bar{s}(t)$.

According to the foregoing discussion we can obtain the maximum possible peak signal to root-mean-square noise at any time t_0 desired, provided that we obtain a physically realizable filter when t_0 is inserted in Eq. (37). In order for the filter specified by Eq. (37) to be physically realizable, its impulse response, according to Sec. 8.2, must vanish for $t < 0$ and must vanish exponentially as $t \to +\infty$. From Eqs. (9), (37), and (24) we have for the impulsive response

$$\begin{aligned} B(t) &= \frac{1}{2\pi}\int_{-\infty}^{+\infty} Y(\omega)\epsilon^{j\omega t}d\omega \\ &= \frac{1}{2\pi}\int_{-\infty}^{+\infty} |S(\omega)|\epsilon^{j[\omega(t-t_0)-\phi_s(\omega)]}d\omega = s(t_0 - t) \end{aligned} \qquad (39)^3$$

It is apparent from Eq. (39) and as shown in Fig. 8-7, that in order for $Y(\omega)$ to be physically realizable, *i.e.*, in order for $B(t)$ to vanish when $t < 0$, the signal $s(t)$ must be entirely past before time $t = t_0$. It is also necessary that $s(t)$ shall vanish exponentially as $t \to -\infty$. This latter condition is always satisfied for practical signals any-

[1] J. H. Van Vleck and David Middleton, "A Theoretical Comparison of the Visual, Aural, and Meter Reception of Pulsed Signals in the Presence of Noise," *Jour. Applied Phys.* vol. 17, November, 1946, pp. 940–971.

[2] This has been pointed out by W. H. Huggins and by M. J. DiToro.

[3] We have here made use of the fact that the Fourier transform of $s(-t)$ is the complex conjugate of the Fourier transform of $s(t)$, since $s(t)$ is real.

way, so that the important restriction is that all of $s(t)$ must have entered the filter before the time t_0 when the filter is expected to give maximum signal-to-noise ratio. This restriction was, of course, to be expected. Provided that this restriction is satisfied, however, there will be a physically realizable $Y(\omega)$ satisfying Eq. (37) and therefore giving maximum signal-to-noise ratio at t_0.

In Fig. 8-8 is shown an example of an $s(t)$ and its corresponding $Y(\omega)$ and $\bar{s}(t)$ to give maximum signal-to-noise ratio at a particular

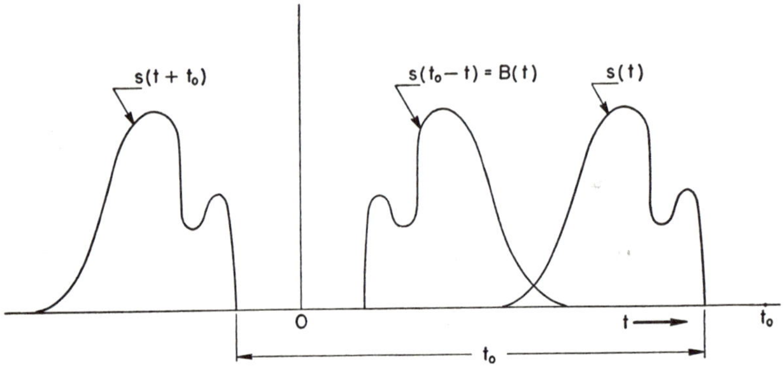

Fig. 8-7. An example of a signal $s(t)$ and the impulsive response, $B(t)$, of the filter for which it will have maximum signal-to-noise ratio at t_0.

t_0. This example was chosen for extreme simplicity, and it is therefore quite artificial. For practical signal functions $s(t)$ one can only expect to get an approximation to the prescribed $Y(\omega)$ with a reasonably simple filter.

Let us next consider for a moment the mechanism by which the filter defined by Eq. (37) succeeds in giving maximum signal-to-noise ratio. The phase part of the mechanism is very simple. When Eq. (37) is satisfied,

$$\bar{s}(t) = \frac{1}{2\pi} \int_{-\infty}^{+\infty} S^*(\omega)\epsilon^{-j\omega t_0} \cdot S(\omega)\epsilon^{j\omega t} d\omega = \frac{1}{2\pi} \int_{-\infty}^{+\infty} |S(\omega)|^2 \epsilon^{j\omega(t-t_0)} d\omega \tag{40}$$

so that all the frequency components of the output signal are in phase at the time $t = t_0$. The signal will therefore have its peak value at this time. The amplitude part of the mechanism does not have such a simple statement. It is, in fact, just an example of the condition required for the Schwarz inequality to become an equality for real functions.

Finally, it is of interest that Van Vleck and Middleton[1] have shown that, in case the functions $s(t)$ and $\bar{s}(t)$ are modulated carriers and $\bar{s}(t)$ enters a detector, then for a wide variety of detectors the

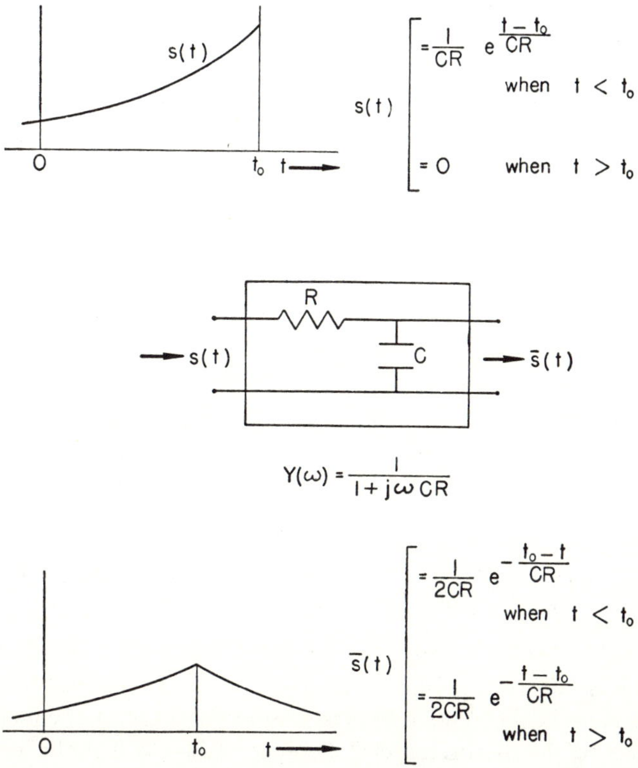

Fig. 8-8. An example of an $s(t)$ and its corresponding $Y(\omega)$ and $\bar{s}(t)$ to give maximum signal-to-noise ratio at t_0.

value of $\left(\dfrac{\text{peak signal}}{\text{root-mean-square noise}}\right)$ is a maximum at the output of the detector, if it is a maximum at the input.

8.4 The correlation functions

(a) *Definitions.* In the last section we mentioned that, when the correct linear filter is used to give maximum peak signal to root-

[1] *Jour. Applied Phys.*, vol. 17, November, 1946, pp. 940–971.

mean-square noise in the output, the desired signal in the output is the finite-auto-correlation function of the desired signal in the input. It is found that the auto-correlation function and its analogue, the cross-correlation function, occur frequently in noise and information analyses, and we will use them in the remainder of this chapter. As an introduction to these functions we wish, briefly, to discuss correlation.

Correlation is a term used in statistical studies as a measure of the dependence between two sets of numbers. If $x_1,x_2, \ldots ,x_n$ are the numbers of one set and $y_1,y_2, \ldots ,y_n$ are the numbers of the other set, then

$$r = \frac{\sum_{i=1}^{n} x_i y_i}{\sqrt{\sum_{i=1}^{n} x_i^2 \sum_{i=1}^{n} y_i^2}} \tag{41}$$

is called the *coefficient of correlation* between the two sets of numbers. It is shown in books on statistics that, regardless of the values of the x's and y's, the coefficient of correlation must lie in the range[1]

$$-1 \leq r \leq +1 \tag{42}$$

If r is equal to either $+1$ or -1, then there is a linear functional relationship between the x and y sets of the form

$$x_i = ay_i \tag{43}$$

where a is a constant independent of i.

Correlation in the general case is a numeral measure of coherence. According to the discussion of coherence in Sec. 3.7, if the x and y sets of numbers in the above example are completely coherent, the x set would be completely determined in all its properties when the y set is known. This would be the case if r in Eq. (41) is ± 1. It should be realized, however, that Eq. (41) is a measure only of the *linear correlation* between two sets of numbers. The two sets may be completely coherent, but if the relation of dependence between the sets is not a linear relation, then Eq. (41) will not measure the extent of the coherence. For example, if

$$y_i = x_i^2 \tag{44}$$

[1] This is closely related to the Schwarz inequality discussed in the preceding section.

for all i, then the x_i's and y_i's will be completely coherent, but $|r| < 1$. In fact, examples can be given where the linear correlation between two sets is zero, but the sets are nevertheless completely coherent.

Despite the foregoing limitation, however, linear correlation is probably the simplest and most important measure of the dependence of sets of numbers. If the sets are monotonically related, even if the relation is not linear, the coefficient of correlation will usually give an approximate measure of the extent of dependence. Moreover, we shall find that the correlation functions, which are closely related to the coefficient of correlation, are valuable tools in dealing with phenomena in linear systems.

The coefficient of correlation, r in Eq. (41), is a normalized measure of the correlation. For many purposes it is more practical to study the average value of the numerator of Eq. (41), namely,

$$\frac{1}{n} \sum_{i=1}^{n} x_i y_i \tag{45}$$

The type of correlation in which we shall be chiefly interested is that between a certain infinite sequence of values and the same or another infinite sequence of values with the index (*i.e.*, subscript) shifted. Since the sequences in our work will usually be the sequences of values of a signal for different values of time, this shift in index will be equivalent to a shift in time. With this introduction, we now define the *auto-correlation function* of the sequence of x numbers as

$$\phi_{xx}(j) = \lim_{N \to \infty} \frac{1}{2N+1} \sum_{k=-N}^{+N} x_k \cdot x_{k+j} \tag{46}$$

We define the *cross-correlation function* between the x and y sequences as

$$\phi_{xy}(j) = \lim_{N \to \infty} \frac{1}{2N+1} \sum_{k=-N}^{+N} x_k \cdot y_{k+j} \tag{47}$$

As the notation indicates, these functions will have the shift index j as their independent variables.

In most of our work, we shall be interested in continuous functions rather than sequences. For this work, we define the *auto-correlation function* of *the continuous function* $g(t)$ as

$$\phi(\tau) = \lim_{T\to\infty} \frac{1}{2T} \int_{-T}^{+T} g(t) \cdot g(t+\tau)dt \tag{48}$$

We define the cross-correlation function between the continuous functions $g_1(t)$ and $g_2(t)$ as

$$\phi_{12}(\tau) = \lim_{T\to\infty} \frac{1}{2T} \int_{-T}^{+T} g_1(t) \cdot g_2(t+\tau)dt \tag{49}$$

The auto-correlation function is thus the time average of $g(t) \cdot g(t+\tau)$. It is a measure of the correlation between the values of $g(t)$ separated by the time τ. It will, of course, be a function of the separation τ. Analogous remarks could be made concerning the cross-correlation function. There is one additional item, however, regarding the cross-correlation function. In the latter case, it is important to know whether g_1 or g_2 is taken at the displaced time $t + \tau$. In general

$$\phi_{21}(\tau) = \lim_{T\to\infty} \frac{1}{2T} \int_{-T}^{+T} g_2(t)g_1(t+\tau)dt \tag{50}$$

need not be the same as $\phi_{12}(\tau)$. However, by identity, we must have

$$\phi_{21}(\tau) = \phi_{12}(-\tau) \tag{51}$$

Applying the same reasoning to the auto-correlation function, we must also have

$$\phi(\tau) = \phi(-\tau) \tag{52}$$

so that we know that the auto-correlation function must be an even function.

In this connection, we also note that

$$\begin{aligned}\phi(0) &= \lim_{T\to\infty} \frac{1}{2T} \int_{-T}^{+T} g(t)g(t)dt \\ &= \text{average value of } |g(t)|^2 \end{aligned} \tag{52a}$$

Furthermore for all values of (τ)

$$\phi(\tau) \leqq \phi(0) \tag{52b}$$

To prove this, we note that

$$0 \leqq [g(t) \pm g(t+\tau)]^2 \equiv [g(t)]^2 + [g(t+\tau)]^2 \pm 2g(t)g(t+\tau) \tag{52c}$$

Therefore

$$\phi(\tau) = \operatorname*{Lim}_{T\to\infty} \frac{1}{2T} \int_{-T}^{+T} g(t)g(t+\tau)dt$$

$$\leq \tfrac{1}{2}\left\{\operatorname*{Lim}_{T\to\infty} \frac{1}{2T} \int_{-T}^{+T} [g(t)]^2dt + \operatorname*{Lim}_{T\to\infty} \frac{1}{2T} \int_{-T+\tau}^{T+\tau} [g(t')]^2dt'\right\}$$

$$= \phi(0) \quad (52d)$$

where

$$t' = t + \tau \quad (52e)$$

The auto-correlation function is one of the measures[1] of the intersymbol influence between values of the function $g(t)$ separated by the time interval τ. As may be seen in the examples in the next section, the right side of Eq. (48) approaches a limit as T approaches infinity in the case of most ordinary functions, so that the function $\phi(\tau)$ exists. In case $g(t)$ is a typical function of an ergodic ensemble, the function $\phi(\tau)$ not only exists but is the same as the auto-correlation function of every other typical function of the ensemble and can be calculated as an ensemble average, for this is the same as the time average. Since the range of the intersymbol influence is finite, in the case of an ergodic ensemble, there is a finite value of τ beyond which the value of $\phi(\tau)$ must be essentially zero in this case. In order to calculate $\phi(\tau)$ as an ensemble average, we write

$$\phi(\tau) = \operatorname*{Lim}_{T\to\infty} \frac{1}{2T} \int_{-T}^{+T} g(t) \cdot g(t+\tau)dt$$

$$= \iint y_1 \cdot y_2 \cdot p_\tau(y_1,y_2)dy_1dy_2 \quad (53)$$

In the last term of Eq. (53)

$$y_1 = g(t_1) \quad (54)$$

where t_1 is some arbitrary value of t, while

$$y_2 = g(t_1 + \tau) \quad (55)$$

The factor $p_\tau(y_1,y_2)dy_1dy_2$ is the probability of occurrence of a function in the ensemble in which $g(t_1)$ lies between y_1 and $y_1 + dy_1$ and $g(t_1 + \tau)$ lies between y_2 and $y_2 + dy_2$. It does not matter what value of t_1 is chosen, for if the ensemble is ergodic, the answer will be the same regardless of the value of t_1.

[1] It measures the linear correlation.

Let us next consider the cross-correlation function $\phi_{12}(\tau)$ between the functions $g_1(t)$ and $g_2(t)$, defined in Eq. (49). If $g_1(t)$ and $g_2(t)$ are each typical members of ergodic ensembles, then we also have

$$\phi_{12}(\tau) = \iint y_1 \cdot y_2 p_\tau(y_1,y_2) dy_1 dy_2 \tag{56}$$

In Eq. (56)

$$y_1 = g_1(t_a) \tag{57}$$

where t_a is some arbitrary value of t, and

$$y_2 = g_2(t_a + \tau) \tag{58}$$

The factor $p_\tau(y_1,y_2)dy_1dy_2$ is the probability of the combined occurrence of a function in the first ensemble in which $g_1(t_a)$ lies between y_1 and $y_1 + dy_1$ along with a function in the second ensemble in which $g_2(t_a + \tau)$ lies between y_2 and $y_2 + dy_2$. In case the two ensembles are independent

$$p_\tau(y_1,y_2) = p_1(y_1) \cdot p_2(y_2) \tag{59}$$

Then

$$\begin{aligned} \phi_{12}(\tau) &= \iint y_1 \cdot y_2 p_1(y_1) \cdot p_2(y_2) dy_1 dy_2 \\ &= \int y_1 p_1(y_1) dy_1 \cdot \int y_2 p_2(y_2) dy_2 = \tilde{y}_1 \cdot \tilde{y}_2 \end{aligned} \tag{60}$$

Here $\tilde{y}_1$ is the ensemble average[1] of $g_1(t_a)$, which is also equal to the time average (d-c component) of $g_1(t)$, since the ensemble is ergodic. Likewise, $\tilde{y}_2$ is the ensemble average of $g_2(t_a + \tau)$, which is also equal to the time average of $g_2(t)$. The cross-correlation function between functions of independent ergodic ensembles is therefore a constant, independent of τ, and is equal to the product of the average values of the respective functions. If either function has an average value of zero, and they belong to independent ensembles, the cross-correlation function is zero for all values of τ.[2]

The correlation functions of sequences of ergodic systems may likewise be expressed as ensemble averages. Thus, for example,

$$\phi_{xx}(j) = \lim_{N\to\infty} \frac{1}{2N+1} \sum_{k=-N}^{+N} x_k \cdot x_{k+j} \tag{61}$$

$$= \sum\sum x_a x_b p(x_a,x_b) \tag{62}$$

[1] The mark $\sim$ over a quantity will indicate an ensemble average in this chapter.

[2] One should not, however, conclude that if the cross-correlation function is zero for all values of τ, that the ensembles are necessarily independent, *i.e.*, incoherent. It is only true, that they are not linearly related, but a different type of dependence may still exist. (See, for instance, the example at the end of Sec. 3.7.)

or
$$= \iint x_a x_b p(x_a,x_b)dx_a dx_b \tag{63}$$

In Eq. (62), $p(x_a,x_b)$ is the probability of the simultaneous occurrence of the value x_a for x_k and of the value x_b for x_{k+j}. The summations are taken over all possible values of x_a and x_b. In case x_a and x_b have a continuous range of possible values, then Eq. (63) must be used.

In case the function $g(t)$ of interest has only a finite total amount of energy, such as a single pulse, or the function

$$\left.\begin{aligned} g(t) &= \frac{1}{CR}\epsilon^{-\frac{t-t_0}{CR}} \quad \text{when } t < t_0 \\ &= 0 \qquad\qquad\quad \text{when } t > t_0 \end{aligned}\right\} \tag{64}$$

of Fig. 8-8 of the last section, then $\phi(\tau)$ would vanish for all values of τ. The correlation in such cases is nevertheless sometimes still a matter of interest. In order to deal with such cases where the total energy is finite, we use a slightly different form for the correlation function, which we shall call the *finite-correlation function.* This is defined as

$$\psi(\tau) = \int_{-\infty}^{+\infty} g(t)g(t+\tau)dt \tag{65}$$

Corresponding to Eq. (41), we can also formulate normalized versions of the correlation functions. If we write $\rho(\tau)$ for the normalized form of $\phi(\tau)$, then the normalized correlation functions for the variational components of $g_1(t)$, $g_2(t)$, and $g(t)$ are defined as

$$\rho_{12}(\tau) = \frac{\overline{[g_1(t) - \overline{g_1(t)}][g_2(t+\tau) - \overline{g_2(t)}]}}{\sqrt{\{\overline{[g_1(t)]^2} - [\overline{g_1(t)}]^2\}\{\overline{[g_2(t)]^2} - [\overline{g_2(t)}]^2\}}} \tag{66a}$$

and
$$\rho(\tau) = \frac{\overline{[g(t) - \overline{g(t)}][g(t+\tau) - \overline{g(t)}]}}{\overline{g(t)^2} - [\overline{g(t)}]^2} \tag{66b}$$

As has been our custom, we use a bar over a quantity for a time average. In case the functions $g(t)$, $g_1(t)$, and $g_2(t)$ involved are typical functions of ergodic ensembles, then this will also be equal to the ensemble average.

It follows from Eq. (66b) that for all functions $g(t)$

$$\rho(0) = 1 \tag{66c}$$

(b) *Relations between Correlation Functions and Power Spectra.* We will now prove Wiener's theorem that the auto-correlation function of $g(t)$ is the Fourier transform of its smoothed power spectrum. This theorem is the basis for the usefulness of the autocorrelation function in the analysis of communication systems.

In our discussion of the correlation functions, we have dealt principally with functions which continued to have finite values even as t approached infinity, such as a random-noise function or $\cos \omega t$. Such functions do not have a finite total energy content. Now it is one of the characteristics of a function which has a Fourier transform in the ordinary sense and for which the Fourier integral energy theorem is applicable that

$$\int_{-\infty}^{+\infty} [g(t)]^2 dt$$

must be finite. In the following analysis, in order to make some of the previously described functions amenable to Fourier analysis, we define an auxiliary function $g_T(t)$ as

$$\left.\begin{aligned} g_T(t) &= g(t) \quad \text{when } -T < t < +T \\ &= 0 \qquad \text{elsewhere} \end{aligned}\right\} \tag{67}$$

$g_T(t)$ will then be subject to Fourier analysis, and as $T \to \infty$, those properties of $g_T(t)$ which approach limiting values will also be properties of $g(t)$.

Let the Fourier transform of $g_T(t)$ be $\Omega_T(\omega)$. Then

$$\Omega_T(\omega) = \int_{-\infty}^{+\infty} g_T(t)\epsilon^{-j\omega t} dt \tag{67a}$$

According to the Fourier integral energy theorem

$$\begin{aligned} \int_{-\infty}^{+\infty} [g_T(t)]^2 dt &= \frac{1}{2\pi} \int_{-\infty}^{+\infty} \Omega_T(\omega) \cdot \Omega_T^*(\omega) d\omega \\ &= \int_{-\infty}^{\infty} |\Omega_T(\omega)|^2 df \end{aligned} \tag{67b}$$

since

$$\omega = 2\pi f \tag{67c}$$

The expression $|\Omega_T(\omega)|^2$ thus is a measure of the energy of $g_T(t)$ in the frequency range f to $f + df$. It may therefore be called the energy spectrum of $g_T(t)$. If we divide the energy spectrum by the effective duration of $g_T(t)$, namely, $2T$, we then get what we may call the power spectrum of $g_T(t)$. Let $W_T(f)$ be the power spectrum of $g_T(t)$. Thus

$$W_T(f) = \frac{1}{2T} |\Omega_T(\omega)|^2 \tag{67d}$$

Finally, we define $W(f)$ as the *power spectrum* of $g(t)$, where

$$W(f) = \lim_{T\to\infty} W_T(f) \tag{68}$$

if this limit exists. $W(f)$ as thus defined is a measure of the frequency distribution of the power in the function $g(t)$ which extends from $t = -\infty$ to $+\infty$. We will soon show that $W(f)$ is itself the Fourier transform of the auto-correlation function of $g(t)$.

Let us next define the auxiliary function

$$C_T(\tau) = \frac{1}{2T} \int_{-\infty}^{+\infty} g_T(t) \cdot g_T(t+\tau)dt \tag{69}$$

Consequently $\qquad C_T(\tau) = 0 \qquad \text{when } |\tau| > 2T \qquad (70)$

and for all functions $g(t)$ of interest, $C_T(\tau)$ will then have a Fourier transform. It follows from Eq. (69) that

$$\lim_{T\to\infty} C_T(\tau) = \phi(\tau) \tag{71}$$

where $\phi(\tau)$ is the auto-correlation function.

Let the Fourier transform of $C_T(\tau)$ be $P_T(\omega)$. Then

$$\begin{aligned} P_T(\omega) &= \int_{-\infty}^{+\infty} C_T(\tau)\epsilon^{-j\omega\tau}d\tau \\ &= \frac{1}{2T}\int_{-\infty}^{+\infty} \epsilon^{-j\omega\tau}\left[\int_{-\infty}^{+\infty} g_T(t)g_T(t+\tau)dt\right]d\tau \\ &= \frac{1}{2T}\int_{-\infty}^{+\infty}\left[\int_{-\infty}^{+\infty} g_T(t)g_T(t+\tau)\epsilon^{j\omega t}\epsilon^{-j\omega(t+\tau)}dt\right]d\tau \end{aligned} \tag{72}$$

We next interchange the order of integration and introduce the variable

$$s = t + \tau \tag{73}$$

Consequently, Eq. (72) becomes

$$\begin{aligned} P_T(\omega) &= \frac{1}{2T}\int_{-\infty}^{+\infty} g_T(t)\epsilon^{j\omega t}dt \cdot \int_{-\infty}^{+\infty} g_T(s)\epsilon^{-j\omega s}ds \\ &= \frac{1}{2T}\Omega_T^*(\omega)\cdot\Omega_T(\omega) = \frac{1}{2T}|\Omega_T(\omega)|^2 \end{aligned} \tag{74}$$

From Eqs. (67d) and (74), we therefore have

$$P_T(\omega) = W_T(f) \tag{75}$$

In the case of a random function which is a typical member of an ergodic ensemble, we found in Sec. 3.6(a) that

$$\frac{1}{2T} |\Omega_T(\omega)|^2$$

does not approach a limit as $T \to \infty$. Referring back to Sec. 3.6(a) and Fig. 3-6, we see that in any frequency range Δf, provided that

$$\Delta f \gg \frac{1}{T} \tag{76}$$

the function $\frac{1}{2T} |\Omega_T(\omega)|^2$ will have approximately $T\,\Delta f$ fluctuations. For a fixed value of Δf, as T becomes larger, the average value of $\frac{1}{2T} |\Omega_T(\omega)|^2$ in the interval will, however, approach a limit. For this case, let us therefore define $W(f)$ as

$$W(f) = \operatorname*{Lim}_{\Delta f \to 0} \operatorname*{Lim}_{T \to \infty} \frac{1}{\Delta f} \int_f^{f+\Delta f} \frac{1}{2T} |\Omega_T(\omega)|^2 df \tag{77}$$

The function $W(f)$ as thus defined may be called the *smoothed power spectrum* and will exist for every random function which is a typical member of an ergodic ensemble. If instead of using Eq. (77), we defined the *ensemble average power spectrum* as the limit as $T \to \infty$ of the ensemble average of $\frac{1}{2T} |\Omega_T(\omega)|^2$ we would also eliminate the fluctuations. Thus, the ensemble average power spectrum and the smoothed power spectrum are equal. We may therefore use as an alternative form for $W(f)$

$$W(f) = \operatorname*{Lim}_{T \to \infty} \frac{1}{2T} \widetilde{|\Omega_T(\omega)|^2} \tag{78}$$

where the $\sim$ stands for an ensemble average.

We now proceed to derive the relation between the power spectrum and the auto-correlation function. To do this let us start with Eq. (44) of Chap. 3, namely,[1]

[1] Note that $2T$ is used here in the sense that T was used in Chap. III.

$$\frac{1}{2T}\left|\Omega\left(\frac{2\pi q}{2T}\right)\right|^2 = \frac{1}{W}\left\{\frac{1}{8WT}\sum_{n=-2TW}^{+2TW}\left[g\left(\frac{n}{2W}\right)\right]^2\right.$$
$$+\frac{1}{4WT}\sum_p \cos\left(\frac{\pi qp}{2TW}\right)\left[\sum_n g\left(\frac{n}{2W}\right)\cdot g\left(\frac{n+p}{2W}\right)\right]$$
$$\text{(over the range of the intersymbol influence)}$$
$$\left.+\frac{1}{4WT}\sum_p \cos\left(\frac{\pi qp}{2TW}\right)\left[\sum_n g\left(\frac{n}{2W}\right)\cdot g\left(\frac{n+p}{2W}\right)\right]\right\} \tag{79}$$
$$\text{(outside the range of the intersymbol influence)}$$

The quantity $\frac{1}{2T}\left|\Omega\left(\frac{2\pi q}{2T}\right)\right|^2$ above is actually the value of $\frac{1}{2T}\,|\Omega_T(\omega)|^2$ at the qth sample point in the frequency domain. In the case of a function which is a typical member of an ergodic ensemble, the range of the intersymbol influence is finite, say approximately equal to the integer K. Therefore, the second term on the right side of Eq. (79) has the order of $K(4TW)$ terms in the double summation while the third term has $(2TW - K)(4TW)$ terms. In the case of a particular typical function of the ensemble, as pointed out in Sec. 3.6, the random-walk analysis then shows that, as T approaches infinity, the second term will approach its ensemble average value, but the third term will not. Specifically the third term will in general remain finite even though its ensemble average value is zero.

Let us however take an ensemble average value of both sides of Eq. (79). Then we have

$$\frac{1}{2T}\widetilde{\left|\Omega\left(\frac{2\pi q}{2T}\right)\right|^2} = \frac{1}{W}\left\{\frac{1}{8WT}\widetilde{\sum_{n=-2TW}^{+2TW}\left[g\left(\frac{n}{2W}\right)\right]^2}\right.$$
$$\left.+\frac{1}{4WT}\widetilde{\sum_p \cos\left(\frac{\pi qp}{2TW}\right)\left[\sum_n g\left(\frac{n}{2W}\right)g\left(\frac{n+p}{2W}\right)\right]}\right\} \tag{80}$$
$$\text{(over the range of the intersymbol influence)}$$

The ensemble average bars on the right side of Eq. (80) are actually superfluous because the value of the summations will be the same for every typical function as for the ensemble average.

The smallest usable value of W in Eq. (80) is limited by the spectral content of $g(t)$, but we can choose W as large as we please and the equation will still be true. If we allow W and T to become very large, the summations in Eq. (80) can be approximated by integrals, where

$$dt = \frac{1}{2W} = d\tau \tag{81}$$

$$t = \frac{n}{2W} \tag{82}$$

$$\tau = \frac{p}{2W} \tag{83}$$

$$\omega = \frac{2\pi q}{2T} \tag{84}$$

and

$$\frac{1}{W}\left\{\frac{1}{8WT}\sum_{n=-2TW}^{2TW}\left[\left(g\,\frac{n}{2W}\right)^2\right]\right\} \rightarrow 0 \tag{85}$$

Thus Eq. (80) becomes

$$\lim_{T\to\infty} \frac{1}{2T}\,\widetilde{|\Omega_T(\omega)|^2} = \lim_{T\to\infty}\left\{\frac{1}{T}\int_0^R \cos\omega\tau\left[\int_{-\infty}^{+\infty} g(t)g(t+\tau)dt\right]d\tau\right\} \tag{86}$$

R in Eq. (86) stands for the range of the intersymbol influence. Since the integral inside the brackets vanishes when $\tau > R$, we can replace R by $+\infty$ in Eq. (86). Thus, the latter may be written

$$\begin{aligned} W(f) &= \lim_{T\to\infty} \frac{1}{2T}\,\widetilde{|\Omega_T(\omega)|^2} \\ &= 2\int_0^\infty \cos\omega\tau\left[\lim_{T\to\infty}\frac{1}{2T}\int_{-T}^{+T} g(t)g(t+\tau)dt\right]d\tau \\ &= \int_0^\infty 2\cos\omega\tau\cdot\phi(\tau)d\tau \end{aligned} \tag{87}$$ [1]

[1] In direct experimental measurement of the power spectrum the spectra for positive and negative frequencies will always be combined. Since $W(f)$ has equal values at the frequencies f and $-f$, and since both represent positive power, the measured power spectrum, since positive and negative frequencies are not distinguished, will be $2W(f)$. For this reason, our formulas involving the power spectrum will differ by a factor of 2 from those of some other writers who define $2W(f)$ as the power spectrum.

Since $\phi(\tau)$ is an even function of τ, Eq. (87) may also be written

$$W(f) = \int_{-\infty}^{+\infty} \epsilon^{-j\omega\tau}\phi(\tau)d\tau \tag{88}$$

Equation (88), or its equivalent Eq. (87), shows that $W(f)$ is the Fourier transform of $\phi(\tau)$. It then follows that we must also have

$$\phi(\tau) = \frac{1}{2\pi}\int_{-\infty}^{+\infty} W(f)\epsilon^{j\omega\tau}d\omega = \int_{-\infty}^{+\infty} W(f)\epsilon^{j2\pi f\tau}df \tag{89}$$

and since $W(f)$ is also an even function

$$\phi(\tau) = \int_{0}^{\infty} 2W(f) \cos 2\pi f\tau \cdot df \tag{90}$$

The auto-correlation function is thus the Fourier transform of the *smoothed power spectrum.*[1] The latter is usually referred to simply as the power spectrum. It should, however, be realized that the fluctuations in $1/2T|\Omega_T(\omega)|^2$ (which fluctuations carry the information content of the power spectrum as pointed out in Sec. 3.6) cannot be obtained from the correlation function. The latter does however give the ensemble average form of the power spectrum and therefore does describe the constraints on the power spectrum, which we will find to be useful knowledge.

By following a similar analysis, which however is much simpler because there is no difficulty with limits, we can find the Fourier transform of the finite-correlation function of a signal of finite total energy. If

$$E(\omega) = |\Omega(\omega)|^2 \tag{91}$$

is called the energy spectrum of $g(t)$, then the finite-correlation function $\psi(t)$ and the energy spectrum $E(\omega)$ are Fourier transforms. Thus

$$\psi(\tau) = \frac{1}{2\pi}\int_{-\infty}^{+\infty} E(\omega)\epsilon^{j\omega\tau}d\omega \tag{92}$$

and

$$E(\omega) = \int_{-\infty}^{+\infty} \psi(\tau)\epsilon^{-j\omega\tau}d\tau \tag{93}$$

The derivation of these equations is left to the reader as an exercise.

[1] This result is sometimes called Wiener's theorem.

The Fourier transforms of the cross-correlation functions also give important spectral information. If we use the notation

$$\Omega_{1T}(\omega) = \int_{-T}^{+T} g_1(t)\epsilon^{-j\omega t}dt \tag{94}$$

and

$$\Omega_{2T}(\omega) = \int_{-T}^{+T} g_2(t)\epsilon^{-j\omega t}dt \tag{95}$$

then the *cross-spectral density* may be defined as

$$W_{12}(f) = \operatorname*{Lim}_{T\to\infty}\left[\frac{1}{2T}\,\overline{\Omega_{1T}^*(\omega)\cdot\Omega_{2T}(\omega)}\right] \tag{96}$$

where the bar indicates an ensemble average. Since

$$\Omega_{1T}(\omega) = \Omega_{1T}^*(-\omega) \tag{97}$$

and

$$\Omega_{2T}(\omega) = \Omega_{2T}^*(-\omega) \tag{98}$$

it follows that

$$W_{12}(f) = W_{12}^*(-f) = W_{21}(-f) \tag{99}$$

By carrying out an analysis similar to that which was performed above for the relations between the power spectrum and the auto-correlation function, one can find the corresponding relation between the cross-spectral density and the cross-correlation function. The analysis will not be given here, but the following results are readily obtained:

$$\phi_{12}(\tau) = \frac{1}{2\pi}\int_{-\infty}^{+\infty} W_{12}(f)\epsilon^{j\omega\tau}d\omega \tag{100}$$

and

$$W_{12}(f) = \int_{-\infty}^{+\infty} \phi_{12}(\tau)\epsilon^{-j\omega\tau}d\tau \tag{101}$$

As a preliminary to the use of the correlation functions, we shall next calculate the correlation functions and power spectra in a few instructive examples.

Problems

1. Derive Eqs. (100) and (101).

2. If two incoherent ergodic ensembles are linearly superimposed, show that their auto-correlation functions are additive and so are their power spectra.

8.5 Examples of correlation functions

(a) *The Auto-correlation Function of Random Noise of Bandwidth W.* Suppose we have random noise of bandwidth W. This is expressed analytically as [Eq. (91) of Chap. III]

$$g(t) = \sum_{n=-\infty}^{+\infty} g_n \frac{\sin (2\pi Wt - n\pi)}{2\pi Wt - n\pi} \tag{102}$$

where the g_n's are the values of $g(t)$ at the sample points in the time domain and are normally distributed with a mean-square value

$$\overline{g_n^2} = \sigma_I^2 = N = \text{average noise power} \tag{103}$$

It would be an involved and difficult matter to substitute Eq. (102) into Eq. (48) and calculate the auto-correlation function by direct integration. We will therefore calculate it, instead, from the ensemble average of the power spectrum. Thus from Eqs. (82) and (85) of Chap. III, we have[1]

$$\begin{aligned} \widetilde{|\Omega_T(\omega)|^2} &= \begin{cases} \rho^2 = \dfrac{T}{W}\sigma_I^2 = \dfrac{T}{W}N & \text{from } -W < \dfrac{\omega}{2\pi} < W \\ 0 & \text{elsewhere} \end{cases} \end{aligned} \tag{104}$$

Therefore

$$\left.\begin{aligned} W(f) = \lim_{T\to\infty} \frac{1}{2T}\widetilde{|\Omega_T(\omega)|^2} &= \frac{N}{2W} \quad \text{from } -W < \frac{\omega}{2\pi} < W \\ &= 0 \quad \text{elsewhere} \end{aligned}\right\} \tag{105}$$

Consequently, by Eq. (89)

$$\phi(\tau) = \int_{-\infty}^{+\infty} \epsilon^{j2\pi f\tau} W(f)\,df = \int_{-W}^{+W} \frac{N}{2W} \epsilon^{j2\pi f\tau}\,df = N \cdot \frac{\sin 2\pi W\tau}{2\pi W\tau} \tag{106}$$

(b) *A Finite-Correlation Function.* In Fig. 8-8 of Sec. 8.3 the signal

$$g(t) = \begin{cases} \dfrac{1}{CR}\epsilon^{(t-t_0)/CR} & \text{when } t < t_0 \\ 0 & \text{when } t > t_0 \end{cases} \tag{107}$$

[1] Note that $2T$ is used here in the sense that T was used in Chap. III.

of finite total energy appeared at the input of a filter. The output was supposed to be its finite-auto-correlation function. Let us now find the finite-auto-correlation function of this $g(t)$. By Eq. (65), we have

$$\psi(\tau) = \int_{-\infty}^{+\infty} g(t)g(t+\tau)dt$$
$$= \begin{cases} \int_{-\infty}^{t_0} \frac{1}{C^2R^2} \epsilon^{(t-t_0+t+\tau-t_0)/CR}dt = \frac{1}{2CR}\epsilon^{+\tau/CR} & (\text{when } \tau < 0) \\ \int_{-\infty}^{t_0-\tau} \frac{1}{C^2R^2} \epsilon^{(t-t_0+t+\tau-t_0)/CR}dt = \frac{1}{2CR}\epsilon^{-\tau/CR} & (\text{when } \tau > 0) \end{cases} \qquad (108)$$

Exercise

Find the finite-auto-correlation function of $g(t)$ in Eq. (107) by obtaining the Fourier transform of the energy spectrum of $g(t)$.

(c) *The Auto-correlation Function of a Periodic Function.* Let us next find the auto-correlation function of

$$A \sin(\omega_0 t + \alpha)$$

Here

$$\phi(\tau) = \lim_{T\to\infty}\left[\frac{1}{2T}\int_{-T}^{+T} A^2 \sin(\omega_0 t + \alpha) \sin[\omega_0(t+\tau) + \alpha]dt\right]$$
$$= \lim_{T\to\infty}\left[\frac{1}{2T}\int_{-T}^{+T} A^2 \sin^2(\omega_0 t + \alpha)\cos\omega_0\tau\, dt\right.$$
$$\left. + \frac{1}{2T}\int_{-T}^{+T} A^2 \sin(\omega_0 t + \alpha)\cos(\omega_0 t + \alpha)\sin\omega_0\tau\, dt\right]$$
$$= \frac{A^2}{2}\cos\omega_0\tau = \frac{A^2}{4}[\epsilon^{j\omega_0\tau} + \epsilon^{-j\omega_0\tau}] \qquad (109)$$

The auto-correlation function in this case is also periodic and has the same period as the original function. We also note that in this case the auto-correlation function does not vanish as $\tau \to \infty$, as it does in the case of the auto-correlation function of a typical function of an ergodic ensemble.

In the present case, $\phi(\tau)$ does not have a proper Fourier transform. However, the power spectrum of $A \sin(\omega_0 t + \alpha)$ is clearly an impulsive-type function, with impulses at $\pm \frac{\omega_0}{2\pi}$. The average power of $A \sin (\omega_0 t + \alpha)$ is $A^2/2$, so that

$$\int_{-\infty}^{+\infty} W(f)df = \frac{A^2}{2} \tag{110}$$

The strength of the impulses at $\pm \frac{\omega_0}{2\pi}$ is therefore $\frac{A^2}{4}\cdot$ Thus

$$W(f) = \frac{A^2}{4}\left[\delta\left(f + \frac{\omega_0}{2\pi}\right) + \delta\left(f - \frac{\omega_0}{2\pi}\right)\right] \tag{111}$$

(d) *The Auto-correlation Function of a Function with a d-c Component.* Suppose that we have a function

$$g(t) = C + g_1(t) \tag{112}$$

where C is the average value or d-c component of $g(t)$. Then

$$\begin{aligned} g(t)g(t+\tau) &= [C + g_1(t)][C + g_1(t+\tau)] \\ &= C^2 + C[g_1(t) + g_1(t+\tau)] + g_1(t)g_1(t+\tau) \end{aligned} \tag{113}$$

Thus

$$\begin{aligned} \phi(\tau) &= \lim_{T\to\infty} \frac{1}{2T}\int_{-T}^{+T} g(t)g(t+\tau)dt \\ &= C^2 + \lim_{T\to\infty} \frac{1}{2T}\int_{-T}^{+T} C[g_1(t) + g_1(t+\tau)]dt \\ &\quad + \lim_{T\to\infty} \frac{1}{2T}\int_{-T}^{+T} g_1(t)g_1(t+\tau)dt \end{aligned} \tag{114}$$

The second term on the right side of Eq. (114) vanishes since the average value of $g_1(t)$ is zero. Therefore

$$\begin{aligned} \phi(\tau) &= C^2 + \lim_{T\to\infty} \frac{1}{2T}\int_{-T}^{+T} g_1(t)g_1(t+\tau)dt \\ &= C^2 + \text{auto-correlation function of } g_1(t) \end{aligned} \tag{115}$$

The power spectrum is then the power spectrum of $g_1(t)$ plus an impulsive-type function at $f = 0$. Since the average power of C is C^2, we have

$$W(f) = C^2\delta(f) + W_1(f) \tag{116}$$

for the power spectrum of $g(t)$, where $W_1(f)$ is the power spectrum of $g_1(t)$.

(e) *An Auto-correlation Function Calculated as an Ensemble Average.* Suppose we have a random function which is a typical member of an ergodic ensemble which consists of a sequence of rectangular pulses of randomly determined height and random duration (see Fig. 8-9).[1]

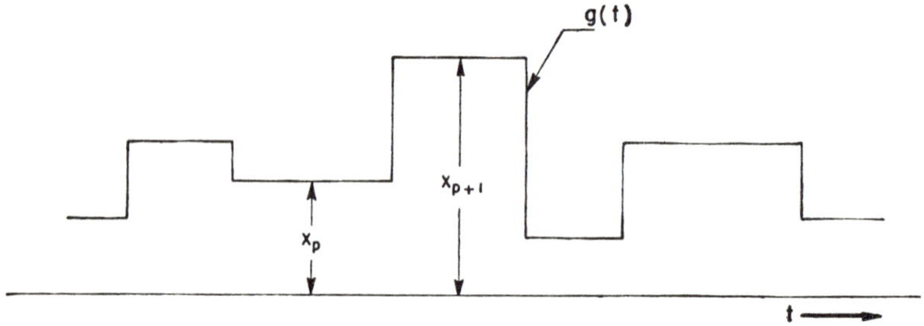

Fig. 8-9.

Let the probability distribution of heights (x) of the pulses be

$$P(x) = \frac{1}{\sqrt{2\pi\sigma^2}} \epsilon^{-(x-g_{dc})^2/2\sigma^2} \tag{117}$$

Let the probability distribution of the lengths l of the pulses be

$$H(l) = \beta\epsilon^{-\beta l} \tag{118}$$

and let us suppose that the height of a pulse is independent of its length. The distribution $H(l)$ will be taken as a fractional distribution over the total number of pulses and not over their total length. In other words

$$\int_0^\infty H(l)dl = 1 \tag{119}$$

and for the ensemble average of l

$$\bar{l} = \int_0^\infty lH(l)dl \tag{120}$$

The fraction of any typical long interval of time (containing a very large number of pulses) which will be filled with pulses of length

[1] Some practical aspects of this type of function are discussed in James, Nichols, and Phillips, *Theory of Servomechanisms*, McGraw-Hill, New York (1947), Sec. 6.12, to which the present derivation is largely indebted.

l between l and $l + dl$ will then be

$$R(l)dl = \frac{lH(l)dl}{\int_0^\infty lH(l)dl} = \frac{l}{\bar{l}} H(l)dl \tag{121}$$

As a preliminary to calculating the auto-correlation function, let us calculate the probability $Q(\tau)$ that $g(t)$ and $g(t + \tau)$ lie in the same pulse. Now the probability that if t is within an arbitrary pulse of length l then $t + \tau$ will also be within the pulse is (see Fig. 8-10)

$$q_l(\tau) = \begin{cases} 1 - \dfrac{|\tau|}{l} & \text{if } |\tau| < l \\ 0 & \text{if } l < |\tau| \end{cases} \tag{122}$$

L = Range in which both t and t+T lie in same interval.

Therefore

$$Q(\tau) = \int_{|\tau|}^\infty \left(1 - \frac{|\tau|}{l}\right) R(l)dl = \int_{|\tau|}^\infty \left(1 - \frac{|\tau|}{l}\right) \frac{l}{\bar{l}} H(l)dl \tag{123}$$

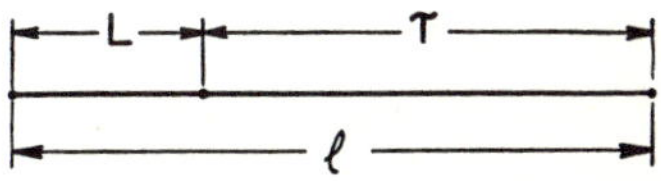

Fig. 8-10.

For the distribution $H(l)$ of Eq. (118)

$$\bar{l} = \int_0^\infty l\beta\epsilon^{-\beta l}dl = \frac{1}{\beta} \tag{124}$$

Thus

$$Q(\tau) = \int_{|\tau|}^\infty \left(1 - \frac{|\tau|}{l}\right) \beta l \cdot \beta\epsilon^{-\beta l}dl = \epsilon^{-\beta|\tau|} \tag{125}$$

Next calculating the auto-correlation function as an ensemble average, we have

$$\phi(\tau) = \widetilde{g(t)g(t + \tau)} = \widetilde{x(t)x(t + \tau)} \tag{126}$$

Since the distributions of heights and lengths are independent,

$$\widetilde{x(t)x(t + \tau)} = \widetilde{[x(t)]^2}Q(\tau) + \widetilde{x(t)(t + T)}[1 - Q(\tau)] \tag{127}$$

where T is used for the value of τ when $x(t)$ and $x(t + \tau)$ are not in the same pulse. Since $x(t)$ and $x(t + T)$ are not in the same pulse their values are uncorrelated, and

$$\widetilde{x(t)x(t + T)} = [\widetilde{x(t)}]^2 = g_{\text{dc}}^2 \tag{128}$$

For the distribution of pulse heights given by Eq. (117),

$$\widetilde{[x(t)]^2} = \sigma^2 \tag{129}$$

Therefore

$$\phi(\tau) = \sigma^2\epsilon^{-\beta|\tau|} + g_{\text{dc}}^2[1 - \epsilon^{-\beta|\tau|}] = (\sigma^2 - g_{\text{dc}}^2)\epsilon^{-\beta|\tau|} + g_{\text{dc}}^2 \quad (130)$$

This is the desired equation for the auto-correlation function. Let us next calculate the power spectrum. For this

$$\begin{aligned} W(f) &= \int_{-\infty}^{+\infty} \phi(\tau)\epsilon^{-j\omega\tau}d\tau \\ &= \int_{-\infty}^{+\infty} [(\sigma^2 - g_{\text{dc}}^2)\epsilon^{-\beta|\tau|} + g_{\text{dc}}^2]\epsilon^{-j\omega\tau}d\tau \\ &= (\sigma^2 - g_{\text{dc}}^2)\frac{2\beta}{\beta^2 + \omega^2} + g_{\text{dc}}^2\delta(f) \end{aligned} \quad (131)$$

It is noteworthy that in this case $\phi(\tau)$ and $W(f)$ do not depend upon the x distribution $P(x)$ in detail but only upon its mean-square value σ^2 and its average value g_{dc}.

(f) *A Cross-correlation Function.* As an example of the calculation of a cross-correlation function, we will calculate the correlation between $\sin \omega_0 t$ and $\cos \omega_0 t$. An example of a cross-correlation function involving a random function will be given in Sec. 8.8.

For the cross-correlation function of $\sin \omega_0 t$ and $\cos \omega_0 t$, we have

$$\begin{aligned} \phi_{12}(\tau) &= \lim_{T\to\infty} \frac{1}{2T}\int_{-T}^{+T} \sin \omega_0 t \cos [\omega_0(t + \tau)]dt \\ &= \lim_{T\to\infty} \int_{-T}^{+T} \frac{1}{2T} \sin \omega_0 t \cos \omega_0 t \cos \omega_0\tau \, dt \\ &+ \lim_{T\to\infty} \int_{-T}^{+T} \frac{1}{2T} \sin^2 \omega_0 t(-\sin \omega_0\tau)dt = -\frac{1}{2}\sin \omega_0\tau \end{aligned} \quad (132)$$

It is interesting to note that even though $\cos \omega_0 t$ and $\sin \omega_0 t$ are 90 deg out of phase, their cross correlation is not zero, but is a periodic function of the same frequency.

The cross-spectral density in this case is

$$\begin{aligned} W_{12}(f) &= \int_{-\infty}^{+\infty} \phi_{12}(\tau)\epsilon^{-j\omega\tau}d\tau \\ &= \int_{-\infty}^{+\infty} -\tfrac{1}{2}\sin \omega_0\tau\epsilon^{-j\omega\tau}d\tau \\ &= -\frac{1}{2}\int_{-\infty}^{+\infty} \frac{\epsilon^{j\omega_0\tau} - \epsilon^{-j\omega_0\tau}}{2j} \cdot \epsilon^{-j\omega\tau}d\tau \\ &= -\frac{1}{4j}\left[\int_{-\infty}^{+\infty} \epsilon^{-j(\omega-\omega_0)\tau}d\tau - \int_{-\infty}^{+\infty} \epsilon^{-j(\omega+\omega_0)\tau}d\tau\right] \end{aligned} \quad (133)$$

Comparing Eq. (133) with Eq. (4), we obtain

$$W_{12}(f) = j\frac{\pi}{2}[\delta(\omega - \omega_0) - \delta(\omega + \omega_0)]$$

$$= \frac{j}{4}[\delta(f - f_0) - \delta(f + f_0)] \qquad (134)^1$$

The j multiplier in Eq. (134) arises from the fact that $\cos \omega_0 t$ and $\sin \omega_0 t$ are 90 deg out of phase. We thus see that the cross-spectral density, unlike the power spectrum, does not need to be real. It gives information regarding the phase relations of the respective frequency components of the correlated functions.[2]

PROBLEMS

1. Find the cross-correlation function between $\sin \omega t$ and $\sin 2\omega t$.

2. Find the cross-correlation function between $\sin \omega_1 t$ and $\sin \omega_2 t$ and discuss the result.

3. Find the cross-correlation function between $f_1(t)$ and $f_2(t)$ where

$$f_1(t) = \text{random noise plus } \sin \omega t$$

and

$$f_2(t) = \cos \omega t$$

4. (Due to C. R. Cahn.) Let $G(t)$ be a function of time. Let

$$G_1(t) = -\frac{1}{\pi}\int_{-\infty}^{+\infty} \frac{G(\tau)}{t - \tau}\, d\tau$$

be its Hilbert transform. Let

$$\phi_{GG}(x) = \int_{-\infty}^{+\infty} G(t)G(t + x)dt$$

[1] To obtain the numerical relation between $\delta(\omega)$ and $\delta(f)$ one can proceed as follows:

$$\int \delta(\omega)d\omega = 1 = \int \delta(f)df = \int \delta(f)\frac{d\omega}{2\pi}$$

Therefore

$$\delta(\omega) = \frac{1}{2\pi}\delta(f)$$

[2] The cross-correlation function does not give phase information if the functions being cross-correlated are random functions, but only if one of them is periodic (with known phases) and does not have the characteristic fluctuations of a random function.

be the finite-auto-correlation function of $G(t)$, and let

$$\phi_{GG_1}(x) = \int_{-\infty}^{+\infty} G(t)G_1(t+x)dt$$

be the finite-cross-correlation function of $G(t)$ and $G_1(t)$. Show that

(a) $\phi_{GG_1}(x) = -\frac{1}{\pi}\int_{-\infty}^{+\infty} \frac{\phi_{GG}(y)}{x-y}\,dy$

(b) $\phi_{GG_1}(0) = 0$

(c) $\phi_{GG}(x) = \phi_{G_1G_1}(x)$

(d) $\phi_{GG}(x) = +\frac{1}{\pi}\int_{-\infty}^{+\infty} \frac{\phi_{GG_1}(y)}{x-y}\,dy$

5. (a) Let $\Omega(\omega)$ be the Fourier transform of $G(t)$ and let $\Omega^*(\omega)$ be the complex conjugate of $\Omega(\omega)$. Define $\psi(\omega')$ as

$$\psi(\omega') = \frac{1}{2\pi}\int_{-\infty}^{+\infty} \Omega(\omega)\Omega^*(\omega+\omega')d\omega$$

Show that the Fourier transform of $\psi(\omega')$ is $[G(-t)]^2$.

(b) For an ergodic ensemble of functions, the average value of $[G(-t)]^2$ is the same for all values of t. Thus for any particular function of the ensemble, $[G(-t)]^2$ fluctuates about a constant average. On the basis of part (a) of this problem, show that there is no linear correlation between the frequency components of the functions of an ergodic ensemble depending upon their frequency separation. Any linear correlation which may exist must therefore depend upon the absolute frequency.

8.6 Linear least-square smoothing and prediction[1]

(a) *Formulation of the Problem.* We are now ready to apply correlation theory to an interesting and important problem. Sup-

[1] The basic material for this section was originally developed in Norbert Wiener's N.D.R.C. report "The Extrapolation, Interpolation and Smoothing of Stationary Time Series," which was later published as a book by John Wiley & Sons. A related theory was independently developed by A. Kolmogoroff, "Interpolation und Extrapolation von Stationären Zufälligen Folgen," *Bull. Acad. Sci.* (*U.S.S.R.*) Ser. Math., vol. 5, 1941, pp. 3–14.

The reader is likely to find difficulty with Wiener's presentation, which is difficult and obscure in vital places. A more readable account for the material here is Norman Levinson, "A Heuristic Exposition of Wiener's Mathematical Theory of Prediction and Filtering," which is published as Appendix C in Wiener's book. Our own presentation is particularly indebted to H. W. Bode and C. E. Shannon "A Simplified Derivation of Linear Least Square Smoothing and Prediction Theory," *Proc. I.R.E.*, vol. 38, April, 1950, pp. 417–425.

pose that a desired signal $s(t)$ and a noise signal $n(t)$ have been superimposed by a linear process so that the signal

$$g(t) = s(t) + n(t) \tag{135}$$

is obtained. Suppose that $g(t)$ is available to us and from it we wish to find $s(t)$. If possible, we would like to find not only $s(t)$, but also $s(t + \alpha)$, some future values of $s(t)$. This is one of the most common types of problems which occur in communication engineering, and closely related problems occur in many other

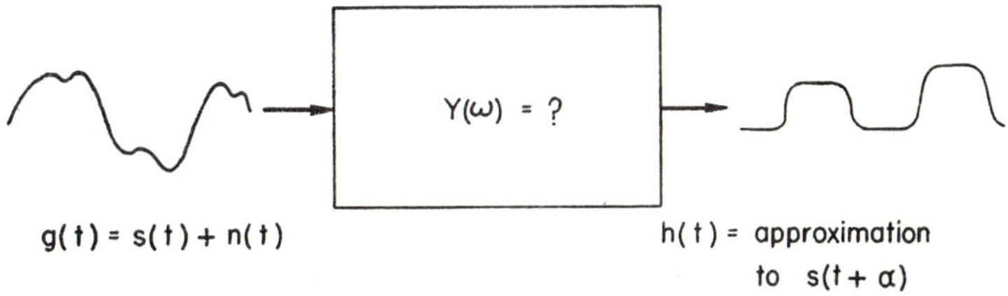

Fig. 8-11.

fields of human activity. We will find that a certain type of solution to a restricted formulation of this problem will be possible.[1]

The restricted problem which we will now attempt to solve is set up with the following conditions:

1. In the first place it is assumed that $s(t)$ and $n(t)$ are typical functions of ergodic ensembles. It is further assumed that we know the power spectra of the ensembles to which $s(t)$ and $n(t)$ belong.
2. The operation to be performed on $g(t)$ will be to send it through a *linear filter*, as shown in Fig. 8-11.
3. The filter $Y(\omega)$ shall be designed to get on the (*ensemble*) *average* the smallest possible *mean-square discrepancy* between the

[1] Conventional filter theory gives an elegant solution to the problem of the separation of two signals or of signal and noise which have essentially non-overlapping spectra. This, however, is rarely the problem which is met in practice, except when carrier frequencies are used. In the more usual case, when the spectra overlap, conventional filtering may still give preferred weighting to the desired signal, if the spectra are sufficiently different. However, this type of filtering uses only the spectral amplitude distribution of signals involved and does not use the phase information or waveshape as a basis of separation. The filter theory developed in the present section is based directly upon the waveshape of the signals involved and therefore deals more completely with the problem.

actual output $h(t)$ and the desired output $s(t + \alpha)$[1]. In other words the filter will be designed for the ensembles of functions $s(t)$ and $n(t)$, and not for particular functions of these ensembles. This procedure of dealing with the ensembles, instead of dealing with particular functions, is the only reasonable method of approach, since we never know with which particular functions $s(t)$ and $n(t)$ we are dealing. Thus our purpose shall be to minimize the ensemble average of the mean-square error, *i.e.*,

$$\widetilde{[h(t) - s(t + \alpha)]^2} \tag{136}$$

Actually, if $s(t)$ and $n(t)$ are typical functions of their ensembles, then if the ensemble average (136) is a minimum, then the expression

$$\frac{1}{2T} \int_{-T}^{T} [h(t) - s(t + \alpha)]^2 dt \tag{137}$$

will also be a minimum if the integral is taken over a sufficiently long time. The reasons for, and the consequences of, designing the filter for the least mean-square discrepancy will be discussed in Sec. 8.6(f).

The problem we will now try to solve is to find the complex transfer function $Y(\omega)$ of the linear filter which will make (136) and (137) a minimum. This has been called the problem of *linear least-square prediction and smoothing of a stationary time series.*[2] The name is

[1] If α is positive, then $h(t)$ attempts to predict the input; if α is negative, then $h(t)$ attempts to tell what the input was α seconds ago. If α is zero, $h(t)$ merely attempts to remove the noise.

[2] Smoothing describes the operation of approximate removal of the noise. The term *stationary time series* is here used in place of *a typical function of an ergodic ensemble.* A stationary time series is actually somewhat more general. It is a random function of time whose average probability distributions are independent of time, but unlike the typical member of an ergodic ensemble, it need not have a finite range of the intersymbol influence. In other words, the way a stationary time series happened to get started may continue to affect its future course forever. In the case of continuous signal functions, such infinite range of the intersymbol influence would give rise to periodic components having energy of comparable amount with that of the complete signal. In other words, there would be δ functions in the power spectrum.

In the method of solution employed here it will be assumed that we are dealing with typical functions of ergodic ensembles. Direct-current and periodic components, when they are present in a practical case, can be completely determined or predicted ahead of time from the statistics of the process, and can therefore be subtracted from the various signals and dealt with separately.

consistent with the three conditions of the problem. The reason for choosing this particular restricted formulation of the problem is that this restricted problem has been solved. At the end of the section we will discuss the theoretical and practical value of the solution as well as the significance of the restrictions.

In Fig. 8-12 are diagrams of $s(t)$, $n(t)$, $g(t)$, and $h(t)$. This figure shows the type of result that can be expected so that the reader may have a better picture of what the problem is about. The figure also illustrates the result which we shall find, that significant prediction cannot be made beyond the range of the intersymbol influence. In other words, predicton is an extrapolation into the future based upon a knowledge of the intersymbol influence. The (smoothed) output, $h_1(t)$ of Fig. 8-12(d), is quite similar in shape to $s(t + \alpha_1)$ when α_1 is negative. The (predicted) output, $h_2(t)$ of Fig. 8-12(e), is still somewhat similar in shape to $s(t + \alpha_2)$, when α_2 is positive but well inside the range of the intersymbol influence.[1] The (predicted) output $h_3(t)$ of Fig. 8-12(f) is unrelated in detailed waveshape to $s(t + \alpha_3)$ when α_3 is positive and greater than the range of the intersymbol influence.

(*b*) *A General Formula for the Mean-square Error.* The difference between the desired output $s(t + \alpha)$ and the actual output $h(t)$ will be called the error $e(t)$. Thus

$$e(t) = s(t + \alpha) - h(t) \tag{138}$$

According to the conditions of the problem, we are interested in finding a $Y(\omega)$ so that the mean-square value of $e(t)$ shall be a minimum.

Let $e_T(t)$ be the function which is equal to $e(t)$ everywhere in the range $(-T < t < +T)$ and is zero outside this range. Let us now choose T so large that $e_T(t)$ may be considered a typical function of an ergodic ensemble. Then

$$\widetilde{[e(t)]^2} = \frac{1}{2T}\int_{-\infty}^{+\infty} [e_T(t)]^2 dt = \frac{1}{2T}\cdot\frac{1}{2\pi}\int_{-\infty}^{+\infty} |E_T(\omega)|^2 d\omega$$
$$= \int_{-\infty}^{+\infty} W_e(f)\,df \tag{139}$$

[1] Even though the output in Fig. 8-13(e) is often less similar to the desired signal than the input is, it may be useful because it predicts the signal before it occurs.

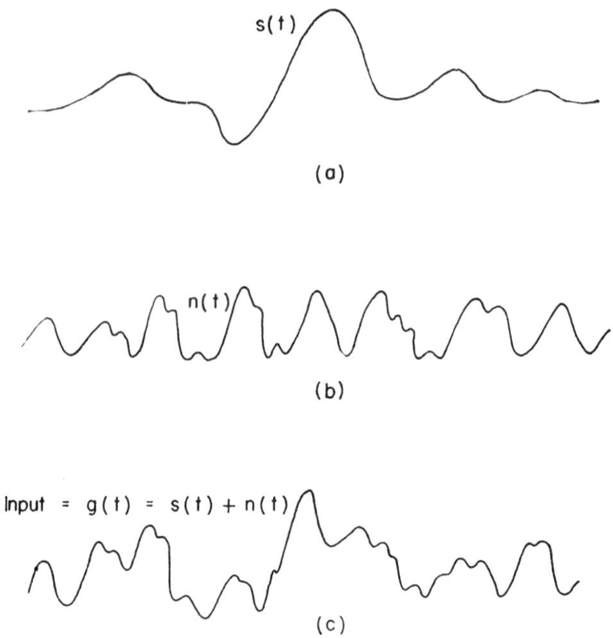

Fig. 8-12 (a), (b), (c). Illustration of smoothing and prediction.

where $E_T(\omega)$ is the Fourier transform of $e_T(t)$ and $W_e(f)$ is the smoothed power spectrum of $e(t)$. Thus $|E_T(\omega)|^2$ is the energy spectrum of $e_T(t)$. If $s_T(t)$, $h_T(t)$, and $n_T(t)$ are given definitions analogous to $e_T(t)$, then

$$g_T(t) = s_T(t) + n_T(t) \tag{140}$$

and

$$e_T(t) = s_T(t + \alpha) - h_T(t) \tag{141}$$

Let, $H_T(\omega)$, $N_T(\omega)$, $S_T(\omega)$, and $S_{T,\alpha}(\omega)$ be the Fourier transforms of $h_T(t)$, $n_T(t)$, $s_T(t)$, and $s_T(t + \alpha)$, respectively. Then if $Y(\omega)$ is the complex transfer function of the filter, regardless of whether this yields the least-square error or not,

$$H_T(\omega) = Y(\omega)S_T(\omega) + Y(\omega)N_T(\omega) \tag{142}$$

$$S_{T,\alpha}(\omega) = S_T(\omega)\epsilon^{j\omega\alpha} \tag{143}$$

Equation (143) is only approximately true for finite values of T, but the relative error approaches zero as T becomes infinite. Then we may write

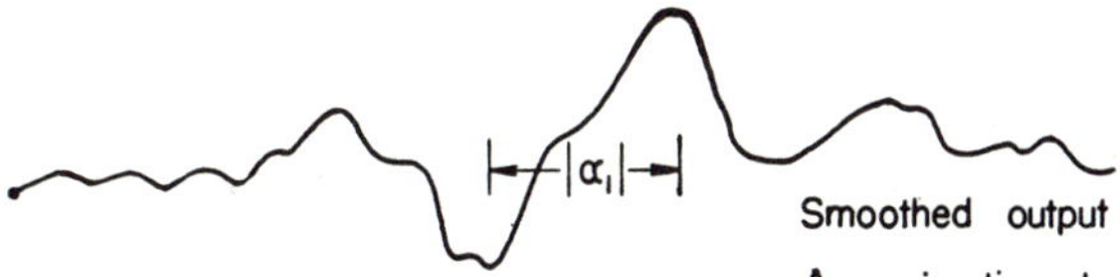

Smoothed output (α_1 negative)
Approximation to $s(t + \alpha_1)$

(d)

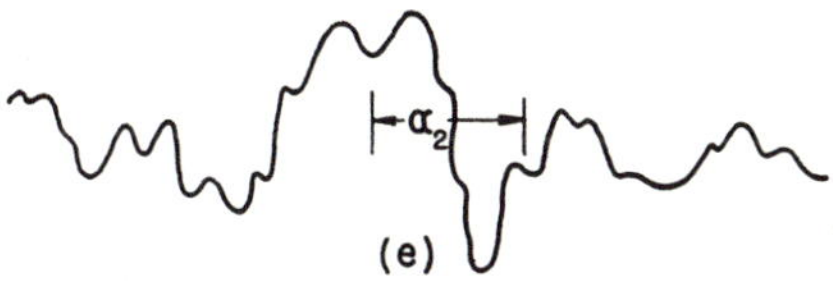

Smoothed output (α_2 positive but less than the range of the intersymbol influence). [Approximation to $s(t + \alpha_2)$] Even though this signal is often less similar to the desired signal than the input is, it may be useful because it predicts the signal before it occurs.

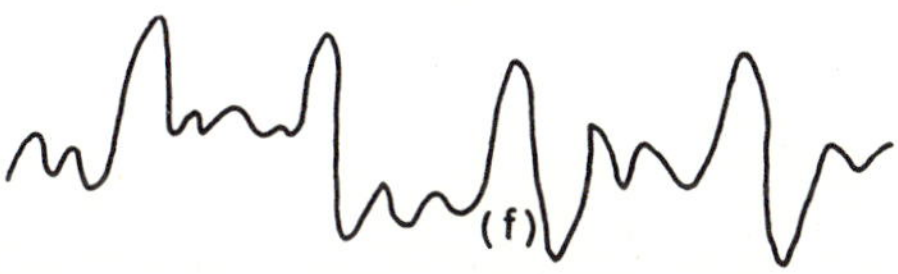

Smoothed output (α_2 positive and greater than the range of the intersymbol influence). This output is not similar to $s(t)$ in waveshape. Only its average properties are similar to those of $s(t)$.

Fig. 8-12 (d), (e), (f).

$$E_T(\omega) = S_{T,\alpha}(\omega) - H_T(\omega) = S_T(\omega)[\epsilon^{j\omega\alpha} - Y(\omega)] - Y(\omega)N_T(\omega) \tag{144}$$

If $s(t)$ and $n(t)$ are incoherent, it follows that

$$W_e(f) = |Y(\omega) - \epsilon^{j\omega\alpha}|^2 W_s(f) + |Y(\omega)|^2 W_n(f) \tag{145}$$

where $W_s(f)$ and $W_n(f)$ are the smoothed power spectra of $s(t)$ and $n(t)$, respectively. Substituting Eq. (145) into Eq. (139) we obtain finally

$$\widetilde{[e(t)]^2} = \int_{-\infty}^{+\infty} [|Y(\omega) - \epsilon^{j\omega\alpha}|^2 W_s(f) + |Y(\omega)|^2 W_n(f)]df \tag{146}$$

Equation (146) is a general expression for the mean-square error $\overline{[e(t)]^2}$. We see that it depends only on $Y(\omega)$, α, and the power spectra of the input signal and noise ensembles. The problem is now to find a physically realizable $Y(\omega)$ which will minimize the mean-square error.

It is interesting to note that the signal and noise enter into Eq. (146) only through their power spectra. This means that the phase constraints of the spectra of the signal and the noise do not affect the mean-square error. Two quite different types of signal, as long as they had the same power spectrum, would lead to the same $Y(\omega)$ for the optimum filter and to the same mean-square error. We shall make frequent use of this fact.

(c) *Determination of the Optimum Filter Without the Limitation of Physical Realizability.* Let us now find the function $Y(\omega)$ which will make the integral on the right side of Eq. (146) a minimum, and thus give the least mean-square error. We first express $Y(\omega)$ in the form

$$Y(\omega) = C(\omega)\epsilon^{j\beta(\omega)} \tag{147}$$

where $C(\omega)$ and $\beta(\omega)$ are real and $C(\omega)$ is also positive. Substituting Eq. (147) into Eq. (146), we obtain

$$\overline{[e(t)]^2} = \int_{-\infty}^{+\infty} \{[C^2 + 1 - 2C\cos(\omega\alpha - \beta)]W_s + C^2W_n\}df \tag{148}$$

For simplicity we are omitting the writing of the independent variables of $C(\omega)$, $\beta(\omega)$, $W_s(f)$, and $W_n(f)$ in Eqs. (148) to (155).

It is apparent from Eq. (148) that, as far as $\beta(\omega)$ is concerned, the integral will be a minimum if $\cos(\omega\alpha - \beta)$ has its maximum possible value. This will occur when

$$\beta = \omega\alpha \tag{149}$$

When Eq. (149) is satisfied, Eq. (148) becomes

$$\overline{[e(t)]^2} = \int_{-\infty}^{+\infty} [(C^2 + 1 - 2C)W_s + C^2W_n]df \tag{150}$$

This may also be expressed in the instructive form

$$\overline{[e(t)]^2} = \int_{-\infty}^{+\infty} \left\{\left[\sqrt{W_s + W_n}\,C - \frac{W_s}{\sqrt{W_s + W_n}}\right]^2 + \frac{W_sW_n}{W_s + W_n}\right\}df \tag{151}$$

The term in brackets under the integral is the square of a real number and is therefore positive or zero. Furthermore, this term is the only part of the integral which involves C. Therefore, as far as C is concerned, the integral will be a minimum, if

$$\sqrt{W_s + W_n}\, C - \frac{W_s}{\sqrt{W_s + W_n}} = 0 \tag{152}$$

which gives

$$C = \frac{W_s}{W_s + W_n} \tag{153}$$

If we substitute Eqs. (153) and (149) into Eq. (147), we then have the function $Y(\omega)$ which will make the mean-square error a minimum. Accordingly, this is

$$Y(\omega) = \frac{W_s}{W_s + W_n} \epsilon^{j\omega\alpha} \qquad (154)^1$$

and when this form of $Y(\omega)$ is used, Eq. (151) for the mean-square error reduces to

$$\widetilde{[e(t)]^2} = \int_{-\infty}^{+\infty} \frac{W_s W_n}{W_s + W_n}\, df \qquad (155)^2$$

If $Y(\omega)$ as expressed in Eq. (154) *represents a physically realizable filter, then Eq.* (154) *represents the solution to the linear least-square smoothing and prediction problem.* Let us therefore examine under what circumstances $Y(\omega)$ will be physically realizable. In the present case, this can be examined most effectively in terms of the

[1] The original solution of Wiener was carried out in the time domain rather than in the frequency domain as we have done in arriving at Eq. (154), following Bode and Shannon. Using the calculus of variations, Wiener's method shows that, for mininum mean square error

$$\phi_s(t + \alpha) - \int_0^\infty [\phi_s(t - \tau) + \phi_n(t - \tau)]B(\tau)d\tau = 0 \quad \text{for } t \geq 0$$

This integral equation is then solved for $B(t)$, giving the Fourier transform of $Y(\omega)$, the optimum predicting filter. In the above equation, $\phi_s(t)$ is the auto-correlation function of the signal $s(t)$, and $\phi_n(t)$ is the auto-correlation function of the noise $n(t)$.

[2] In the absence of noise, $W_n = 0$. Then $Y(\omega) = \epsilon^{j\omega\alpha}$ and the output is precisely $s(t + \alpha)$, so that there is no error. If α is a negative quantity, then $Y(\omega) = \epsilon^{j\omega\alpha}$ represents a physically realizable filter and thus represents an exact solution to the smoothing problem in the absence of noise.

impulse response $B(t)$ of the filter $Y(\omega)$. According to Eq. (9), we have

$$B(t) = \frac{1}{2\pi} \int_{-\infty}^{+\infty} Y(\omega)\epsilon^{j\omega t} d\omega$$

$$= \frac{1}{2\pi} \int_{-\infty}^{+\infty} \frac{W_s(f)}{W_s(f) + W_n(f)} \epsilon^{j\omega(t+\alpha)} d\omega \qquad (156)^{1}$$

In order for $Y(\omega)$ to be physically realizable, it is necessary that $B(t)$ should vanish for negative values of t. Except in very special cases, this will not usually be true. Therefore $Y(\omega)$ as defined in Eq. (154) will not usually represent a physically realizable filter. However, it does represent the transfer function, among all complex transfer functions whether physically realizable or not, which would give the minimum mean-square error. When the $Y(\omega)$ of Eq. (154) is not physically realizable, the filter which gives the minimum mean-square error among physically realizable filters will therefore give a greater error than that given by the $Y(\omega)$ of Eq. (154), *i.e.*, greater than the error given by Eq. (155).

(d) *Solution of the Pure Prediction Problem.* We will now try to find the linear filter which will give the least mean-square error among physically realizable filters in the case of pure prediction, *i.e.*, when no noise is present. In other words, we wish to find the physically realizable filter whose output will be the best approximation to $s(t + \alpha)$ for positive α, when we know $s(t)$ from $t = -\infty$ up to $t = 0$ and no perturbing noise is present.

In Eq. (146) we found that the mean-square error will depend only on the smoothed power spectrum W_s of $s(t)$ and not upon its specific waveform. We may therefore replace $s(t)$ by any other signal, say $s_1(t)$, having the same power spectrum and the solution for the optimum filter will be the same. Now Bode and Shannon[2] have found an $s_1(t)$ for which the optimum filter can be found in a remarkably simple and instructive manner. In order to obtain the $s_1(t)$ in question, random noise $r_1(t)$ of very wide bandwidth is sent through a filter $Y_s(\omega)$ for which

$$|Y_s(\omega)| = \sqrt{W_s} \qquad (157)$$

[1] In Eq. (156), we have again introduced the independent variable f in the notation for W_s and W_n to emphasize that these power spectra are functions of frequency and therefore are functions of ω.

[2] *Proc. I.R.E.*, vol. 38, April, 1950, pp. 417–425.

The output of this filter is the desired $s_1(t)$, and it will clearly have the power spectrum W_s. In order for $Y_s(\omega)$ to be a physically realizable filter, it is necessary that W_s should have a frequency characteristic which will allow $|Y_s(\omega)|$ as defined in Eq. (157) to satisfy the Paley-Wiener criterion. In practically all cases of interest the criterion will be satisfied. It is then possible to choose a $Y_s(\omega)$ which has neither poles nor zeros in the lower half of the complex ω plane such that

$$Y_s(\omega) \cdot Y_s^*(\omega) = W_s \tag{158}$$

We will now introduce a new notation by writing

$$Y_s(\omega) = [W_s]^+ \tag{159}$$

This will signify that all the poles and zeros of $Y_s(\omega)$ are in the upper half of the complex ω plane and that $|Y_s(\omega)| = \sqrt{W_s}$. The complex conjugate of $Y_s(\omega)$ will then be

$$Y_s^*(\omega) = [W_s]^- \tag{160}$$

whose magnitude is also $\sqrt{W_s}$ but all of whose poles and zeros are in the lower half of the complex ω plane. If follows from Eqs. (158) to (160) that

$$W_s = [W_s]^+ \cdot [W_s]^- \tag{161}$$

According to the above specification, it follows that $Y_s(\omega)$ is a physically realizable filter of minimum-phase-shift type. From Sec. 8.2, we then know that $Y_s(\omega)$ has a physically realizable inverse. It is through the $Y_s(\omega)$, as thus specified, that we send the wideband random noise $r_1(t)$, in order to get $s_1(t)$. Since $Y_s(\omega)$ has a physically realizable inverse of minimum phase shift type $Y_s^{-1}(\omega)$, we can send $s_1(t)$ through the inverse filter to recover $r_1(t)$ exactly, without a time delay. [see Fig. 8-13(a)].

According to the conditions of the problem, we may consider that we know $s_1(t)$ from $t = -\infty$ up to the present time, $t = t_0$. This is equivalent to knowing $r_1(t)$ from $t = -\infty$ up to the present. We now want to know what physically realizable filter $Y_0(\omega)$ having $r_1(t)$ as its input will give the best approximation to $s_1(t_0 + \alpha)$ in the least-square sense when averaged over all possible values of t_0.

According to Chap. III the random noise $r_1(t)$ consists of the superposition of a large number of impulsive-type signals of sampling function shape. We are at liberty to choose the bandwidth of $r_1(t)$ so wide that it may be considered a succession of impulses at its sampling points. These impulses have a normal distribution in height with no intersymbol influence. Each of these impulses causes an output of $Y_0(\omega)$ of the form $B_0(t - t')$, where t' is the time of occurrence of the impulse and $B_0(t)$ is the impulsive response of $Y_0(\omega)$. The actual output signal $h_1(t)$ is the superposition of these impulsive responses. Our problem is to design $Y_0(\omega)$ so that $h_1(t_0)$ is the best approximation to $s_1(t_0 + \alpha)$.

We can see from Fig. 8-13(a) that $r_1(t)$ is the output of $Y_s^{-1}(\omega)$ when the input is $s_1(t)$. Therefore if $r_1(t)$ were the input to the filter

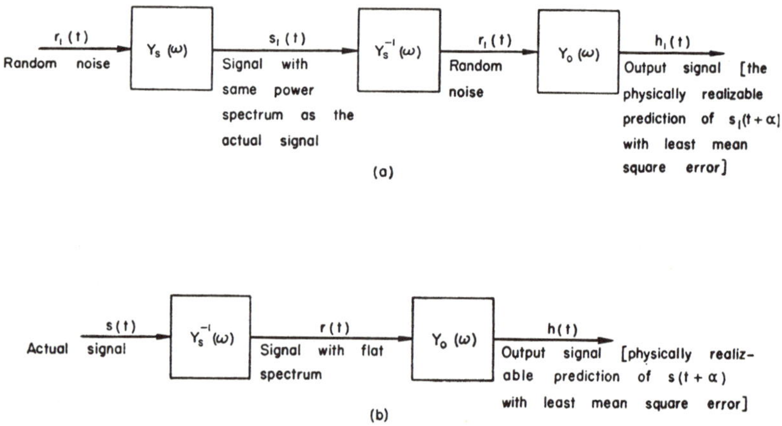

Fig. 8-13.

$Y_s(\omega)$, the output would be $s_1(t)$. In this output, the future signal $s_1(t_0 + \alpha)$ is made up of two parts: the first part being due to the responses to the impulses of $r_1(t)$ that occur prior to the present time t_0, and the second part being due to the responses to the impulses which occur after t_0. The first part is completely predictable at time t_0, the second part is completely unpredictable, since $r_1(t)$ has no intersymbol influence. We will now show how to obtain the predictable part of $s_1(t_0 + \alpha)$ and show what to do about the unpredictable part.

Consider a filter $Y_0(\omega)$ whose impulsive response is (see Fig. 8-14)

$$B_0(t) = \begin{cases} B_s(t + \alpha) & \text{for } t > 0 \\ 0 & \text{for } t < 0 \end{cases} \tag{162}$$

This means that, if identical impulses enter the filters Y_0 and Y_s, the response of Y_0 at any time t_0 will be the same as the response of Y_s at time $t_0 + \alpha$. From the discussion in the preceding paragraph, it then follows that, if $r_1(t)$ were the input to the filter Y_0, its output would be the part of $s_1(t_0 + \alpha)$ which is exactly predictable at time t_0.

The unpredictable part of $s_1(t_0 + \alpha)$ corresponds to impulses in $r_1(t)$ which occur between t_0 and $t_0 + \alpha$. Since the impulses in $r_1(t)$ have a normal distribution in heights, the arithmetic average value of the contribution of these future impulses to $s_1(t_0 + \alpha)$ averaged over the ensemble will be zero. Therefore, the arithmetic average

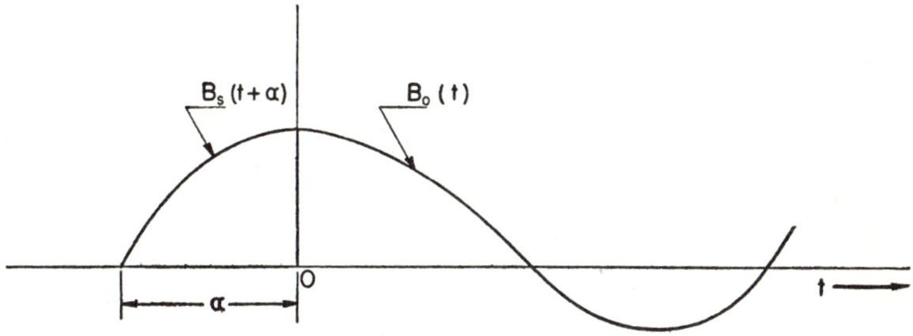

Fig. 8-14.

of the possible values of $s_1(t_0 + \alpha)$, when $s_1(t)$ is known prior to t_0 and is unknown thereafter, is the output of Y_0 when its input is $r_1(t)$. Since the least mean-square error occurs about the arithmetic average of any distribution,[1] it follows that *the output of* Y_0 *itself* when $r_1(t)$ is its input *will predict* $s_1(t_0 + \alpha)$ *with the least mean-square error*. This mean-square error can be a mean taken over all possible functions of the ensemble for a particular t_0, or over all possible values of t_0 for a particular function of the ensemble. Since the ensemble is ergodic, the result will be the same either way. No physically realizable filter can do better than Y_0. A non-physically realizable filter could do better than Y_0, in fact it could make a perfect prediction of $s_1(t_0 + \alpha)$ in the absence of noise as indicated by Eq. (155), but this would depend upon obtaining responses to impulses which had not yet occurred.

Referring once more to Fig. 8-13(a) we see that the output of a filter whose transfer characteristic is

$$Y_s^{-1}(\omega) \cdot Y_0(\omega) \tag{163}$$

[1] Cramer, *Mathematical Methods of Statistics*, p. 175.

will in the absence of noise and when its input is $s_1(t_0)$ give $s_1(t_0 + \alpha)$ with the least mean-square error averaged over the ensemble. Since this is true for every t_0, we may say that $Y_s^{-1}(\omega) \cdot Y_0(\omega)$ will give the best prediction of $s_1(t + \alpha)$ for the input $s_1(t)$.

Equation (159) gives us the value of $Y_s^{-1}(\omega)$, namely,

$$Y_s^{-1}(\omega) = \frac{1}{[W_s]^+} \tag{164}$$

On the other hand, $Y_0(\omega)$ is the Fourier transform of $B_0(t)$ given by Eq. (162). Thus

$$Y_0(\omega) = \int_{-\infty}^{+\infty} B_0(t)\epsilon^{-j\omega t}dt = \int_0^{\infty} B_s(t + \alpha)\epsilon^{-j\omega t}dt \tag{165}$$

where $B_s(t)$ is the Fourier transform of $Y_s(\omega)$.

As mentioned earlier, we know from Eq. (146) that the mean-square error depends only upon the smoothed power spectrum of the ensemble of input signals, so that the design of the optimum predictor will also depend only on the power spectrum. Since $s_1(t)$ and $s(t)$ have the same power spectrum, *it follows that* (163) *also gives the transfer characteristic of the optimum predictor for* $s(t + \alpha)$ *when the input is* $s(t)$. We have thus solved the problem of the optimum[1] predictor in the absence of noise.

The ensemble average of the mean-square error in the prediction of $s(t + \alpha)$ can also be calculated. This can be done by calculating the error in the prediction of $s_1(t + \alpha)$, since these average errors are the same. The error in predicting $s_1(t_0 + \alpha)$ is due to the impulses which occur between t_0 and $t_0 + \alpha$. Since these impulses are uncorrelated, the total mean-square error is the sum of the individual mean-square errors. Now an impulse of strength M at time t between t_0 and $t_0 + \alpha$ will contribute an [error]^2 in the estimate of $s(t_0 + \alpha)$ of amount

$$[MB_s(t_0 + \alpha - t)]^2 \tag{166}$$

The total mean-square error in the estimate of $s_1(t_0 + \alpha)$ will therefore be

$$\widetilde{[e(t)]^2} = \widetilde{m^2} \int_{t_0}^{t_0+\alpha} [B_s(t_0 + \alpha - t)]^2dt = \widetilde{m^2} \int_0^{\alpha} [B_s(\tau)]^2d\tau \tag{167}$$

where

$$\tau = t_0 + \alpha - t \tag{168}$$

[1] For the purpose of least mean-square error.

and $\widetilde{m^2}$ is the mean-square impulse strength per unit of time[1] in $r_1(t)$. By a similar argument, we would get for the mean-square value of $s_1(t_0 + \alpha)$

$$\widetilde{[s_1(t_0 + \alpha)]^2} = \widetilde{m^2} \int_{-\infty}^{t_0+\alpha} [B_s(t_0 + \alpha - t)]^2 dt$$

$$= \widetilde{m^2} \int_0^{\infty} [B_s(\tau)]^2 d\tau \tag{169}$$

Let us define the relative error of prediction by the optimum filter as

$$\text{relative error of prediction} = \left\{ \frac{\widetilde{[e(t)]^2}}{\widetilde{[s_1(t + \alpha)]^2}} \right\}^{\frac{1}{2}}$$

$$= \left\{ \frac{\int_0^{\alpha} [B_s(t)]^2 dt}{\int_0^{\infty} [B_s(t)]^2 dt} \right\}^{\frac{1}{2}} \tag{170}$$

It follows from Eq. (170) that the prediction will be good if the area under the curve $[B_s(t)]^2$ out to α is small compared to the total area under the curve between zero and infinity. If the area under the curve out to α is relatively large, the prediction will be poor. The relative error starts at zero for $\alpha = 0$ and increases monotonically to the value of unity as α approaches infinity.

(e) *Prediction and Smoothing in the Presence of Noise.* We will now extend the analysis of the noiseless case to solve the problem of finding the optimum filter for prediction and smoothing when noise is present. We require the best estimate, in the least mean-square-error sense, of $s(t + \alpha)$ when $g(t) = s(t) + n(t)$ is the input. Proceeding as in the noiseless case, we pass $g(t)$ through a filter of minimum-phase type to obtain an output $r(t)$ which has a flat spectrum. This filter will be

$$Y_1^{-1}(\omega) = \frac{1}{[W_s + W_n]^+} \tag{171}$$

The filter Y_1^{-1} will exist as a physically realizable filter according to the Paley-Wiener criterion, provided that

$$\int_0^{\infty} \frac{\log \left| \frac{1}{W_s + W_n} \right|}{1 + \omega^2} d\omega = \int_0^{\infty} \frac{-\log |W_s + W_n|}{1 + \omega^2} d\omega \tag{172}$$

[1] The value of $\widetilde{m^2}$ depends only on the energy per kilocycle of $r_1(t)$ and not upon its bandwidth.

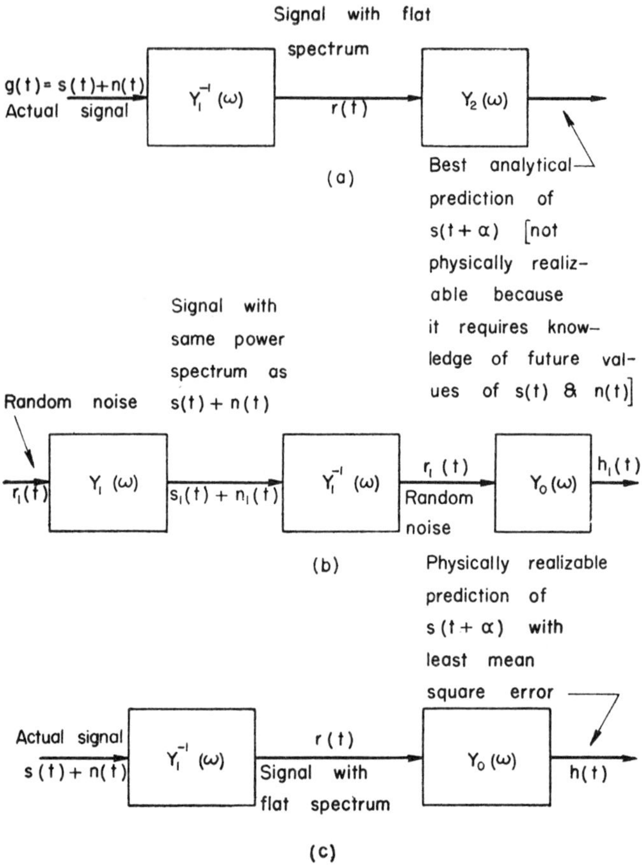

Fig. 8-15.

is finite. Since $Y_1(\omega)^{-1}$ has a physically realizable inverse

$$Y_1(\omega) = [W_s + W_n]^+ \tag{173}$$

knowledge of its output is equivalent to knowledge of its input. Furthermore, the best linear operation on the output of $Y_1^{-1}(\omega)$ to approximate $s(t + \alpha)$ will give the same prediction as the corresponding best linear operation on the input.

If we were not limited to physically realizable filters, the best filter into which to send $s(t) + n(t)$ to predict $s(t + \alpha)$ would be given by Eq. (154) as we have previously shown. This result would be obtained analytically, even though it is not realizable, by sending the output of $Y_1^{-1}(\omega)$ through a filter $Y_2(\omega)$, where

$$Y_2(\omega) = \frac{W_s}{[W_s + W_n]^-} \epsilon^{j\omega\alpha} \tag{174}$$

This is shown diagrammatically in Fig. 8-15(a). The impulsive response $B_2(t)$ corresponding to this $Y_2(\omega)$ is shown in Fig. 8-16.

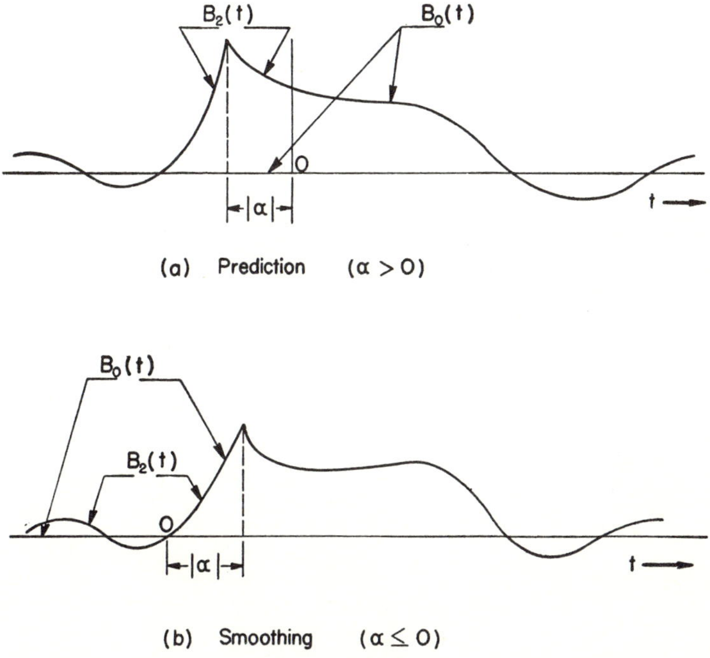

Fig. 8-16.

Since $Y_2(\omega)$ is not physically realizable, $B_2(t)$ will have tails extending to both positive and negative infinity.

As pointed out earlier in this section, the same filter

$$Y_1^{-1}(\omega) \cdot Y_2(\omega)$$

which is the optimum predictor for $s(t + \alpha)$ will also be the optimum predictor for any other signal $s_1(t + \alpha)$ which has the same power spectrum provided that $s_1(t)$ is originally mixed with noise $n_1(t)$ having the same power spectrum as $n(t)$. Let us therefore consider an $s_1(t) + n_1(t)$ of this type which is derivable from random noise as shown in Fig. 8-15(b); and let us send this $s_1(t) + n_1(t)$ through $Y_{\overline{1}}^{-1}(\omega)$ to obtain the random-noise signal $r_1(t)$. We now wish to determine the *physically realizable* filter $Y_0(\omega)$, which will predict $s_1(t + \alpha)$ with the least mean-square error.

As before, the random-noise signal $r_1(t)$ may be considered a succession of impulsive-type signals between which there is no intersymbol influence. Suppose that the present time is t_0. Then the signal $h_1(t)$, which is the output of $Y_0(\omega)$ when $r_1(t)$ is its input, is the superposition of the weighted effects of the impulses in $r_1(t)$ which have occurred before time t_0, each impulse being weighted in proportion to the value of $B_0(\tau)$ at its time of occurrence. To be precise

$$h_1(t_0) = \int_{-\infty}^{t_0} r_1(t)B_0(t_0 - t)dt = \int_0^{\infty} r_1(t_0 - \tau)B_0(\tau)d\tau \tag{175}$$

where

$$\tau = t_0 - t \tag{176}$$

On the other hand, if instead of using $Y_0(\omega)$ we used the optimum analytical but not physically realizable filter $Y_2(\omega)$, the output would be

$$\begin{aligned}\text{output of } Y_2(\omega) \text{ at } t_0 \text{ when } r_1(t) \text{ is its input} &= \int_{-\infty}^{+\infty} r_1(t)B_2(t_0 - t)dt \\ &= \int_{-\infty}^{+\infty} r_1(t_0 - \tau)B_2(\tau)d\tau \\ &= \int_{-\infty}^{0} r_1(t_0 - \tau)B_2(\tau)d\tau + \int_0^{\infty} r_1(t_0 - \tau)B_2(\tau)d\tau \end{aligned} \tag{177}$$

In the case of the non-physically realizable filter $Y_2(\omega)$, the integral must extend over future impulses as well as past impulses, since the weighting function $B(\tau)$ in this case extends into the future as well as the past.

Equation (177) gives the best approximation to $s_1(t_0 + \alpha)$ which is analytically possible with a linear filter as already pointed out. It is therefore better than $h_1(t)$ as given by Eq. (175). Now the total output of $Y_2(\omega)$ in Eq. (177) is divided into two parts. The first part

$$\int_{-\infty}^{0} r_1(t_0 - \tau)B_2(\tau)d\tau \tag{178}$$

is derived from impulses in $r_1(t)$ which occur after the present time t_0. The second part

$$\int_0^{\infty} r_1(t_0 - \tau)B_2(\tau)d\tau \tag{179}$$

is derived from impulses in $r_1(t)$ which occur before the present time t_0. A physically realizable filter like $Y_0(\omega)$, on the other hand, can only derive signals from impulses in $r_1(t)$ which occur before the present time. Equation (177) will now give us important information regarding the optimum physically realizable filter. It follows from the last paragraph in Appendix XI that, since there is no intersymbol influence between the impulses in $r_1(t)$, the same weighting function should be applied to the past impulses of $r_1(t)$ to get the optimum prediction, in the least mean-square-error sense, regardless of whether the weighting of future impulses is controllable or not. Equation (177) therefore gives us the best weighting function for the past impulses even in the physically realizable case, when the future impulses must be given zero weighting. Consequently $B_0(t)$ is completely determined, for

$$B_0(t) = 0 \quad \text{when } t < 0 \tag{180}$$

since $Y_0(\omega)$ is physically realizable, and

$$B_0(t) = B_2(t) \quad \text{when } t > 0 \tag{181}$$

according to the foregoing argument. Since $Y_0(\omega)$ is the Fourier transform of $B_0(t)$, we have

$$Y_0(\omega) = \int_{-\infty}^{+\infty} B_0(t)\epsilon^{-j\omega t}dt = \int_0^{\infty} B_2(t)\epsilon^{-j\omega t}dt \tag{182}$$

where

$$B_2(t) = \frac{1}{2\pi}\int_{-\infty}^{+\infty} Y_2(\omega)\epsilon^{j\omega t}d\omega = \frac{1}{2\pi}\int_{-\infty}^{+\infty} \frac{W_s}{[W_s + W_n]^-}\epsilon^{j\omega(t+\alpha)}d\omega \tag{183}$$

with the aid of Eq. (174). The optimum physical realizable filter to predict $s_1(t + \alpha)$ is thus

$$Y_1^{-1}(\omega) \cdot Y_0(\omega) \tag{184}$$

where $Y_0(\omega)$ is given in Eq. (182) and $Y_1^{-1}(\omega)$ is given by Eq. (171) Since this is the optimum filter to predict $s_1(t + \alpha)$, it is also the optimum filter to predict $s(t + \alpha)$, so that our problem is solved [(see also Fig. 8-15(c)].

In case α is less than zero, we are dealing with a smoothing problem rather than a prediction problem. The same solution (184) will still apply, but it would appear from Fig. 8-16(b) that the error will be considerably less in this case, because the optimum weighting of the unused future impulses will be small.

(f) *Bode and Shannon's Critique of the Basic Assumptions.* Bode and Shannon[1] have given a remarkably clear and incisive discussion of the basic assumptions of the foregoing theory. Rather than paraphrase their remarks, the most important parts of them are quoted here from the original.

> A result in applied mathematics is only as reliable as the assumptions from which it is derived. The theory developed above is especially subject to misapplication because of the difficulty in deciding, in any particular instance, whether the basic assumptions are a reasonable description of the physical situation. Anyone using the theory should carefully consider each of the three main assumptions with regard to the particular smoothing or prediction problem involved.
>
> The assumption that the signal and noise are stationary is perhaps the most innocuous of the three, for it is usually evident from the general nature of the problem when this assumption is violated. The determination of the required power spectra W_s and W_n will often disclose any time variation of the statistical structure of the time series. If the variation is slow compared to the other time constants involved, such nonstationary problems may still be solvable on a quasi-stationary basis. A linear predictor may be designed whose transfer function varies slowly in such a way as to be optimal for the local statistics.
>
> The least square assumption is more troublesome, for it involves questions of values rather than questions of fact. When we minimize the mean-square error we are, in effect, paying principal attention to the very large errors. The prediction chosen is one which, on the whole, makes these errors as small as possible, without much regard to relatively minor errors. In many circumstances, however, it is more important to make as many very accurate predictions as possible, even if we make occasional gross errors as a consequence. When the distribution of future events is Gaussian, it does not matter which criterion is used since the most probable event is also the one with respect to which the mean-square error is the least.

[1] H. W. Bode and C. E. Shannon, "A Simplified Derivation of Linear Least Square Smoothing and Prediction Theory," *Proc. I.R.E.*, vol. 38, April, 1950, pp. 424–425.

With lopsided or multimodal distributions, however, a real question is involved.

As a simple example, consider the problem of predicting whether tomorrow will be a clear day. Since clear days are in the majority, and there are no days with negative precipitation to balance days when it rains, we are concerned here with a very lopsided distribution. With such a curve, the average point, which is the one given by a prediction minimizing the mean-square error, might be represented by a day with a light drizzle. To a man planning a picnic, however, such a prediction would have no value. He is interested in the probability that the weather will really be clear. If the picnic must be called off because it in fact rains, the actual amount of precipitation is of comparatively little consequence.

The third assumption, that of linearity, is neither a question of fact, nor of evaluation, but a self-imposed limitation on the types of operations or devices to be used in prediction. The mathematical reason for this assumption is clear; linear problems are always much simpler than their nonlinear generalizations. In certain applications the linear assumption may be justified for one or another of the following reasons:

(1) The linear predictor may be an absolute optimal method, as in the Gaussian time series mentioned above.

(2) Linear prediction may be dictated by the simplicity of mechanization. Linear filters are easy to synthesize and there is an extensive relevant theory, with no corresponding theory for nonlinear systems.

(3) One may use the linear theory merely because of the lack of any better approach. An incomplete solution is better than none at all.

How much is lost by restricting ourselves to linear prediction? The fact that nonlinear effects may be important in a prediction can be illustrated by returning to the problem of forecasting tomorrow's weather. We are all familiar with the fact that the pattern of events over a period of time may be more important than the happenings taken individually in determining what will come. For example, the sequence of events in the passage of a cold or warm front is characteristic. Moreover, the significance of a given happening may depend largely upon the intensity with which it occurs. Thus, a sharp dip in the barometer may mean that moderately unpleasant weather is coming. Twice as great a drop in the same time, on the other hand, may not indicate that the weather will be merely twice as unpleasant; it may indicate a hurricane.

As a final point, we may notice that the requirement that the prediction be obtained from a linear device and the objective of minimizing the mean-square error are not, in all problems, quite compatible with one another. The absolute best mean-square prediction (ignoring the assumption of linearity) would, of course, always pick the mean of the future distribution, *i.e.*, the "center of gravity," since in any case this minimizes the mean-square error. In general, however, the position of this center of gravity will be a nonlinear function of the past history. When we require that the prediction be a linear operation on the past history, the mathematics is forced to compromise among the conflicting demands of various possible past histories.

(g) *Practical Aspects of the Design of Optimum Filters.* The early parts of this section have given us formal solutions to the optimum filtering and prediction problems. For example, the optimum predictor in the presence of noise consists of the two filters in cascade

$$Y_1^{-1}(\omega) = \frac{1}{[W_s + W_n]^+} \tag{185}$$

and

$$Y_0(\omega) = \int_0^\infty B_2(t)\epsilon^{-j\omega t}dt \tag{186}$$

where

$$B_2(t) = \frac{1}{2\pi}\int_{-\infty}^{+\infty} \frac{W_s}{[W_s + W_n]^-}\epsilon^{j\omega(t+\alpha)}d\omega \tag{187}$$

In order to carry through this solution in a particular case it is first necessary to determine the smoothed power spectra W_s and W_n of the signal and noise, respectively. It is then necessary to find the linear factors of $[W_s + W_n]$ in order to determine $[W_s + W_n]^+$ and $[W_s + W_n]^-$. Once this is done, $B_2(t)$ is next calculated and then $Y_0(\omega)$. In actual cases, it is rarely possible to carry through these operations except by approximation methods.

In practice W_s and W_n are usually obtained from their respective auto-correlation functions, or $[W_s + W_n]$ can be obtained from the auto-correlation function of the combined signal plus noise. In order for $[W_s + W_n]$ to be expressible as a product of linear factors, the correlation function

$$\phi_{(s+n)}(\tau) = \phi_s(\tau) + \phi_n(\tau) \tag{188}$$[1]

must be expressible as the sum of terms of the form

$$A\epsilon^{a_1\tau} \quad A_2\tau^{m_2}\epsilon^{a_2\tau} \quad A_3\epsilon^{a_3\tau}\cos(b_3\tau + \theta_3)$$

and

$$A_4\tau^{m_4}\epsilon^{a_4\tau}\cos(b_4\tau + \theta_4)$$

It is therefore customary to start the approximation procedure by fitting the curve of the correlation function approximately with the sum of a series of terms of the above type. On the other hand,

[1] Equation (188) follows from the fact that the signal and noise are incoherent.

if $[W_s + W_n]$ is determined directly, then the curve fitting is done on $[W_s + W_n]$ with a function of f having linear factors. After $[W_s + W_n]$ is factored, the integration in Eqs. (186) and (187) can be approximated graphically. The actual design of an optimum filter in any practical case is invariably laborious.[1]

(h) *General Remarks.* In Weiner's original work he considered more general cases than we have considered here. In the first place, he considered the case when the signal and noise are correlated. Our own analysis could be generalized in this manner without too much trouble. Wiener also considered the problem in which additional information could be obtained from signals correlated with $s(t)$ in order to approximate $s(t + \alpha)$. This becomes quite involved. Finally Wiener considered the problem of approximating the derivative of $s(t + \alpha)$ instead of $s(t + \alpha)$ itself. This can, in effect, be reduced to the ordinary problem of the optimum predictor with the aid of differentiating circuits.

It is worthy of remark that the Wiener optimum filter takes advantage of the phase as well as the amplitude of the frequency spectrum in removing noise. In this way it is superior to the ordinary frequency filter. In other words it filters on the basis of waveshape, not just the spectrum. It is a bit surprising under the circumstances that the design formulas for the optimum Wiener filter turn out to depend only on the power spectra of the signal and the noise, respectively. However, this is a consequence of the use of the least mean-square criterion and does not indicate that phase considerations have been neglected.

A few characteristics of Wiener predictors may also be noted. A Wiener predictor does not predict the future of a signal all at once. It predicts the future at the same rate at which the signal is unfolded, but is a time α ahead of it (see Fig. 8-12). The prediction of a signal depends on the intersymbol influence in the ensembles of which the signal and the noise are, respectively, typical. The auto-correlation function is a measure of this intersymbol influence if it is of a linear

[1] Chapter VII of the Radiation Laboratory Series Vol. 25, on "Theory of Servomechanisms" by R. S. Phillips discusses an actual case. An excellent discussion is also given by Robert Cohen in Technical Report No. 69 of the Research Laboratory of Electronics at M. I. T. on "Some Analytical and Practical Aspects of Wiener's Theory of Prediction." An extensive, but more formal discussion of the factoring problem is also given in Wiener, *The Extrapolation, Interpolation and Smoothing of Stationary Time Series.*

type. This accounts for its importance and that of its Fourier transform, the power spectrum, in prediction theory.[1]

We note in passing that the behavior of the auto-correlation function in the immediate vicinity of $\tau = 0$ gives the most important information for predicting a signal (since it appears in the weighting of all elements of the past signal). The form of the auto-correlation function in the neighborhood of $\tau = 0$ must therefore be approximated with great accuracy.

8.7 The use of random noise in determining transfer functions[2]

A novel and interesting use of correlation theory is the determination of transfer functions with the aid of random noise. Suppose we have a linear transmission system as shown in Fig. 8-17. Let $g_i(t)$, a typical function of an ergodic ensemble, be the input, and let $g_o(t)$ be the output. As pointed out in Chap. III, $g_o(t)$ will then also be a typical function of an ergodic ensemble. The cross-correlation function between input and output will then be

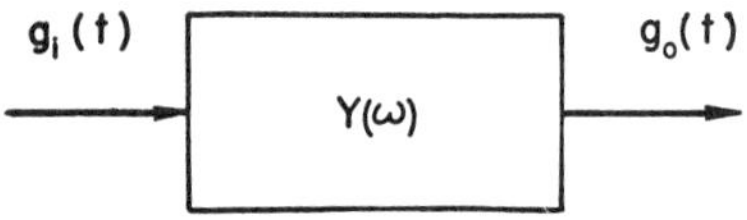

Fig. 8-17.

$$\phi_{io}(\tau) = \lim_{T\to\infty} \frac{1}{2T} \int_{-T}^{T} g_i(t) g_o(t + \tau) dt \tag{189}$$

The output
$$g_o(t) = \int_{-\infty}^{+\infty} B(\sigma) g_i(t - \sigma) d\sigma \tag{190}$$

Substituting Eq. (190) in (189), we obtain

$$\phi_{io}(\tau) = \lim_{T\to\infty} \frac{1}{2T} \int_{-T}^{+T} g_i(t) \left\{ \int_{-\infty}^{+\infty} B(\sigma) g_i(t + \tau - \sigma) d\sigma \right\} dt \tag{191}$$

Inverting the order of integration, which is allowable in this case, we have

$$\begin{aligned} \phi_{io}(\tau) &= \int_{-\infty}^{+\infty} B(\sigma) \left[\lim_{T\to\infty} \frac{1}{2T} \int_{-T}^{+T} g_i(t) g_i(t + \tau - \sigma) dt \right] d\sigma \\ &= \int_{-\infty}^{+\infty} B(\sigma) \phi_{ii}(\tau - \sigma) d\sigma \end{aligned} \tag{192}$$

[1] In connection with the discussion of predictors, it is of interest to note that a Taylor expansion is a (non-statistical) prediction operation.

[2] This section is based upon a discussion by Y. W. Lee, "Application of Statistical Methods to Communication Problems," Technical Report No. 181, Research Laboratory of Electronics at M. I. T. This report also gives references to papers describing experimental methods for measuring correlation functions.

where ϕ_{ii} is the auto-correlation function of the input. Equation (192) is a fundamental equation for any linear transmission system whose input is a typical function of an ergodic ensemble.

Equation (192) is of the form of a convolution integral. It may therefore be expressed in the frequency domain as

$$W_{io}(f) = Y(\omega)W_{ii}(f) \tag{193}$$

Thus

$$Y(\omega) = \frac{W_{io}(f)}{W_{ii}(f)} \tag{194}$$

We note in passing, that since $W_{ii}(f)$ is real, the phase of $W_{io}(f)$ is the same as that of $Y(\omega)$ at all frequencies.

In case the input signal $g_i(t)$ is random noise of wide bandwidth, Eqs. (194) and (192) become very simple. In this case

$$W_{ii}(f) = K^2 \tag{195}$$

in the frequency band of the random noise, and

$$\phi_{ii}(\tau - \sigma) = K^2\delta(\tau - \sigma) \quad \text{approximately} \tag{196}$$

where K^2 is a positive constant. Thus Eq. (194) becomes

$$Y(\omega) = \frac{1}{K^2} W_{io}(f) \tag{197}$$

and Eq. (192) becomes

$$\phi_{io}(\tau) = \int_{-\infty}^{+\infty} B(\sigma)K^2\delta(\tau - \sigma)d\sigma = K^2B(\tau) \tag{198}$$

Equations (197) and (198) give a novel means of determining the transfer function and impulsive response of a linear system. This method has the significant advantage that any extraneous noise, either internal or external, which may be present when the measurements are made will not appreciably affect the results. The reason for this is that any extraneous noise will be incoherent with the input random noise used for measurement, so that the cross correlation between them will be zero.

8.8 The detection of repetitive signals in the presence of noise by correlation methods[1]

Correlation methods have proved very useful in the detection of repetitive signals upon which a strong ergodic noise signal has been superimposed. Radar signals are repetitive signals of this type.

[1] This section, like Sec. 8.7, is based upon the report of Lee, Technical Report No. 181, Research Laboratory of Electronics, at M. I. T.

Let $s(t)$ be a repetitive signal and let the ergodic noise $n(t)$ be superimposed upon it. The resultant input signal is thus

$$g_i(t) = s(t) + n(t) \tag{199}$$

The auto-correlation function of $g_i(t)$ is then

$$\begin{aligned}\phi_{ii}(\tau) &= \phi_{ss}(\tau) + \phi_{nn}(\tau) + \phi_{sn}(\tau) + \phi_{ns}(\tau) \\ &= \phi_{ss}(\tau) + \phi_{nn}(\tau)\end{aligned} \tag{200}$$

The cross correlation between the signal and noise vanishes because they are incoherent. A diagram showing $g_i(t)$, $s(t)$, $n(t)$, and $\phi_{ii}(\tau)$ is shown in Fig. 8-18. We see that auto-correlation shows

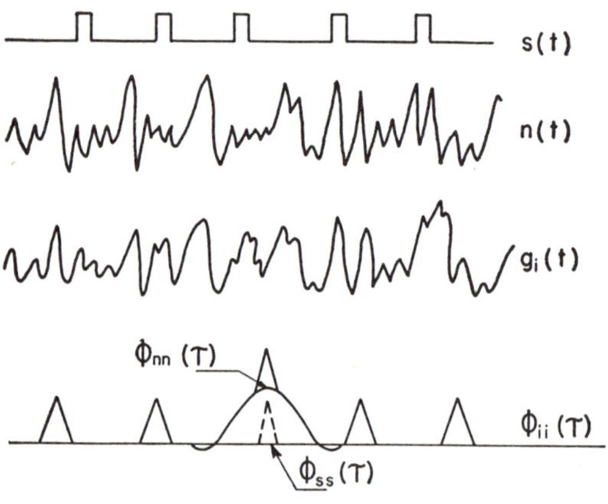

Fig. 8-18. Use of auto-correlation to discover a signal in a strong background of noise.

the presence of a repetitive signal despite the superposition of a strong noise signal. Unfortunately it generally distorts the signal and fails to give its location in time, a matter which is important in radar. The accuracy in determining $\phi_{ii}(\tau)$ in practical correlators increases with the number of sampling values.

The amount of improvement in signal-to-noise ratio in practical correlation equipment therefore increases with the time cycle of operation of the correlator. Descriptions of correlation equipment and its operation may be found elsewhere.[1]

[1] For a bibliography, see the above article of Lee. It may also be noted that for the case where $s(t)$ is a sine wave, so that $\phi_{ii}(\tau)$ is of the same waveshape, Lee gives the formula for the signal-to-noise ratio in the output of an auto-correlator

In case the period of repetition of the signal is known, an even greater improvement in signal-to-noise ratio can be obtained by means of cross correlation. Using a local signal of the same repetition period for cross correlation, we have

$$\begin{aligned}\phi_{12}(\tau) &= \lim_{T\to\infty} \frac{1}{2T}\int_{-T}^{+T} [s_1(t) + n(t)]s_2(t+\tau)dt \\ &= \phi_{s_1s_2}(\tau) + \phi_{ns_2}(\tau) = \phi_{s_1s_2}(\tau) \qquad (201)\end{aligned}$$

Here $s_1(t)$ is the desired incoming signal, $n(t)$ is the noise, and $s_2(t)$ is the locally supplied signal. Since $n(t)$ and $s_2(t)$ are incoherent, $\phi_{ns_2}(\tau)$ vanishes. Unless the repetition frequencies of $s_1(t)$ and $s_2(t)$

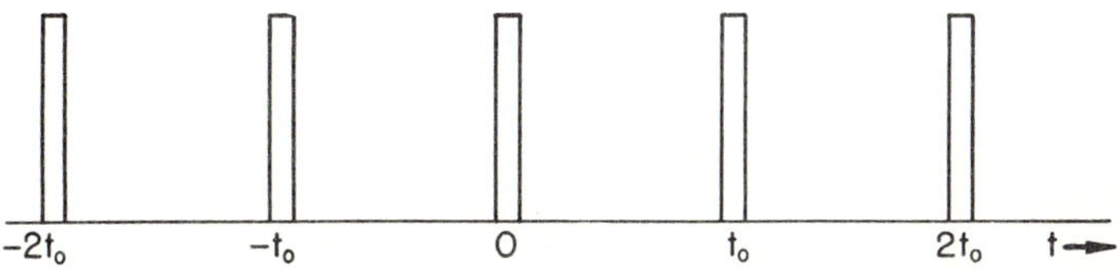

Fig. 8-19. A periodic sequence of impulses to be used for cross-correlation.

are determined by the same ultimate source, $\phi_{s_1s_2}(\tau)$ also vanishes. When $s_1(t)$ and $s_2(t)$ are incoherent, but of approximately the same frequency of repetition, if the cross correlation sampling points are taken over a relatively short total time, a moderate amount of signal-to-noise improvement may still be possible, by using the short-time approximation to $\phi_{s_1s_2}(\tau)$. The short-time approximation will not vanish. However, to get the great improvements of signal-to-noise ratio to be expected by the elimination of $\phi_{ns_2}(\tau)$ in Eq. (201), the cross-correlation points used must extend over a long period of time. In this case $s_1(t)$ and $s_2(t)$ must be coherent.[1]

$$R_{oa} = 10 \log_{10} \frac{N}{1 + 4\rho_i^2 + 2\rho_i^4} \quad db$$

where N is the number of sample products which the correlator takes for a point on the output curve, and ρ_i is the root-mean-square noise-to-signal ratio at the input.

[1] For the detection of a sine wave in random noise by cross correlation with a sine wave, Lee quotes the formula

$$R_{oC} = 10 \log_{10} \frac{N}{1 + 2\rho_i^2} \quad db$$

for the output signal-to-noise ratio. This assumes a coherent $s_1(t)$ (see also Exercise).

There is one particular local signal which gives very impressive results when used for cross correlation. This consists of a series of sharp impulses (see Fig. 8-19) whose period of repetition is coherent with that of the desired incoming signal. This coherence is easily arranged in the case of radar. When the signal of Fig. 8-19 is used for cross-correlation $\phi_{s_1s_2}(\tau)$ has the same wave-shape as $s_1(t)$, and furthermore, the phase of $s_1(t)$, which determines radar range, for example, is also not lost. Let us now analyze this case.

Since $s_1(t)$ is a periodic function, it may be expressed as a Fourier series in the form

$$s_1(t) = \sum_{n=-\infty}^{+\infty} C_n \epsilon^{jn\left(\frac{2\pi}{t_0}\right)t} \tag{202}$$

where t_0 is the period of repetition and

$$C_n = \frac{1}{t_0} \int_0^{t_0} s_1(t) \epsilon^{-jn\left(\frac{2\pi}{t_0}\right)t} dt \tag{203}$$

We may likewise express $s_2(t)$ as a Fourier series. In this case we have

$$s_2(t) = \sum_{n-\infty}^{+\infty} c_n \epsilon^{jn\left(\frac{2\pi}{t_0}\right)t} = \frac{K}{t_0} \sum_{n=-\infty}^{+\infty} \epsilon^{jn\frac{2\pi}{t_0}(t-t_1)} \tag{204}$$

since

$$\begin{aligned} c_n &= \frac{1}{t_0} \int_0^{t_0} s_2(t) \epsilon^{-jn\left(\frac{2\pi}{t_0}\right)t} dt \\ &= \frac{1}{t_0} \int_0^{t_0} K\delta(t - t_1) \epsilon^{-jn\left(\frac{2\pi}{t_0}\right)t} dt \\ &= \frac{K}{t_0} \epsilon^{-jn\left(\frac{2\pi}{t_0}\right)t_1} \end{aligned} \tag{205}$$

where t_1 indicates the location of the impulse in the cycle and K is its strength. Let us choose our origin of time so that $t_1 = 0$. Then

$$s_2(t) = \frac{K}{t_0} \sum_{n=-\infty}^{+\infty} \epsilon^{jn\frac{2\pi t}{t_0}} \tag{206}$$

Now suppose that the input to the correlation equipment is

$$h(t) = s_1(t) + n(t) \tag{207}$$

where, as usual, $n(t)$ is a noise signal. Then for the cross-correlation function between $s_2(t)$ and the signal (207), we have

$$
\begin{aligned}
\phi_{s_2h}(\tau) &= \phi_{s_2s_1}(\tau) + \phi_{s_2n} = \phi_{s_2s_1}(\tau) \\
&= \lim_{T\to\infty} \frac{1}{2T} \int_{-T}^{+T} s_2(t)s_1(t+\tau)dt \\
&= \frac{1}{t_0} \int_0^{t_0} s_2(t)s_1(t+\tau)dt \\
&= \frac{1}{t_0} \int_0^{t_0} s_2(t-\tau)s_1(t)dt
\end{aligned}
\tag{208}
$$

Now $s_2(t)$ is an even function, so that

$$
s_2(t-\tau) = s_2(\tau - t) \tag{209}
$$

Therefore, Eq. (208) becomes

$$
\begin{aligned}
\phi_{s_2h}(\tau) &= \frac{1}{t_0} \int_0^{t_0} s_2(\tau - t)s_1(t)dt \\
&= \frac{1}{t_0} \int_0^{t_0} s_1(t) \left[\frac{K}{t_0} \sum_{n=-\infty}^{+\infty} \epsilon^{jn\frac{2\pi}{t_0}(\tau - t)} \right] dt \\
&= \frac{K}{t_0^2} \sum_{n=-\infty}^{+\infty} \epsilon^{jn\frac{2\pi}{t_0}\tau} \left[\int_0^{t_0} s_1(t)\epsilon^{-jn\frac{2\pi}{t_0}t}\, dt \right] \\
&= \frac{K}{t_0} s_1(\tau)
\end{aligned}
\tag{210}
$$

The output of the cross correlator is thus precisely the desired input signal $s_1(t)$, except for the magnitude factor K/t_0, which can be adjusted to any desired value by changing K. If the cross-correlation equipment is able to give $\phi_{s_2h}(\tau)$ with a high degree of accuracy, it

(1) gives the desired signal without distortion

(2) removes the noise

(3) gives the location in time of $s_1(t)$, without any unknown displacement of the origin.

Signal-to-noise improvement in radar by cross-correlation means is remarkably effective for stationary targets in case $s_1(t)$ is truly periodic and coherence between $s_1(t)$ and $s_2(t)$ can be maintained over long periods of time. Figure 8-20 (after Lee) shows how radar echoes can be located by cross correlation in the presence of noise which is so strong that it would normally completely mask the signal.

EXERCISE

Derive the formulas for the signal-to-noise ratios in the output of an auto-correlater and of a cross correlator given in the footnotes of Sec. 8.8.

8.9 Further discussion of noise reduction by correlation methods

Linear correlation methods for detecting signals in the presence of noise are a special type of generalized selectivity. It is therefore of

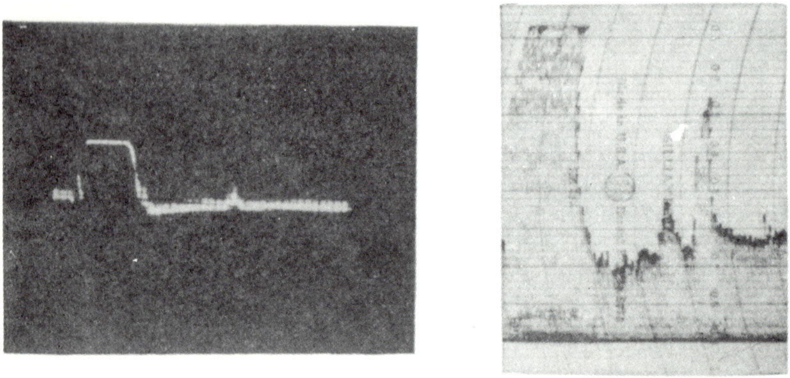

Fig. 8-20. Left, radar A-scope presentation; right, cross-correlated A-scope presentation. (*From Y. W. Lee,* Technical Report No. 181, *Research Laboratory of Electronics, M.I.T.*)

interest to discuss them from the same point of view as was used in Chap. VII. A special feature of these correlation methods is that they use extra time rather than extra bandwidth for noise reduction. We will start with a discussion of auto-correlation.

Auto-correlation is effectively a type of time separation filter. It is sharply selective for highly periodic signals. Like a passive linear frequency filter, it has no threshold. Its selectivity depends only upon an absolute measure of time separation. Also like a passive linear filter, its effective selectivity is rather small. For large improvements in signal-to-noise ratio, the repetitive character of the signal and the period of repetition must hold constant for long periods of time. This makes the signal highly redundant, like a frequency component; and as in the case of frequency selectivity, from an information theory point of view, this redundancy is not used efficiently for noise reduction. However, for signals which are intrinsically highly periodic, auto-correlation may be a practical way to separate signal information from a background of noise.

As a preliminary to considering cross correlation, we point out that the cross correlation between two signals is the scalar product between their vectors in signal space. When the number of dimensions of signal space is large, every signal vector is essentially at right angles to all but a vanishingly small percentage of the others. This means that the cross correlation of any correlating signal with other signals is zero except for a vanishingly small percentage of the other signals. Those which do not show zero cross correlation must be coherent with the correlating signal; but as already pointed out, even coherent signals can have zero cross-correlation functions.

Cross correlation is also a type of generalized selectivity. It gives greatly increased response to those signals which are not at right angles in signal space to the correlating signal. Its standard of coherence is contained in the correlating signal.

In the case of a periodic cross-correlating signal such as shown in Fig. 8-19, the standard of coherence is the location of the periodic synchronizing time (the pulse location in Fig. 8-19). The latter is subject to a random fluctuation whose average value tends to increase as the square root of the total correlating time. This is analogous to the increase of r.m.s noise as the square root of the bandwidth in systems using extra bandwidth for noise reduction. The noise-reduction threshold in this type of cross correlation is reached when the probable error in the correlating pulse location exceeds the desired resolution time in the incoming signal.

The correlation which we have so far discussed is linear correlation. Linear correlation cannot approach the ultimate noise-reducing possibilities indicated by information theory, namely, a noise reduction which increases exponentially with the correlating time used; since linear correlation does not effectively distinguish and use all of the possible permutations in the signal.

Exercise

Show why linear least-square smoothing and prediction as discussed in Sec. 8.6 does not involve a noise-reducing threshold.

CHAPTER IX

VARIOUS ASPECTS OF INFORMATION THEORY

9.1 A survey of our results and their significance

With the many and sometimes complicated analyses that have occupied our attention in the foregoing pages, it is easy to allow oneself to become so absorbed in the details of the procedure that one may lose sight of the over-all picture. Let us therefore, take stock of what we have done by surveying our more important results, without attempting to summarize everything.

(a) *The Two Foundation Stones of Information Theory.* In order to transmit a message (an idea) in a given transmission system, it is necessary to translate the message into a signal which can be carried by the transmission system. This signal must be expressed in terms of a code or language which is appropriate for the transmission system. The fact that it is possible[1] to translate messages pertaining essentially to all items and ideas of interest, no matter how different, into a common language is one of the two foundation stones of information theory. The other is the exponential law of choice in the number of possible messages of a given long duration in a language. It would also be reasonable to call the "law of large numbers" of probability theory (upon which the exponential law of choice depends) the second foundation stone, instead of the exponential law itself.

(*b*) *Exponential Law of Choice.* If a code has an alphabet of L symbols, we found that the number of different possible messages having a duration of n symbols is L^n, provided that the code has no constraints which rule out possible combinations of symbols. The number of different possible messages thus increases exponentially

[1] Information theory does not deal with ideas which cannot be communicated, if there are such things.

with the length of the message in this simple case. In the more general case, in which constraints of the code rule out certain combinations of symbols as not being allowed, and when the symbols of the alphabet are not required to be of equal duration, we still found that for sufficiently long messages the number of allowed possible messages increased exponentially with the duration of the message.

Finally when a code has probability constraints imposed upon it so that it becomes a language,[1] the probability constraints prescribing the relative probabilities of the various combinations of symbols, we still found that for sufficiently long messages the number of allowed reasonable probable messages still increases exponentially with the message duration.

(c) *Intersymbol Influence.* The probability constraints of a language are the relative probabilities of occurrence of the different symbols of the alphabet in the language, and the probabilities of occurrence of the different possible digrams, trigams, and other groups of symbols. The probabilities of occurrences of groups of symbols specify the intersymbol influences, namely, the effect of the presence of one symbol or group of symbols on the probability of occurrence of other symbols. The *range* of the intersymbol influence is the maximum separation between symbols so that the presence of one symbol *appreciably* affects the probability of occurrence of the other. A symbol which is separated from another by a distance greater than the range of the intersymbol influence has a negligible but not necessarily a zero effect on the probability of occurrence of the other. The range of the intersymbol influence is thus an approximate, not an exact, quantity.

(d) *The Definition of Information.* Conventions can be set up identifying the symbol groups of one language with those of another. This is called translation. A fundamental requirement of the definition of information is, of course, that any message must contain the same amount of information regardless of the particular language (say spoken French or written German) in which it is expressed.

The number of allowed possible messages of a given duration may be called the *available choice* in messages of any given duration. It follows from preceding subsections that the *available choice*

[1] For the reader of this chapter who has not read the rest of the book, this method of defining a language is likely to appear strange.

in any language increases exponentially with the message duration.[1] The logarithm of the available choice in a language is, therefore, proportional to the message duration. According to one method of defining information, the logarithm of the available choice per message divided by the message duration (in number of symbols) is called the *average information* (*or entropy*) per symbol of a language.[2] This definition is in accord with the requirement of the conservation of information in translation. It also makes the information of independent messages additive. The definition, however, has the drawback that it does not specify how much information there is in a short message, but deals only with messages which are so long that average conditions prevail.

An alternative definition of information which has all the advantages of the foregoing and which reduces to the foregoing for very long messages, but which also specifies the information in short messages, is

$$\begin{matrix}\text{information received}\\ \text{with message}\end{matrix} = \log \left[\frac{\begin{matrix}\text{probability at the receiver}\\ \text{of the event after the message}\\ \text{is received}\end{matrix}}{\begin{matrix}\text{probability at the receiver of}\\ \text{the event before the message}\\ \text{is received}\end{matrix}}\right]$$

As shown in the early pages of the book, this definition actually corresponds pretty closely to the usual idea of information among people in general. Furthermore, it deals equally well with correct or erroneous messages.

(e) *Entropy.* A great boon to the development of a mathematical theory of information transmission is the existence of a simple mathematical expression, namely,

$$-\sum_i p_i \log p_i$$

which is a measure of choice of messages as well as of average information in long messages in communication systems. This

[1] These remarks apply strictly only to messages which are sufficiently long so that average conditions prevail.

[2] For example, in a system in which average number of possible allowed messages in a long time "t" is 2^{Rt}, the rate of transmission of information is R binits per unit of time.

expression is the same as that used by Boltzmann and Gibbs for entropy in statistical mechanics. Because of this identity, the mathematical methods of statistical mechanics can be used as a guide in the development of information theory. The name entropy itself for the above expression is also used in information theory.

In information theory, we find that entropy is a maximum when there are no constraints. In statistical mechanics, entropy is a maximum when there is complete disorder. The point of view is different, but the mathematics is the same.

(f) *Information and Energy.* It was our avowed purpose as we started the book to develop a quantitative theory of information. Even before developing our quantitative theory; we had certain ideas as to the nature of information. We considered it to be a sort of essence or pertinent part of all messages, such as, for example, those which come to the brain from the outside world. We believed that information was the fundamental material used in thinking and that it formed the basis for intelligent action. We had an intuitive feeling that information as thus described really did exist. One of the most important results of the book is that we have shown that an entity exists which has the properties which we have ascribed to information and which can be measured quantitatively.

There is an interesting analogy between "information" in communication theory and "energy" in physics. We all know that there are many natural phenomena, as well as man-made devices, which transform one type of mechanical activity into another, or transform mechanical activity into heat or electrical activity, or chemical change. A unifying concept which greatly simplifies all considerations of these transformations is the idea of energy. The discovery that a measurable entity exists, namely, the energy which can measure mechanical, electrical, thermal, or any other kind of physical activity, and which can measure potential as well as actual activity, greatly simplifies thinking about physical phenomena. In the study of any complicated physical or physicochemical transformation the guiding principle today is usually to see what happens to the energy.

In a similar manner, the discovery that a measurable entity, information, exists, which can measure what is transmitted by messages, regardless of their form, is a unifying step in communication theory. When the content of a scene in the outside world is transformed into a spoken message, or a written message, or from one language into another, probably the simplest way of studying what is going on from a communication point of view is to determine what happens to the information.[1]

(g) *Redundancy.* The average information per symbol in a language is always considerably less than its maximum possible value allowed by the L^n formula. In effect this means that parts of messages usually tell things which are already partly known. Thus the intersymbol influences (including interword influences) can predict the nearby succeeding parts of a message to a considerable extent. The actual reception of the message then gives partly a verification or correction of the prediction in addition to any completely new ideas. This partial or complete repetition of message content which occurs in languages is called *redundancy.* By coding messages to eliminate redundancy, it is possible to increase the information per symbol. However, despite the loss in rate of transmission of information which it causes, redundancy is a very useful property of languages, for it allows individual errors in the transmission of messages to be recognized easily and corrected. Redundancy is the "noise-reducing" mechanism of languages.

(h) *Noise and Negative Information.* If the same message is not always received as that which is sent, we say that there is noise in the transmission. In general, when we say that there is noise in a message, we mean that the message is not certain to be correct. The effect of noise is to reduce the information content of messages. Individual noisy messages whose reception actually results in less knowledge of a given situation than existed before the messages were received contain negative information. A liar tries to give negative information. However, once he becomes known as a liar, his statements no longer have much effect on probabilities. Consequently, a liar's effectiveness in giving negative information decreases rapidly.

(i) *Noise Reduction and Reliability.* The reliability of a transmission system is a measure of the probability that the messages which it transmits are true. In certain types of transmissions, the

[1] The analogy between information and energy appears to go beyond what has already been stated. Thus, for example, in relativity theory, neither information (considered as a signal) nor energy can travel with a velocity greater than that of light.

reliability may also be a measure of the accuracy of messages. We know according to information theory that the amount of information in a message cannot be increased anywhere along the transmission path, once the message has left its source. Furthermore, in a noiseless transmission system (one in which all received messages are sure to be correct) no additional information is obtained at the receiver, if a message is repeated. However, for a transmission path of a given noisiness, the reliability (or accuracy) of transmission can be increased by means of redundancy. This increase in reliability by means of redundancy is called *noise reduction.* Information theory gives a quantitative analysis of the mechanisms of noise reduction and the amounts of noise reduction possible in different systems.

In its simplest form, the use of redundancy for noise reduction is accomplished by the simple means of just repeating the message. This method is frequently used in human conversations and communications. For a given amount of redundancy, however, this is a relatively inefficient method of improving reliability. More efficient methods of using redundancy, which may be called planned-pattern redundancy, are obtained by using intersymbol constraints. This is done in such cases as noise reduction in frequency modulation and in the case of the use of intersymbol (including interword) constraints in a language to avoid misinterpretation.

The use of logic combined with probability theory in thinking in order to arrive at conclusions is essentially a noise-reduction process. Logic sets up certain constraints (relations) between messages which must hold in order for the messages to be true, simultaneously. These constraints are then used to obtain conclusions of high reliability.

It is likely that the redundancy of a language is used quite efficiently to improve reliability since it has been developed by a feedback method (trial and error) with this as a primary objective.

(j) *Effect of the Human Attention Span.* The exponential law of choice in allowed messages holds only if different meanings are assigned to all permutations among the allowed messages. In human communication systems, the exponential rise in choice with time is effective for only small durations of time, because the maximum span in human beings of the required concentrated attention to distinguish the permutations is relatively short, being a matter of seconds. As the duration of messages increases beyond this value, the percentage of permutations with different assigned meanings decreases rapidly in all languages.[1]

The effect of this situation is equivalent to redundancy in long messages. This increases their reliability for human communication, since otherwise the messages would be likely to be misunderstood. This redundancy, therefore, decreases random errors originating in the human thinking mechanism. We may, therefore, consider the human thinking mechanism as starting to get noisy after a concentrated attention span of a few seconds.

(k) *Continuous Information and Bandwidth.* In many types of communications, the signals do not come as a series of discrete symbols but rather as continuous functions. Since there is an infinite number of points on a line, if we knew nothing ahead of time about the values of the signal at future points, a continuous signal would contain an infinite amount of information. Actually this is never the case, for the signal values in a continuous signal change gradually, and the value of a signal at a point can be predicted with considerable accuracy from the signal values at immediately preceding points. An approximate value of the range of the intersymbol influence in a continuous signal, *i.e.*, an approximate value of the time between independent signal values, is determined by the frequency bandwidth. If the bandwidth is great (wide), the time between independent signal values is small. If the bandwidth is narrow, the time between independent signal values is large.

(l) *Frequency and Time Representation.* The idea of the frequency bandwidth of a signal depends upon Fourier analysis. Fourier analysis allows a function of time (or any other independent variable) to be expressed in terms of periodic (frequency) components. The frequency components are over-all properties of the entire signal. By means of Fourier analysis one can express the value of a signal at any point in terms of the over-all frequency properties of the signal; or vice versa, one can obtain these over-all properties from the values of the signal at its various points. In Chap. II and Appendix VI we were able to show that a signal has the same number of independent frequency components as it has independent time values.

[1] As an example of permutations having the same meaning we cite, "She wore a red dress and black shoes" and "she wore black shoes and a red dress."

With the aid of studies of frequency and time components, information theory has given us a more fundamental understanding of random noise, and of the relation between the correlation function and the power spectrum.

(m) *Prediction, Smoothing, and Filtering.* A most important aspect of the use of information is to predict what is going to happen on the basis of past experience. In order to predict what is going to happen in a particular signal, we study the past of the signal in order to determine its intersymbol influences, *i.e.*, its constraints.

We next determine the value of the signal at the present time and its values back into the past for as long a time as the range of the intersymbol influence. From the past and present values of a signal we can then predict the future of the signal as far as the influence runs. This prediction will decrease in accuracy the further we get into the future. If we go further into the future than the range of the intersymbol influence, we can predict only average values. A piece of equipment which automatically carries out the prediction process is called a *predictor*.

In case we are dealing not only with a desired signal, but also with noise superimposed upon the signal, then the separation of the signal from the noise is another problem. In this case we first try to determine the average values and the intersymbol constraints of the signal and of the noise from the past history of the system. We then determine the values of the combination of desired signal plus noise during the period of interest. On the basis of the differences between the average properties of the desired signal and the noise, we then estimate the best approximation which we can for the signal during the period of interest. The process of separation of signal and noise is called *smoothing*. If the separation is performed automatically by a piece of equipment, the smoothing process is called *filtering*, and the equipment which performs the separation is called a *filter*.

In Chap. VIII, we found it possible, by ascribing sufficiently simple properties to a system, to design networks which would perform both prediction and filtering operations.

(n) *Introduction to the Succeeding Sections of This Chapter.* In the preceding chapters of this book, we have developed principally the information theory of one-dimensional *communication* systems. We were able to develop a reasonably well-rounded quantitative theory in this simple case. The methods which we have found useful can, however, be generalized, and they are certainly suggestive in dealing with several more complex fields of human interest. These fields are still practically virgin territory for the application of information theory, but certain elementary ideas in the application of information theory to an analysis of thinking, knowledge, and scientific procedures are outlined in the remainder of the present chapter. We shall start out by discussing

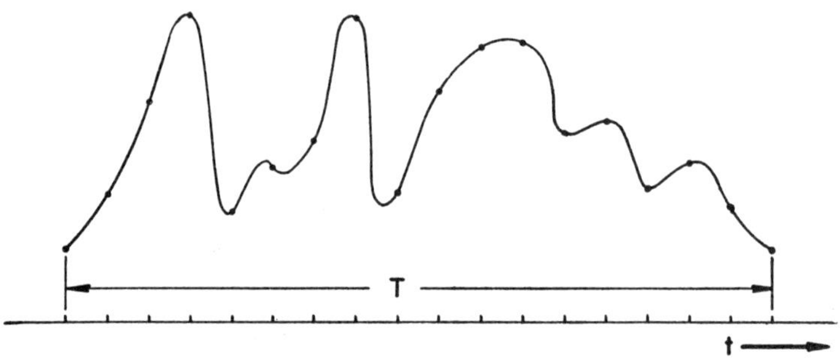

Fig. 9-1. A signal of duration T and bandwidth W, showing its sample points in the time domain.

some properties of multidimensional information, since this is the type of information with which we shall be dealing.

9.2 The dimensionality of information

(a) *Mapping Dimensions and Alphabet Dimensions.* In Fig. 9-1 is shown a signal as a function of time. If this signal is a member of an ensemble having a bandwidth W and a duration T, the signal is completely specified if we know its values at $2TW$ sample points. The signal therefore has $2TW$ degrees of freedom and can be specified by a single point in signal space of $2TW$ dimensions as discussed in Chaps. VI and VII.

Consider next the picture in Fig. 9-2. Consider this first as a stationary portrait in black and white. This may be considered as a two-dimensional signal from the point of view of information theory. We will say that it has *two mapping dimensions*. The resolving power of the eye (or the reproduction process) has the same effect as a limitation in bandwidth in Fig. 9-1. If D is the distance between

Fig. 9-2. J. Willard Gibbs (1839–1903), whose ergodic hypothesis is the forerunner of fundamental ideas in information theory. (*Photograph courtesy Dr. Ralph Gibbs Van Name.*)

resolvable points in Fig. 9-2, then the picture is completely specified if we know the optical density at about

$$\left(\frac{A}{D}\right)\left(\frac{B}{D}\right) = \frac{AB}{D^2} \tag{1}$$

sample points of the picture, A and B being its height and width.[1]

The picture therefore has AB/D^2 degrees of freedom and can be specified by a single point in signal space of AB/D^2 dimensions.

[1] This statement assumes that there is a sampling theorem in two dimensions, similar to the one already derived for one dimension.

As the next step in complication suppose we consider a signal which specifies a section of three-dimensional space in terms of

$$\frac{A}{D} \cdot \frac{B}{D} \cdot \frac{C}{D} = \frac{ABC}{D^3} \tag{2}$$

sample points. Suppose that at each sample point the density, pressure, and temperature are specified. In this case we may consider the alphabet at each sample point as being three dimensional, *i.e.*, as being a three-dimensional array.[1] In addition, the space itself has three mapping dimensions. We thus see that information has both mapping dimensionality and alphabet dimensionality which should not be confused with each other. For example if there are L_1 specifiable values of density, L_2 of pressure and L_3 of temperature, the alphabet has

$$(L_1 \cdot L_2 \cdot L_3) \tag{3}$$

symbol triplets. The total number of different possible signals is

$$(L_1 \cdot L_2 \cdot L_3)^{(ABC/D^3)} \tag{4}$$

Finally consider the external world itself. This has three dimensions of extensions in space and one in time. These four dimensions may usually be considered the mapping dimensions of the world. Furthermore, at any point in the space-time map there may be not only a density, but a color,[2] a temperature, a pressure, a chemical identity, an electric potential, a gravitational potential, a magnetic potential, and other properties, some of which may be unknown to man. It is also true that, similar to the intersymbol constraints which can exist within the one-dimensional mapping of Fig. 9-1 and which reduce its degrees of freedom below the $2TW$ value, there may be additional inter-dimensional constraints between the apparent mapping dimensions and/or alphabet dimensions of the external world which reduce the over-all number of independent dimensions as well as the number of degrees of freedom. For

[1] For example the element (1.26; 723; 320) standing for density = 1.26, pressure = 723 mm Hg, temperature = 320°K might be one element of this three-dimensional array.

[2] Color alone is often considered to have a three-dimensional alphabet consisting of possible intensities of three primary colors.

example, the chemical identity and temperature pretty well determine the color.

Consequently, prior to taking constraints into account, there are

$$(L_1 \cdot L_2 \cdot \ . \ . \ . \ \cdot L_q)^{(ABC/D^3)\cdot 2TW} \qquad (4.1)$$

possible signals any one of which might give the history of q different properties of the world throughout the volume ABC and for the time T. Only one of these signals gives the actual history of the real external world.

(b) *Information Which Is Not Mapped in the Space-Time World.* In discussions of abstract matters of the real world, the messages involved are frequently not signals mapped in the four-dimensional space-time world. Thus if one is told that George is kind and generous, intelligent but lazy, a married man and an attorney-at-law, the message will carry information if it represents something not already known. The message as a language communication is a sequence of language symbols and is thus mapped in time. However, the idea which the language message conveys is also a message. This latter message is not mapped in the four-dimensional space-time world. This message has, in effect, only one "sample point," namely, George. He has a multidimensional alphabet of possible characteristics. One dimension of this alphabet covers his possible occupations, another dimension his marital status, a third his mental competence, and so on. The actual message is thus chosen from a multidimensional alphabet and has only one sample point.

In messages of the foregoing type, the sample point, or sample points, may be quite abstract. Thus modern art or adolescence or philosophy may be the sample point of a message. On the other hand, the Republican party and the Democratic party may be the two sample points of a message.

Studies of ethics, sociology, politics, tastes, and prejudices, the relations between individuals, and similar topics all deal with information which is fundamentally not mapped in terms of sample points in the four-dimensional space-time world. Nevertheless certain aspects of information theory will be applicable in dealing with these studies. Whether or not they will be useful, only the future will tell.

(c) *The Information in Quantitative Data.* Multidimensional alphabets can be used in studying the amount of information in quantitative data. Suppose we have a multidimensional alphabet having

$$L = L_1 \cdot L_2 \cdot \ . \ . \ . \ \cdot L_q \tag{5}$$

symbols, where L_1 is the number of sub-symbols in the first dimension of the alphabet, with analogous meanings for $L_2, \ . \ . \ . \ ,L_q$. Suppose there are no constraints between the different dimensions of the alphabet. Then if s_1 is a particular sub-symbol in dimension (1), and in general s_q is a particular sub-symbol in dimension (q), and s is the symbol consisting of the set of sub-symbols

$$s_1,s_2, \ . \ . \ . \ ,s_q$$

then

$$\log [p(s)] = \log [p(s_1)] + \log [p(s_2)] + \ . \ . \ . \ + \log [p(s_q)] \tag{6}$$

It follows that, when there are no constraints between the dimensions of the symbol alphabet, the amounts of information received with the sub-symbols of the different dimensions of the alphabet are additive.

Let us now apply the foregoing to a consideration of quantitative data. Suppose that an actual temperature of

$$T = 246.2° \tag{7}$$

is measured as

$$T = 241.7° \tag{8}$$

How much information is in this measurement?

Let us make the assumption that

$$L = \frac{\text{maximum possible temperature value}}{\text{smallest significant difference in value}} \tag{9}$$

This formula implicitly assumes that the "smallest significant difference in value" is the same regardless of the size of the reading. If this is not the case, the formula must be corrected accordingly, so that L represents the number of significantly different possible readings.

Suppose we now consider the alphabet of possible values as being multidimensional, the hundreds of degrees value being one dimension, the tens value another dimension, and so forth. After the measurement is made, the measurer has approximately complete knowledge ($p = 1$) of the hundreds value but uncertain knowledge ($p < 1$) of the tens, units, and tenths values. Suppose, in general, we write the multidimensional alphabet as

$$L = L_{a1} \cdot L_{a2} \cdot \ . \ . \ . \ \cdot L_{an} \cdot L_{b1} \cdot L_{b2} \cdot \ . \ . \ . \ L_{bm} \tag{10}$$

where $L_{a1}, \ . \ . \ . \ ,L_{an}$ represent the respective number of symbols in the $(p = 1)$ group of dimensions, and $L_{b1}, \ . \ . \ . \ ,L_{bm}$ represent the number of symbols in the $(p < 1)$ group.

Then the amount of information in the measurement is

$$B = \log \frac{1}{p_{a1}} + \log \frac{1}{p_{a2}} + \ . \ . \ . \ + \log \frac{1}{p_{an}} + \log \frac{p'_{b1}}{p_{b1}} + \ . \ . \ . \ + \log \frac{p'_{bm}}{p_{bm}} \tag{11}$$

In Eq. (11), p_{a1} is the measurer's estimate of the probability that the actual correct value of the $(a1)$ dimension is the correct value before the measurement is made, with analogous meanings for $p_{a2}, \ . \ . \ . \ ,p_{an},p_{b1}, \ . \ . \ . \ ,p_{bn}$. The p' values are estimates of the same probability after the measurement is made. Equation (11) actually applies to any multidimensional alphabet, not just to quantitative data.

In order to know how many binits of information there are in the measurement

$$T = 241.7° \tag{12}$$

one must know the a priori probabilities and the a posterieri probabilities of the digits in the correct value

$$T = 246.2° \tag{13}$$

Reasonable values of these may be

$$\begin{aligned} p_{100}(2) &= 1 & p'_{100}(2) &= 1 \\ p_{10}(4) &= 0.2 & p'_{10}(4) &= 0.8 \\ p_{1}(6) &= 0.1 & p'_{1}(6) &= 0.15 \\ p_{0.1}(2) &= 0.1 & p'_{0.1}(2) &= 0.1 \end{aligned} \tag{14}$$

With these values, the amount of information in the measurement is

$$\begin{aligned} &\log \frac{p'_{100}(2)}{p_{100}(2)} + \log \frac{p'_{10}(4)}{p_{10}(4)} + \log \frac{p'_{1}(6)}{p_{1}(6)} + \log \frac{p'_{0.1}(2)}{p_{0.1}(2)} \\ &= \log 1 + \log 4 + \log 1.5 + \log 1 \\ &= 0 + 2 + 0.6 + 0 = 2.6 \text{ binits} \end{aligned} \tag{15}$$

Information theory tells how to measure the *quantity* of information, not its *value*. It does not, therefore, tell when it is worthwhile to expend the prodigious effort sometimes necessary to increase the accuracy of experimental results.

EXERCISES

1. Derive a two-dimensional form for the sampling theorem.

2. Discuss the analogue in the three-dimensional case of the sampling theorem in the time domain for which the bandwidths do not start at zero frequency. Does this have any practical application? How about the four-dimensional case and its relation to the symmetrical and antisymmetrical wave functions of quantum mechanics.

3. A school teacher has a six-year-old pupil. Estimate the amount of information she gets when the pupil tells her that his mother is 31 years old? 21 years old? 51 years old?

9.3 Some applications of information theory to an analysis of thinking[1]

Let us next try to examine certain aspects of thinking from the point of view of information theory. After a human being becomes alive, he receives a continual series of messages from the outside world by way of his various senses. As a result of these messages he gradually builds up a fairly detailed idea of what the outside world is like. It is probable that vision provides the basic three-dimensional spatial world for most people. Certainly, vision gives more detailed information about the external world than any of the other senses. It is actually not logically necessary that the messages brought to an individual by his various senses should all be coming from the same external world. It is probable, however, that he believes this instinctively. In any case, the intersymbol influence between the messages brought by the different senses at particular instants and at succeeding instants is very great, and it turns out that this intersymbol influence can be most simply understood and will normally involve no contradictions if the messages brought by the different senses are interpreted as coming from the same external world.

Furthermore changes in the external world appear to occur slowly,

[1] The remainder of this chapter deals with certain topics related to information theory. However, parts of it involve a good deal of speculative theory. Such material, although perhaps interesting and instructive does not have the same solid mathematical foundation as the rest of the book.

by which we mean that the external world appears to have continuity in time. This means that the bandwidth of the messages coming to our perception mechanism is apparently much less than the maximum bandwidth that this mechanism can handle (*i.e.*, its bandwidth as a transmission system).[1] It thus appears that things in the external world preserve their identity in time. We therefore arrive at the conclusion that there is a single real external world. This belief is further reinforced by the fact that other people

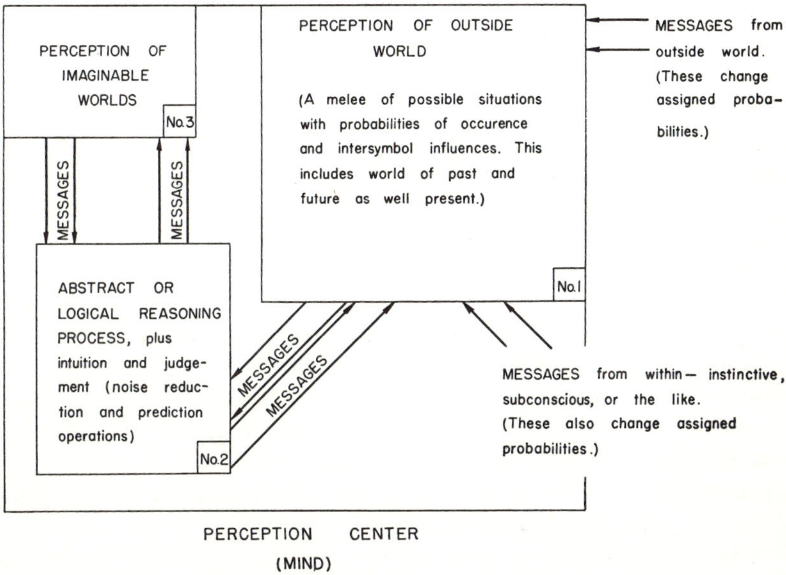

Fig. 9-3. Idealized block diagram of part of the human mind.

have the same belief and the worlds of different people appear to correspond.

In Fig. 9-3 we have a highly idealized block diagram of part of the human mind according to the point of view presented here. Block 1 represents the individual's perception of the "real" external world in which he lives. Messages from the outside world are continually coming in to build up and modify this picture. This

[1] A magician, when he can operate so that "the hand is quicker than the eye" makes the bandwidth of our perception mechanism deficient. This may lead to doubts about our usual ideas concerning the continuity of identity in time.

picture represents the individual's impression of what the external world is like based partly upon the messages he has received from it. The picture is, however, also based in part upon messages which arise within the individual. By this is meant that inherited (instinctive) impressions of what the world is like may alter the individual's picture. He furthermore may alter the picture (subconsciously) in the direction of the way in which he wishes it were.

All the messages involved, whether they arise in the outside world or within the individual are then subjected to a noise-reducing process, represented by block 2. This represents the logical reasoning processes of the mind plus what may be called intuition and judgment. For our purposes, intuition may be described as the subconscious recognition and estimation of intersymbol influence, while judgment is the use of both conscious and subconscious estimates of intersymbol influence for smoothing and prediction.

In addition to the individual's perception of the real world, he can also perceive imaginary[1] worlds. These may be purely imaginary worlds of fiction or daydreams. On the other hand they may be alternative possible future real worlds contemplated for the purpose of deciding on the individual's course of action. In this case the imaginary worlds are subjected to the noise-reducing and prediction operation of block 2, to determine the probable consequences of particular initial conditions. As another possibility, the imaginary worlds perceived by an individual may be purely abstract, and may be coded into logical or mathematical symbols for the purpose of abstract reasoning.

Block 2 of Fig. 9-3 may be described as the reasoning process of the mind. In the language of Chap. VIII, this is a smoothing and prediction operation. However, the simplifying assumptions which made it possible to solve the smoothing and prediction problem in Chap. VIII will not hold in this case.

In Fig. 9-4 is a block diagram indicating some of the over-all operations in this smoothing and prediction process. The first objective of this process is to determine the truth as nearly as possible from the incoming signal combined with previous knowledge. This is the smoothing part of the operation. The next objec-

[1] "Imaginary" here is used with its everyday meaning, not its mathematical meaning.

tive of the process is to predict future events on the basis of this truth combined with knowledge already on hand. Frequently, this objective has an additional complication. It is desired to predict future events in alternative cases where the possible courses of action by the individual supply the alternatives, which are to be added to the incoming signals plus knowledge already on hand.

The situation in Fig. 9-4 is very different from the simple case discussed in Chap. VIII. In the first place, the superposition of signal and noise in Fig. 9-4 is not linear. Secondly, Fig. 9-4 uses cross correlation with signals (knowledge) already on hand to

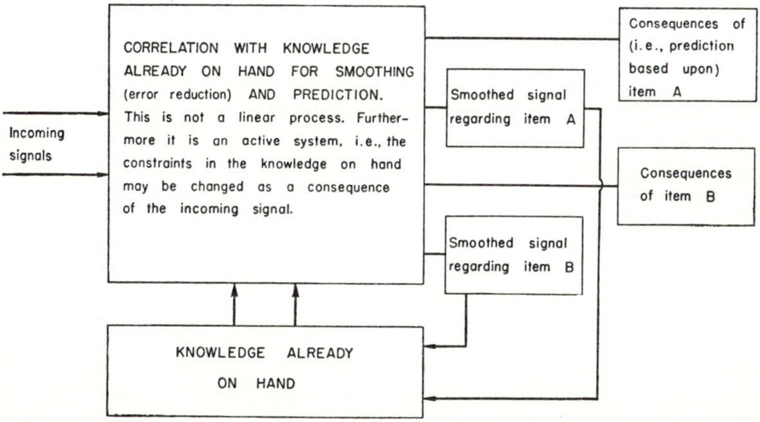

Fig. 9-4. Idealized block diagram of human reasoning process.

eliminate the noise. Next, the correlation used for both smoothing and prediction is not of a linear type. Furthermore, the system is an active system, *i.e.*, the constraints (in the knowledge on hand) may be changed as a consequence of the incoming signal. The optimum prediction is not in general of a least-square type. In fact, it is not even clear what would be meant by a least-square error.

With all these differences, the question then arises as to what are the similarities which make it worth while to study thinking from an information theory point of view. Let us now consider this matter.

Thinking is part of a communication system and therefore has a language with an alphabet. In the conscious thinking of an individual this alphabet consists of a fairly large number of simple concepts,[1] which may be called thought symbols. The knowledge of an individual then consists of his cognizance of the intersymbol constraints between these thought symbols. Simple-minded people are cognizant of only simple intersymbol constraints between thought symbols taken a few at a time. People of greater intellect think in terms of a larger number of thought symbols and are aware of not only the simpler intersymbol constraints, but also the more complicated ones. People of lesser intellect, besides not being aware of the more complicated intersymbol constraints, cannot think in terms of a large number of thought symbols. They therefore tend to identify a number of things having one or a few characteristics in common with the same thought symbol and then in effect assume that these things are identical in all details. It is probable that all people, regardless of their mental caliber, tend to do this upon first association with a new environment, say a new group of people or a new group of concepts. However, those of greater intellect will become aware of more distinctions and will become aware of distinctions in general more quickly. It should also be remarked that the fact that an individual has great (or small) intellect in one field of knowledge is no assurance of a corresponding intellect in another field of knowledge.

Human thinking actually has many languages. For example, there is a visual language and a vocal language. Translations are constantly being made between them. The visual language consists of the mental pictures of things, and the vocal language consists of their names. Thinking may be done in terms of any language, but each particular individual probably has one or more preferred languages in which he does most of his thinking. The various noise-reduction and prediction processes of any individual each probably has a particular language in which it is carried out. Therefore the thoughts in question must be translated into the required language before carrying out any particular noise-reduction process. It is also probably true that there is a certain amount of loss of information in translating from one thought language to another because of alphabet deficit.

It is interesting to note that in human society one individual considers other individuals (and also groups of individuals) as sets of

[1] In the subconscious thinking of an individual, the alphabet probably consists of a much smaller number of more elementary concepts.

constraints. In order to determine how a second person will react to various situations the first person will observe his reactions in certain particular situations. This is illustrated in Fig. 9-5. In the case of an interview, the first person may ask the second certain questions to get his reactions. It is very much like testing a network to determine its system function. The important point to be noted is that human beings assume that, if one has determined the constraints of an individual by observing his reactions to certain situations, these same constraints can be assumed to determine his reactions to other situations which may arise. In other words, it is assumed that the factors which determine the reactions of an individual may be represented as a set of constraints. This set of constraints is the personality of the individual. It is a fact of human experience that these constraints usually change only slowly with time.

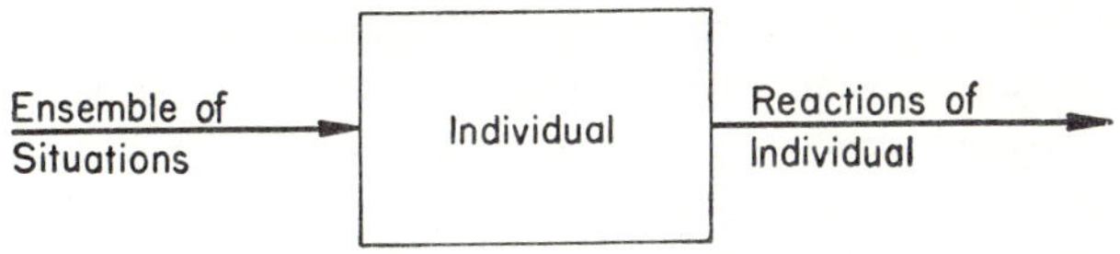

Fig. 9-5.

The attitude of human beings in judging other human beings as sets of constraints is only a special case of their attitude toward the world in general. A human being will, in general, judge practically all things in the real world as sets of constraints. He assumes that their activity and reactions in the future can be predicted from past behavior. This involves the implicit assumption that the constraints of the external world will change only slowly (if at all) with time. In a general way, this assumption appears to agree with human experience. Most human thinking, argument, and discussion is an effort to determine the constraints of the world or to convince others as to the nature of the constraints, and to use these constraints for prediction.

In ascertaining the constraints which determine what the world is like, there is abundant evidence to indicate that human thinking makes extensive use of the facts of probability theory, particularly sampling methods and the law of large numbers. If situation A leads to consequence B in several apparently independent trials, human thinking concludes that it will practically always do so.

If situation A is associated with situation B in several apparently independent samplings of the universe, then human thinking concludes that their association is regularly to be expected. From infancy on, an individual learns about the characteristics (constraints) of things in the world by such sampling methods. Because of the great value and habitual use of this sampling method, an individual is slow to abandon it in cases in which it does not apply because these samples are not independent. A weakness of human thinking appears to be its reluctance to believe that different samplings are not necessarily independent and to realize that, if samplings are not independent, the repetition of a situation has no significance. From time immemorable, certain wily individuals have taken advantage of their fellowmen, by bringing to their attention a large number of presumably independent situations leading to a particular (usually false) conclusion when the situations were not independent at all. In modern times, this method of deception has become an organized activity under the name of propaganda and appears to a regrettable extent in advertising.

The application of the points of view of information theory in the study of psychology and thinking could be greatly expanded. The detailed mechanisms of thinking in translating from one thought language to another, in noise reduction and prediction, in reduction of alphabet size, in "matching" messages and knowledge to the available "apparatus" of the brain in terms of attention span, alphabet size, and duration of conceived intersymbol influence are all worthy of extensive investigation. Psychologists already have considerable knowledge in these fields, but information theory will add a new point of view which should aid in the design of new types of tests and experiments, and should indicate new ways of analyzing the results.

9.4 Some applications of information theory to an analysis of scientific work

In the preceding section we indicated how the thinking processes of man can be described in the language of information theory. It was hoped that some of the points of view already developed or still to be developed in information theory could thereby be exploited in psychology, and might lead to useful results. In this section we describe certain scientific facts and scientific procedures in the language of information theory. It is hoped that some worthwhile ideas may thereby be born.

The laws of science are an expression of constraints (symbol probabilities as well as intersymbol influences). These can be used for prediction. Some of the objectives of science are

(1) To determine the constraints of the universe, *i.e.*, to determine the laws of nature.

(2) To use the constraints for prediction. An example of this would be the use of the law of gravitation to predict the motion of the planets.

(3) To set up initial conditions which with the known constraints

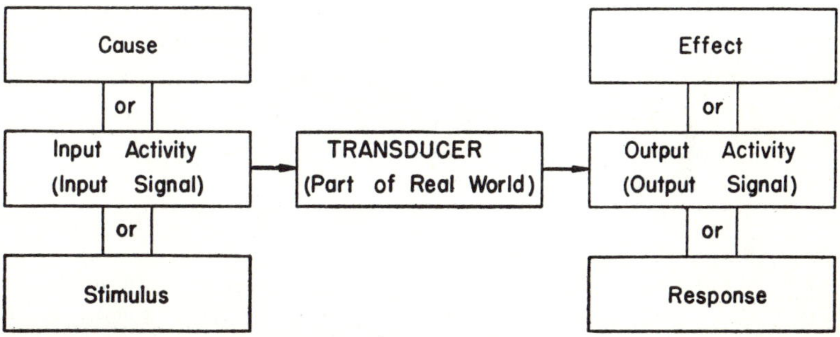

Fig. 9-6. Block diagram of the real world as a transducer between cause and effect.

will give a predictable desired final result. Engineering and medicine are examples of this type of activity.

Let us now consider those parts of science which involve *cause* and *effect.* In such cases, we consider the particular part of the real world under consideration as a transducer between *cause* and *effect.* This is shown diagrammatically in Fig. 9-6. Here the "cause" which may also be called the "stimulus" or the "input activity" acts upon a part of the real world which may be considered a transducer between cause and effect. The result may be called the "effect" or the "response" or "output activity." The input activity may be considered an input signal (usually multidimensional), and the output activity may be considered the corresponding output signal. A study of the real world thus becomes a study of transducers. Scientific laws may then be described as the properties of transducers. Information theory does throw some light on the relations between the inputs and output of transducers. It should therefore be of value in the development of science.

It is outside the objectives of this book to give an extended discussion of science in general or of any particular science from the point of view of information theory. However, a few very general remarks will be made.

In the physical sciences when dealing with the things of *macroscopic* size, experience shows that a given cause acting upon a non-living transducer gives rise to a completely determined effect. For example, if a body is dropped in a vacuum in the earth's gravitational field, its subsequent location is completely determined by the laws of mechanics. In the language of information theory, we may make the general statement that non-living things of macroscopic size, as transducers, add only *fixed constraints* in transforming cause into effect. We limit our statement to non-living things of macroscopic size; for non-living things of microscopic size, according to one interpretation of quantum mechanics, are believed to add probability constraints in transforming cause into effect.

Let us turn now to the biological sciences. Here experience shows that living things as transducers add probability constraints in transforming cause into effect. In other words the result of a biological experiment is, in general, predictable only on the average: it cannot be predicted for certain in the individual case. A large school of biologists believes that the laws of biology are ultimately derivable from the laws of the physical sciences and that if *all* the conditions of a biological experiment were known, as they cannot be in practice, the result would be completely determined. We, of course, do not know whether this is true or not. All we do know is that in the present state of knowledge, the laws of biology are equivalent to probability constraints.

We now wish to give an example[1] of a particular case in which the point of view of information theory has given rise to a new idea in biology. This idea suggests a reason for the division of large organ-

[1] This example is taken from a paper by Goldman, *Proc. I.R.E.*, vol. 36, May, 1948, 594. It is too early to say whether this idea will lead to anything of value in biology.

isms into cells.[1] According to current biological theory, the characteristics of an organism are determined by the chromosomes in its cells, and there are essentially identical sets of chromosomes in each of the cells of the organism (excepting the germ cells). Now the life history of the organism is regulated by the interactions of the various parts of the organism which is living in an environment subject to a considerable amount of random fluctuation. These interactions, whether they are secretions by glands, the sending of nerve impulses, the process of digestion, or any other type of physiological action, represent perfectly good signals in our general theory of communication. The activity of the chromosomes, whatever it may be, may likewise be considered a set of signals.

In the case of a large organism, there is a large amount of structure built according to a detailed plan which must be kept running according to plan during the life cycle, in spite of the general disintegrating processes of the physical world and the random fluctuations of the environment. The large amount of characteristic detail in the organism is the equivalent of a large amount of signal detail in communication theory. The communication-theory equivalent of retaining all the characteristic individual detail despite the presence of the fluctuating environment is the maintenance of a large signal-to-noise ratio. We recall from Chap. VII that in technical communication systems, when it is desired to obtain a large signal-to-noise ratio, such as in frequency modulation, improvement in the signal-to-noise ratio is obtained by repetition of the signal (usually in the frequency domain) according to some predetermined plan. It is, therefore, not surprising to find that, in the case of large organisms, the fundamental knowledge contained in the structure or activity of the chromosomes, which regulate the life history of the organism, is set up in such a way that the knowledge is repeated coherently, *i.e.*, according to a planned layout. Thus the division of the organism into cells, each of which has an essentially identical set of chromosomes, is a way in which nature can keep a large organism operating in a planned life cycle despite the disintegrating effects of the physical world.

The foregoing example has, of course, not been worked out in

[1] There are, of course, also other important reasons for the division of a large organism into cells besides the one suggested here.

detail. It does show, however, that information theory can be suggestive in biology. All the aspects and phenomena which we have noted in information theory, such as modulation, detection, sample points, bandwidth, transformations between time and frequency domains, filtering, noise reduction, power, entropy, information, random noise, and several others, have counterparts in the natural sciences. We may therefore expect information theory to afford a new method of expression of many scientific ideas and perhaps to lead to new scientific points of view and new ideas.[1]

[1] In this connection see *Essays on the Use of Information Theory in Biology* edited by Henry Quastler, University of Illinois Press, Urbana, Ill., 1953.

APPENDIX I

The Growth of the Number of Possible Sequences in a Transmission Channel with a Finite Number of Fixed Constraints[1]

If a transmission channel without constraints has an alphabet having a finite number, L, of symbols, then the number of different sequences which are possible having a duration of n symbols is

$$L^n \tag{1}$$

To show that this is true, we note that there are L different sequences of duration of one symbol, namely, the L symbols of the alphabet. Each of these symbols when followed by another symbol will give a different sequence having a duration of two symbols. There are thus L times L sequences of duration of two symbols. This procedure can be continued at will, thus giving formula (1).

Let us next determine the number of possible sequences in a given duration of time. We will not assume that the durations of the L symbols are necessarily equal, but we will say that these durations are $t_1,t_2,\ldots,t_L$, respectively. The individual symbols will be called $S_1,S_2,\ldots,S_L$, respectively. Let us write $N(T)$ for the function of T which represents the total number of different sequences of duration T.[2] The number of these of duration $(T - t_1)$ is thus $N(T - t_1)$. To each of these $N(T - t_1)$ sequences an S_1 can be added to give a sequence of duration T. It therefore follows that

$$N(T) = N(T - t_1) + N(T - t_2) + \ldots + N(T - t_L) \tag{2}$$

provided that all zero terms in the right side of Eq. (2) with non-negative arguments are set equal to unity. Thus on the right side of Eq. (2), substitute zero for $N(-1)$, $N(-2)$, etc., but substitute unity for $N(0)$, $N(1)$, etc., if their value is zero.

[1] The principal results of Appendices I and III are due to C. E. Shannon, *Bell System Tech. Journal*, vol. 27, July, 1948, pp. 379–423.

[2] More specifically, $N(T)$ is the number of different possible sequences of duration T or less, provided that none of these counted sequences will still be of duration T or less if some allowed symbol is added to it.

As an example, consider an alphabet consisting of the three symbols A, B and C, where

$$t_A = 2 \qquad t_B = 3 \qquad t_C = 6$$

Then

	Sequences
$N(1) \rightarrow N(-1) + N(-2) + N(-5) \rightarrow 0$	(none)
$N(2) \rightarrow N(0) + N(-1) + N(-4) \rightarrow 1 + 0 + 0 = 1$	A
$N(3) \rightarrow N(1) + N(0) + N(-3) \rightarrow 1 + 1 + 0 = 2$	A, B
$N(4) \rightarrow N(2) + N(1) + N(-2) \rightarrow 1 + 1 + 0 = 2$	AA, B
$N(5) \rightarrow N(3) + N(2) + N(-1) \rightarrow 2 + 1 + 0 = 3$	AA, BA, AB
$N(6) \rightarrow N(4) + N(3) + N(0) \rightarrow 2 + 2 + 1 = 5$	AAA, BA, AB, BB, C
$N(7) \rightarrow N(5) + N(4) + N(1) \rightarrow 3 + 2 + 1 = 6$	AAA, BAA, ABA, AAB, BB, C
$N(8) = N(6) + N(5) + N(2) = 5 + 3 + 1 = 9$	AAAA, BAA, ABA, BBA, CA, AAB, BAB, ABB, AC

and so on.

Equation (2) is a linear *difference equation*. Certain general methods of solving linear difference equations have been developed which are similar in a way to the methods of solving linear differential equations with constant coefficients. For example, if we have the linear difference equation with constant coefficients for $y(t)$

$$y(t + 2) = ay(t + 1) + by(t) \tag{3}$$

it is known that the solution for integral values of time will be

$$y(t) = C_1r_1^t + C_2r_2^t \tag{4}$$

where r_1 and r_2 are the roots of

$$r^2 - ar - b = 0 \tag{5}$$

and C_1 and C_2 are arbitrary constants. To prove that Eq. (4) is a solution, we need only substitute it into Eq. (3). Thus according to Eq. (4)

$$y(t + 2) = C_1r_1^{(t+2)} + C_2r_2^{(t+2)} \tag{6}$$

$$y(t + 1) = C_1r_1^{(t+1)} + C_2r_2^{(t+1)} \tag{7}$$

Substituting Eqs. (4), (6), and (7) into Eq. (3)

$$y(t+2) - ay(t+1) - by(t) = 0$$
$$= C_1 r_1^t[r_1^2 - ar_1 - b] + C_2 r_2^t[r_2^2 - ar_2 - b] \tag{8}$$

If r_1 and r_2 are roots of Eq. (5), then the right side of Eq. (8) will vanish for all values of C_1 and C_2, so that Eq. (4) is a solution of Eq. (3).

The same method will now be used to find $N(T)$ from Eq. (2). Assume that

$$N(T) = C_1 r_1^T + C_2 r_2^T + \dots + C_n r_n^T \tag{9}$$

Then, in general,

$$N(T - t_p) = C_1 r_1^{(T-t_p)} + C_2 r_2^{(T-t_p)} + \dots + C_n r_n^{(T-t_p)} \tag{10}$$

Substituting these values into Eq. (2), we have

$$\begin{aligned} &C_1 r_1^T[1 - r_1^{-t_1} + r_1^{-t_2} - \dots r_1^{-t_L}] \\ &+ C_2 r_2^T[1 - r_2^{-t_1} - r_2^{-t_2} - \dots r_2^{-t_L}] \\ &+ \cdot \cdot \cdot \cdot \cdot \cdot \cdot \cdot \cdot \cdot \cdot \cdot \cdot \cdot \cdot \cdot \cdot \cdot \cdot \cdot \cdot \\ &+ C_n r_n^T[1 - r_n^{-t_1} - r_n^{-t_2} \dots r_n^{-t_L}] = 0 \end{aligned} \tag{11}$$

The general solution to Eq. (2) is thus Eq. (9) where $r_1, r_2, \dots, r_n$ are the roots to

$$1 - r^{-t_1} - r^{-t_2} - \dots - r^{-t_L} = 0 \tag{12}$$

and $C_1, C_2, \dots, C_n$ are arbitrary constrants.

$N(T)$ must be real and non-negative for all positive values of T. For sufficiently large values of T, (9) then reduces to

$$N(T) = C_1 r_1^T \tag{13}$$

where r_1 is the largest positive real root of Eq. (12), since the other terms in Eq. (9) become negligible by comparison. In Eq. (13), we have thus found the general solution of Eq. (2) for sufficiently large values of T. This is the number of different possible sequences of long duration T in the system, and we see that it grows exponentially with T, when T is large.

In deriving Eq. (13), we assumed in writing down Eq. (2) that any symbol, say S_p, could be used anywhere in a sequence. Now in some codes and transmission channels, this is not the case. For example, in telegraphy we have the symbols: dot, dash, letter space, word space. A dot or a dash may be used anywhere in a sequence, but a letter space or a word space can only follow a dot or a dash. A letter space or a word space cannot follow either a letter space or a word space. If a particular language is coded into telegraphy, there are additional constraints, but in its simplest aspect, telegraphy has only the constraints mentioned above. These constraints may also be fully described by saying that a generator of telegraphic symbols has two *states*. In the first state, it can generate any of the four symbols listed above; in the second state it can generate only a dot or a dash. The generator will be in state 1 if the last symbol generated was a dot or a dash. It will be in state 2 if the last symbol generated was a word space or a letter space. This situation is depicted in Fig. A-1. The directed lines show which symbols take the system from state to state.

We will next derive the analogue of Eq. (13) for a rather general type of transmission system having fixed[1] constraints. Let us suppose the system has the alphabet of symbols $S_1, S_2, \ldots, S_L$. Let us further suppose that the system has a number of states which we may designate as $1, 2, \ldots, m$, respectively. States, in general, are a means of specifying the constraints

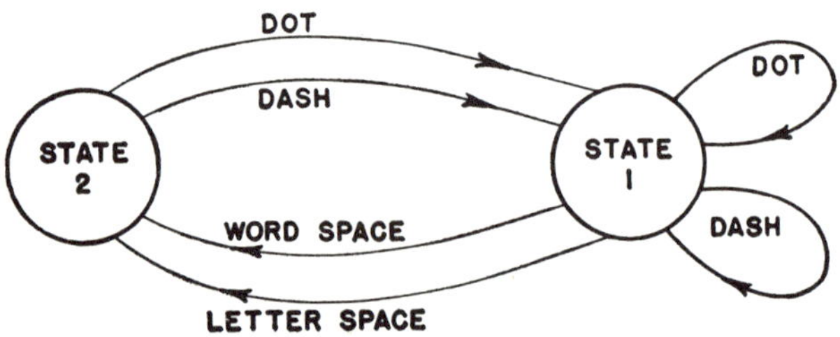

Fig. A-1. Graphical representation of the constraints in telegraphy.

of a sequence, whether the latter are fixed or are probability constraints. In its most general form, a state is an expression of the past history of the sequence as far back as the intersymbol influence goes. Anything in this past history which changes the possibility or probability of what may follow changes the state of the sequence at the location of interest. In the present case, we will consider the states to be specified by the fixed constraints alone, *i.e.*, those regulating possibilities, and we will ignore the probability constraints.

In any system of the above type, let $N_i(T)$ be the number of different possible sequences of duration T ending in state i. Then we may write

$$N_j(T) = \sum_i \left[\sum_s N_i(T - b_{ij}^{(s)})\right] \tag{14}$$

where $b_{ij}^{(1)}, b_{ij}^{(2)}$ are the lengths of the symbols which are possible in state i and lead to state j. The $\sum_s$ in Eq. (14) is taken over all allowable symbols

[1] Fixed constraints are specific restrictions of a transmission channel which are always in force. They specify what is possible and what is not possible. These are to be distinguished from *probability constraints* which affect only average values. Generally speaking, codes have fixed constraints while languages have probability constraints.

which lead to state j from the particular state i, and $\sum_i$ then is taken over all possible states. Equation (14) actually represents a set of linear difference equations, there being a different member of the set for each different state j. By methods similar to those used in deriving Eq. (13), it may be shown that for large T the set of solutions to Eq. (14) may be written as

$$N_j(T) = A_j W^T \tag{15}$$

where W is the same constant for all states j but A_j varies from state to state. We will now substitute Eq. (15) in Eq. (14) and solve for W. Thus

$$A_j W^T = \sum_i [\sum_s A_i W^{(T - b_{ij}^{(s)})}] = \sum_i [A_i \sum_s W^{(T - b_{ij}^{(s)})}] \tag{16}$$

Dividing the equation through by W^T, we obtain

$$A_j = \sum_i A_i [\sum_s W^{-b_{ij}^{(s)}}] \tag{17}$$

If we introduce the symbol

$$\begin{aligned} \delta_{ij} &= 1 \quad \text{when } i = j \\ \delta_{ij} &= 0 \quad \text{when } i \neq j \end{aligned} \tag{18}$$

then Eq. (17) may be written

$$\sum_i A_i [\sum_s (W^{-b_{ij}^{(s)}}) - \delta_{ij}] = 0 \tag{19}$$

Equation (19), of course, represents a different equation for each different value of j. This is a set of homogeneous linear equations in the A_i's, and the set will have a non-trivial solution if and only if the determinant of the coefficients vanishes. Thus

$$\begin{vmatrix} \sum_s (W^{-b_{11}^{(s)}}) - \delta_{11} & \sum_s (W^{-b_{21}^{(s)}}) - \delta_{21} & \cdots \\ \sum_s (W^{-b_{12}^{(s)}}) - \delta_{12} & \sum_s (W^{-b_{22}^{(s)}}) - \delta_{22} & \cdots \\ \vdots & \vdots & \end{vmatrix}$$

$$= \begin{vmatrix} \sum_s (W^{-b_{11}^{(s)}}) - 1 & \sum_s (W^{-b_{21}^{(s)}}) & \cdots \\ \sum_s (W^{-b_{12}^{(s)}}) & \sum_s (W^{-b_{22}^{(s)}}) - 1 & \cdots \\ \vdots & \vdots & \end{vmatrix} = 0 \tag{20}$$

If the largest real root of Eq. (20) is substituted in Eq. (15), we have the solution of Eq. (14).

Equation (20) reduces to Eq. (12) when there is only one state. In this case, Eq. (20) becomes

$$\sum_s (W^{-b_{11}^{(s)}}) - 1 = 0$$

$$= W^{-b_{11}^{(1)}} + W^{-b_{11}^{(2)}} + \cdots + W^{-b_{11}^{(L)}} - 1 = 0 \tag{21}$$

If we then set
$$W = r \tag{22}$$
and
$$\left.\begin{aligned} b_{11}^{(1)} &= t_1 \\ b_{11}^{(2)} &= t_2 \\ b_{11}^{(L)} &= t_L \end{aligned}\right\} \tag{23}$$
we have Eq. (12).

We will now illustrate the use of Eqs. (15) and (20) by finding W in the case of electrical telegraphy. In this case, if we designate "voltage on for unit time" as (+) and "voltage off for unit time" as (−), the four telegraphic symbols have the following specifications

dot	+ −	letter space	− − −
dash	+ + + −	word space	− − − − − −

As mentioned above, telegraphy has two states. In state 1, the next symbol can be any of the above four symbols; in state 2, the next symbol must be either a dot or a dash. Referring to Fig. A-1, we see that the symbols which take the system from state 1 to state 1 are the dot and dash. The symbols which take the system from state 1 to state 2 are the letter space and word space. The symbols which take the system from state 2 to state 1 are the dot and dash. There are no symbols which take the system from state 2 to state 2.

Eq. (20) thus becomes

$$\begin{vmatrix} W^{-4} + W^{-2} - 1 & W^{-4} + W^{-2} \\ W^{-6} + W^{-3} & -1 \end{vmatrix} = 0 \tag{24}$$

or
$$1 - W^{-2} - W^{-4} - W^{-5} - W^{-7} - W^{-8} - W^{-10} = 0 \tag{25}$$

The largest real root of Eq. (25) is

$$W = 1.453 = 2^{0.539} \tag{26}$$[1]

The total number of different possible telegraphic signals of duration T is

$$N(T) = \sum_j N_j(T) = \sum_j A_j W^T = W^T \sum_j A_j \tag{27}$$

[1] $\log_2(x) = \dfrac{\log_{10} x}{\log_{10} 2} = 3.32 \log_{10} x.$

The code capacity of telegraphy is thus

$$C = \lim_{T\to\infty}\left[\frac{1}{T}\log N(T)\right]$$

$$= \lim_{T\to\infty}\frac{1}{T}[T\log W + \log \sum_j A_j] = \log W = 0.539 \text{ binit/unit time} \quad (28)$$

using 2 as the logarithmic base.

APPENDIX II

Derivation of the Number of Different Ergodic Sequences of Duration T

Let us consider an ergodic sequence consisting of m symbols. Let us divide this sequence into

$$r = m/q \tag{1}$$

groups of successive symbols, each group consisting of q symbols. The range of the intersymbol influence is finite in an ergodic sequence, and q will be chosen sufficiently greater than the effective range of the intersymbol influence so that we can assume that there is effectively no intersymbol influence between the r groups of symbols.[1] If L is the number of symbols in the alphabet of the ergodic sequence, then by Eq. (11) of Chap. I there are

$$s = L^q \tag{2}$$

different groups of symbols of duration q. The total effect of the symbol frequencies and intersymbol influences can then be expressed in terms of the probabilities p_i of occurrence of the L^q different symbol groups each having a duration of q symbols.

We will specify the different symbol groups as group 1, group 2, . . . , group s, and we will let $N_1, N_2, \ldots, N_s$ be the respective number of each of the different groups in the original sequence. Then we have

$$r = \frac{m}{q} = N_1 + N_2 + \cdots + N_s \tag{3}$$

By a well-known formula in the theory of combinations, the total number of ways in which these groups of symbols can be arranged is then

$$M(m) = \frac{r!}{N_1!N_2! \ldots N_s!} \tag{4}$$

$M(m)$ is then the total number of different possible arrangements of the m symbols in the original ergodic sequence subject to the constraint that the sequence contains

$$N_1 = p_1 r \tag{5a}$$

[1] There is, of course, some intersymbol influence between the symbols near the neighboring edges of adjacent groups. We are here assuming that, if q is chosen sufficiently large, this intersymbol influence will not appreciably affect the analysis which follows.

groups of type 1,

$$N_2 = p_2 r \tag{5b}$$

groups of type 2, and so on. The probabilities $p_1, \ldots, p_s$ then express the effects of the symbol frequencies and intersymbol influences.

From Eqs. (4) and (5) it then follows that

$$\begin{aligned} \log M &= \log (r!) - \log (N_1!) - \log (N_2!) - \ldots \log (N_s!) \\ &= \log (r!) - \log [(p_1 r)!] - \log [(p_2 r)!] \ldots \log [(p_s r)!] \end{aligned} \tag{6}$$

In an ergodic sequence, $N_1, N_2, \ldots, N_s$ and r are all large numbers. For the factorial of a large number, we have the highly accurate Stirling approximation, namely,

$$\log (N!) = (N + \tfrac{1}{2}) \log N - N + \tfrac{1}{2} \log 2\pi \tag{7}$$ [1]

Using this formula, Eq. (6) becomes

$$\begin{aligned} \log M(m) \\ &= (r + \tfrac{1}{2}) \log r - r + \tfrac{1}{2} \log 2\pi - \sum_{i=1}^{s} [(p_i r + \tfrac{1}{2}) \log (p_i r) \\ &\quad - p_i r + \tfrac{1}{2} \log 2\pi] = r \log r - r + \tfrac{1}{2} \log r + \tfrac{1}{2} \log 2\pi \\ &\quad - \sum_{i=1}^{s} [p_i r \log p_i + p_i r \log r + \tfrac{1}{2} \log p_i + \tfrac{1}{2} \log r \\ &\qquad - p_i r + \tfrac{1}{2} \log 2\pi] \end{aligned} \tag{8}$$

Now

$$\sum_{i=1}^{s} p_i r \log r = r \log r \sum_{i=1}^{s} p_i = r \log r \tag{9}$$

and

$$\sum_{i=1}^{s} p_i r = r \sum_{i=1}^{s} p_i = r \tag{10}$$

since $\sum_{i=1}^{s} p_i$ must equal unity. Therefore, after cancellation of equal terms of opposite sign in Eq. (8), we obtain

$$\begin{aligned} \log M(m) = -r \Big[\sum_{i=1}^{s} p_i \log p_i \Big] - \left(\frac{s-1}{2}\right) \log r - \frac{1}{2} \Big[\sum_{i=1}^{s} \log p_i \Big] \\ - \left(\frac{s-1}{2}\right) \log 2\pi \end{aligned} \tag{11}$$

[1] See, for example, Goldman, *Transformation Calculus and Electrical Transients*, p. 255. The logarithmic base in Eq. (7) is ϵ. For simplicity, ϵ will be used as the logarithmic base throughout this appendix, even though we are dealing with discrete signals.

If r is extremely large, as it will be in an ergodic sequence, Eq. (11) becomes approximately

$$\log M(m) = -r \sum_{i=1}^{s} p_i \log p_i \tag{12}$$

since the other terms on the right side of Eq. (11) become negligible in comparison with the first. Let us now define g, h', and A by the equations

$$\log g = h' = -\frac{1}{q} \sum_{i=1}^{s} p_i \log p_i \tag{13}$$

and $$\log A = -\left(\frac{s-1}{2}\right) \log r - \frac{1}{2}\left[\sum_{i=1}^{s} \log p_i\right] - \left(\frac{s-1}{2}\right) \log 2\pi \tag{14}$$

or $$A = \frac{1}{r^{\left(\frac{s-1}{2}\right)} \cdot p_1^{\frac{1}{2}} \cdots p_s^{\frac{1}{2}} (2\pi)^{\left(\frac{s-1}{2}\right)}} \tag{15}$$

Then from Eq. (11), it follows that

$$M(m) = Ag^m = A\epsilon^{h'm} = A2^{hm} \tag{16}$$

where A is a multiplier which changes slowly with m in comparison with an exponent.

Equation (16) can also be expressed on a time basis. Thus if b is the average symbol duration and T is the total duration of a sequence, then

$$m = \frac{T}{b} \tag{17}$$

The total number of different possible ergodic sequences m symbols long may then be called the total number of different possible ergodic sequences of duration T. From Eqs. (16) and (17), we then have

$$M(T) = M(m) = A2^{(h/b)T} \tag{18}$$

where $M(T)$ is the number of different possible ergodic sequences of duration T.

APPENDIX III

Adjustment of the Probability Constraints to the Fixed Constraints for Maximum Language Transmission Capacity

In Appendix I, we found a general formula, Eq. (27), for the total number of different possible sequences of duration T in a code channel having a rather general type of fixed constraints. If such a channel is used to encode messages which are already expressed in the form of a language, it is often an important matter to "match" the probability constraints of the language to the fixed constraints of the code channel so that despite the additional constraints the language transmission capacity is as nearly as possible equal to the code capacity. In the present appendix we will develop a criterion for matching of this type.[1]

Suppose we have a rather general type of transmission channel. This channel has an alphabet with symbols $S_1, S_2, \ldots, S_L$ and has a number of states which may be designated by number as state 1, state 2, . . . , up to state m. At any location in a sequence of symbols in this transmission channel, the sequence is in a particular state which is specified by the past history of the sequence as far back as the intersymbol influence goes. The past history of the sequence in conjunction with the fixed constraints then determine what succeeding symbols or combinations of symbols are possible. The past history in conjunction with the probability constraints determine the relative probabilities of these succeeding symbols or combinations of symbols. Let the probability constraints be specified as follows:

$$P_i = \text{probability of being in state } i \tag{1}$$

$$p_{ij}^{(s)} = \begin{bmatrix} \text{probability, if in state } i\text{, of going from} \\ \text{state } i \text{ to state } j \text{ with symbol } s \end{bmatrix} \tag{2}$$

On the basis of Eq. (2), we may also define the quantity

$$p_{ij} = \begin{bmatrix} \text{total probability, if in state } i\text{, of going} \\ \text{from state } i \text{ to state } j \text{ with any symbol} \end{bmatrix}$$

$$= \sum_s p_{ij}^{(s)} \tag{3}$$

[1] The use of the shortest code groups in Morse code for the most frequently occurring English letters is an approximation to this type of match.

As in Appendix I, we also use the definition

$$b_{ij}^{(s)} = \text{length of symbol } s \text{ taking system from state } i \text{ to state } j \tag{4}$$

From Eqs. (1) and (2), we have

$$P_i p_{ij}^{(s)} = \text{probability of going from state } i \text{ to state } j \text{ with symbol } s \tag{5}$$

The average length of symbol in a long ergodic sequence in the channel is then

$$\sum_i \{P_i \sum_j [\sum_s p_{ij}^{(s)} b_{ij}^{(s)}]\} = \text{average length of symbol}$$
$$= X \quad \text{say} \tag{6}$$

Now, if the system is known to be in state i, the information received with the symbol s taking the system into state j is, according to Eq. (2) of Chap. 1,

$$\text{information received with symbol } s = -\log p_{ij}^{(s)} \tag{7}$$

in the noiseless case. The average information received per symbol in an ergodic sequence is thus

$$\text{average information per symbol} = -\sum_i \{P_i \sum_j [\sum_s p_{ij}^{(s)} \log p_{ij}^{(s)}]\}$$
$$= Y \quad \text{say} \tag{8}$$

Thus

$$\text{average information per unit length of sequence} = \frac{Y}{X} = R \quad \text{say} \tag{9}$$

Now if the values of $p_{ij}^{(s)}$ can be adjusted so that R equals the code capacity of the fixed constraints alone, then the probability constraints will not be responsible for more than a vanishingly small fraction of the disallowed sequences, and R will have its maximum possible value for the given fixed constraints.[1] We will now show that a solution which makes R equal to the code capacity of the fixed constraints is

$$p_{ij}^{(s)} = \frac{B_j}{B_i} W^{-b_{ij}^{(s)}} \tag{10}$$

where the B_i are solutions of the set of simultaneous equations

$$B_i = \sum_j [\sum_s B_j W^{-b_{ij}^{(s)}}] \tag{11}$$

[1] If the values of the $p_{ij}^{(s)}$ probabilities are determined, this automatically determines the probabilities P_i since

$$\sum_i P_i p_{ij} = P_j \quad \text{and} \quad \sum_i P_i = \sum_j P_j = 1$$

Equations (11) may also be written

$$\sum_j \{B_j[\sum_s (W^{-b_{ij}^{(s)}}) - \delta_{ij}]\} = 0 \tag{12}$$

The symbols W and δ_{ij} are defined in Appendix I. The set of equations, (12), will have a non-trivial set of solutions, if the determinant of the coefficients vanishes, *i.e.*, if

$$\begin{vmatrix} \sum_s (W^{-b_{11}^{(s)}}) - \delta_{11} & \sum_s (W^{-b_{12}^{(s)}}) - \delta_{12} & \cdots \\ \sum_s (W^{-b_{21}^{(s)}}) - \delta_{21} & \sum_s (W^{-b_{22}^{(s)}}) - \delta_{22} & \cdots \\ \vdots & \vdots & \end{vmatrix} = 0 \tag{13}$$

This is the same as Eq. (20) of Appendix I with columns and rows interchanged. Equation (12) therefore has a set of solutions for the B_i's (which is, of course, the same as the set of the B_j's) with W having the same significance as in Appendix I.

In order that the $p_{ij}^{(s)}$ can represent probabilities, it is necessary that

$$\sum_j [\sum_s p_{ij}^{(s)}] = 1 \tag{14}$$[1]

We will now check this for the $p_{ij}^{(s)}$ defined in Eq. (10). Thus

$$\sum_j \left[\sum_s p_{ij}^{(s)}\right] = \sum_j \left[\sum_s \frac{B_j}{B_i} W^{-b_{ij}^{(s)}}\right] = \frac{1}{B_i} \sum_j \left[\sum_s B_j W^{-b_{ij}^{(s)}}\right]$$

$$= \frac{B_i}{B_i} = 1 \tag{15}$$

with the aid of Eqs. (10) and (11). This checks Eq. (14) for the $p_{ij}^{(s)}$, so that they may be a possible set of probabilities.

Using these values for $p_{ij}^{(s)}$, let us calculate R. From Eqs. (6), (8), and (9)

$$R = \frac{-\sum_i \{P_i \sum_j [\sum_s p_{ij}^{(s)} \log p_{ij}^{(s)}]\}}{\sum_i \{P_i \sum_j [\sum_s p_{ij}^{(s)} b_{ij}^{(s)}]\}} \tag{16}$$

From Eq. (10)

$$\log p_{ij}^{(s)} = \log B_j - \log B_i - b_{ij}^{(s)} \log W \tag{17}$$

[1] It is also necessary that all the values of $p_{ij}^{(s)}$ called for by Eq. (10) lie in the range $0 \leq p_{ij}^{(s)} \leq 1$.

Thus the numerator of Eq. (16) is

$$-\sum_i \{P_i \sum_j [\sum_s p_{ij}^{(s)} \log p_{ij}^{(s)}]\}$$
$$= -\sum_i \{P_i \sum_j [\sum_s p_{ij}^{(s)} \log B_j]\} + \sum_i \{P_i \sum_j [\sum_s p_{ij}^{(s)} \log B_i]\} + \sum_i \{P_i \sum_j [\sum_s p_{ij}^{(s)} b_{ij}^{(s)} \log W]\} \tag{18}$$

Now the first two terms on the right side of Eq. (18) cancel. To show this, note that

$$\sum_i \{P_i \sum_j [\sum_s p_{ij}^{(s)} \log B_j]\} = \sum_i \{P_i \sum_j p_{ij} \log B_j\}$$

[with the aid of Eq. (3)] $= \sum_j \sum_i P_i p_{ij} \log B_j$

[by just adding the same terms in a different order]

$$= \sum_j \log B_j \sum_i P_i \cdot p_{ij} = \sum_j \log B_j \cdot P_j \tag{19}$$

since

$$\sum_i P_i p_{ij} = P_j \tag{20}$$

Furthermore

$$\sum_i \{P_i \sum_j [\sum_s p_{ij}^{(s)} \log B_i]\}$$
$$= \sum_i \{P_i \sum_j p_{ij} \log B_i\}$$
$$= \sum_i P_i \log B_i \sum_j p_{ij} = \sum_i P_i \log B_i \tag{21}$$

since

$$\sum_j p_{ij} = 1 \tag{22}$$

Now the right side of Eq. (21) is identically equal to the right side of Eq. (19) since they both represent summations of the same terms. Thus the first two terms on the right side of Eq. (18) cancel, and Eq. (16) reduces to

$$R = \frac{\log W \sum_i \{P_i \sum_j [\sum_s p_{ij}^{(s)} b_{ij}^{(s)}]\}}{\sum_i \{P_i \sum_j [\sum_s p_{ij}^{(s)} b_{ij}^{(s)}]\}} = \log W \tag{23}$$

In Appendix I, we showed that $\log W$ is equal to the code capacity. It therefore follows from Eq. (23) and the discussion in Sec. 1.7 that the values of $p_{ij}^{(s)}$ given by Eq. (10) will cause no appreciable decrease in the number of different possible sequences of time duration T from the value determined by the fixed constraints alone. This represents a type of "match" between the probability constraints of the language and the fixed constraints of the code channel.

The fundamental mathematical situation which makes such a match possible may be seen in Table III of Chap. I, as an example. The total of column 3 shows that there are 1,048,576 possible sequences of 20 symbols duration if the alphabet is limited to A's and B's. Column 3 itself shows that the distribution is sharply peaked around the range having approximately equal numbers of A's and B's. The peaking would actually become very much sharper if the sequence were longer. Therefore, if the probability constraints called for approximately equal numbers of A's and B's, only a negligible number of the possible sequences would be ruled out by the probability constraints, if very long sequences were considered. On the other hand, if the probability constraints called for $\frac{3}{4}$ A's and $\frac{1}{4}$ B's, the allowed (ergodic) sequences would constitute only a small fraction of the possible sequences, and this fraction would decrease exponentially as the sequence considered became longer and longer. The condition of equal percentages of A's and B's in this case corresponds to the probability constraints being matched to the fixed constraints and thus not causing R to be less than $\log W$. The condition of $\frac{3}{4}$ A's and $\frac{1}{4}$ B's corresponds to a mismatched case. In the latter case the probability constraints cause R to be less than $\log W$.

As a matter of interest, let us show that Eq. (10) gives the proper matching condition in the case depicted in column 3 of Table III. In this example

$$N(T) = 2^T \tag{24}$$

We therefore see by inspection that

$$W = 2 \tag{25}$$

without the necessity of solving Eq. (13). There is only one state, since there are no disallowed symbols anywhere in a sequence. There are two symbols, A and B. We therefore have only two probabilities to determine, namely,

$$p_{11}^{(\mathrm{A})} = \frac{B_1}{B_1} W^{-b_{11}^{(\mathrm{A})}} \tag{26}$$

and

$$p_{11}^{(\mathrm{B})} = \frac{B_1}{B_1} W^{-b_{11}^{(\mathrm{B})}} \tag{27}$$

In this case, we note also that

$$b_{11}^{(\mathrm{A})} = 1 = b_{11}^{(\mathrm{B})} \tag{28}$$

Therefore, the conditions for match are

$$p_{11}^{(\mathrm{A})} = W^{-1} = \tfrac{1}{2} \tag{29}$$

$$p_{11}^{(\mathrm{B})} = W^{-1} = \tfrac{1}{2} \tag{30}$$

This checks the values already found.

Let us consider a little further some of the practical aspects of the matching we have been talking about. Ordinarily we are supplied with a language or with messages of given probabilities, and we are also supplied with equipment which will supply a code channel having a certain code capacity. This code capacity determines the maximum average rate at which language information can be sent through the equipment. The practical encoding problem with which we are faced is to determine the best way of encoding the language symbols or messages into symbol groups of the code channel. If time delays in interpreting the received messages were of no consequence, a complete solution would be possible, in theory at least. Thus we know from Theorem A of Sec. 1.7 that it is possible to encode all the combinations of symbol groups allowed by the probability constraints of the language into (long) symbol groups of the code channel in such a way as to get a language transmission capacity equal to the code capacity of the code channel alone. Such encoding, however, besides causing delays would also create great complication in the encoding and decoding processes. It is, therefore, usually necessary in practice to encode the language symbols (or at least messages) individually. The best we can say from Eq. (10) is then:

(a) The more frequently a letter or letter combination is used in a language, the shorter should be the code symbol group which is equivalent to it.

(b) Within the limits imposed by statement (a), each code symbol group which is used for a letter or combination of letter should be as short as possible. Thus no short code symbol groups should be unused.

The reason for statement (a) is obvious from the form of Eq. (10). The reason for statement (b) is to prevent any short code symbol groups which are possible in the code channel from not having a language equivalent and thereby having zero probability.

Equation (10) is actually of more importance in dealing with certain theoretical problems in information theory than it is in the solution of practical problems.

APPENDIX IV

Alphabets, Constraints, Choice, and Order

The Expansion of Alphabets

In Appendix I we showed that there are L^n different possible sequences n symbols long if the alphabet consists of L symbols. For example, as mentioned in Sec. 1.4, there are 10^6 different six-digit numbers from 000,000 to 999,999. In this case there are 10 symbols in the alphabet, namely, 0, 1, 2, 3, 4, 5, 6, 7, 8, and 9.

It is sometimes of value when analyzing communication problems to consider the symbols in groups, and thereby to expand the alphabet. Thus in the above example, we can consider the alphabet to consist of the 100 two-digit symbols between 00 and 99. There are then

$$(100)^3 = 10^6$$

different sequences having a duration of three of these double symbols, and these are, of course, exactly the same sequences as mentioned in the first case.

If there is no intersymbol influence, and all symbols are equally probable, the amount of language information in each sequence in the single-symbol sequences is (in the noiseless case)

$$\text{language information/sequence} = \log \frac{1}{(1/L^n)} = \log L^n = n \log L \quad (1)$$

since there are L^n sequences. In other words there are $\log L$ binits per symbol of language information in the sequence. If, on the other hand, these sequences are considered on the basis of the double-symbol alphabet, there are

$$(L^2)^{n/2} = L^n \quad (2)$$

different sequences, each of probability $1/(L^2)^{n/2}$. The language information per sequence is

$$\text{language information/sequence} = \log \frac{1}{1/(L^2)^{n/2}} = \frac{n}{2} \log L^2 \quad (3)$$

In other words there are

$$\log L^2 = 2 \log L \quad (4)$$

binits per double symbol in the sequence. Fundamentally, of course, this is just what we already found in the single-symbol analysis.

One of the cases in which it is of value to use a multiple-symbol analysis occurs where there is intersymbol influence of finite range. In this case it is possible to express the intersymbol influence in terms of the probability of occurrence of multiple-symbol sequences. Thus if the intersymbol influence can be expressed in terms of the probabilities of four-symbol groups in the above digit alphabet case, we could express the intersymbol influence by assigning probabilities to the 10,000 different four-symbol groups between 0000 and 9999. Those eliminated by the fixed constraints would have zero probability. The probabilities of the others would be determined by the probability constraints. The foregoing situation is, however, simpler than that usually met in practice. In practical cases, the intersymbol influence is not confined to isolated small symbol groups such as the four-symbol groups considered above. Nevertheless, we can for theoretical purposes simplify the actual case by an approximation method. If the intersymbol influence can be considered to be of finite range, say less than a duration of q symbols, then the intersymbol influence may be expressed by assigning probabilities to the

$$L^q = s \quad \text{(say)} \tag{5}$$

different sequences having a duration of q symbols. We may consider it a matter of the definition of the finite range of intersymbol influence that there is a finite number q such that the intersymbol influences can be expressed to any required degree of approximation by assigning probabilities to all sequences q symbols long.

Let us now consider a very long sequence m symbols long, made up of

$$\frac{m}{q} = r \quad \text{(say)} \tag{6}$$

of these groups of q symbols. Then if the probabilities of these L^q different symbol groups are $p_1, p_2, \ldots, p_s$, respectively, the probability of the long sequence is, by the law of large numbers,

$$P = (p_1)^{p_1 r}(p_2)^{p_2 r} \ldots (p_s)^{p_s r} \tag{7}$$

if the sequence is ergodic, and this probability is the same for all ergodic sequences.

If the sequence is not ergodic, its probability is vanishingly small and may be, considered zero. If $\bar{M}$ is the number of ergodic sequences which are possible, then

$$\bar{M} = \frac{1}{P} \tag{8}$$

In Appendix II we showed that if P is of the form given by Eq. (7)

$$\log \overline{M} = -\log P = -r \sum_i p_i \log p_i + \text{a constant} \tag{9}$$

The so-called "constant" in Eq. (9) is not actually independent of r but varies so slowly that it may be considered a constant in calculation. Thus

$$\overline{M} = A2^{-r \sum_i p_i \log p_i} \tag{10}$$

Equation (10) as here developed shows that for long sequences the number of possible sequences rises exponentially with the length of the sequence even when both fixed and probability constraints are present, provided that all constraints are of finite range.

The number $s = L^q$ is, for all practical purposes, the number of "symbols" in a large alphabet. The probabilities of the symbols in this alphabet express all the effects of the constraints, both of the fixed and probability types. The number $\overline{M}$ in turn is the number of symbols in a gigantic alphabet. In this latter alphabet all symbols are of equal probability. The elimination of certain possible symbols, namely, the non-ergodic sequences from this alphabet has taken care of all the constraints. From one point of view we can therefore say that the analysis of systems with constraints can be simplified by the expansion of alphabets.

Choice and Order

In any of the ergodic sequences m symbols long, considered in the preceding paragraphs, the total language information is $-\log P$, and is thus

$$-r \sum_i p_i \log p_i \quad \text{approximately}$$

Therefore

$$\text{language information/symbol} = \frac{-r}{m} \sum_i p_i \log p_i = \frac{-1}{q} \sum_i p_i \log p_i \tag{11}$$

From Eqs. (8), (9), and (11), it is clear that the amount of language information (*i.e.*, the amount of entropy) in an ergodic sequence is equal to the logarithm of the number of different equally probable ergodic sequences of the same duration. The amount of language information (*i.e.*, entropy) in the sequence is thus a measure of the *choice* that was available when the sequence was selected.

When we say that the language information in a sequence is (say) I binits/symbol, we are assuming that all possible use is being made of the order of occurrence of the symbols in the sequence. If, however, only the symbols' occurrence is important and not their order, then the language information per symbol is greatly reduced. Thus, for example if 813, 831,

138, 183, 318, and 381 all have the same meaning, and likewise all other numbers whose digits are the same are assigned the same meaning, then the number of different three-digit sequences between 000 and 999 is about 200 instead of being 1000. As the sequences become longer, the percentage of used sequences would decline rapidly. We thus conclude that the values for language information per symbol in the various formulas we have derived throughout this book assume that advantage is taken of all possible permutations of the order of the symbols in setting up codes.

In the actual transmission of practical messages the order of symbols in a sequence is significant for only a very limited range. Thus the statement "John is 69 inches tall and weighs 150 pounds" usually has the same significance as "John weighs 150 pounds and is 69 inches tall." The human brain is apparently accustomed to deal fully with only relatively short sequences at one time and does not (perhaps cannot) take full advantage of the language transmission capacities of systems for long sequences. Of course, it should also be remembered that long sequences, if fully utilized and not protected by planned redundancy, are very susceptible to great harm by small amounts of noise. This is undoubtedly part of the reason why long sequences are not fully utilized in human communication systems.

Entropy and Choice vs. Probability in Information Theory

Throughout the present text, we have used the formula

$$\text{information received} = \log \left[\frac{\text{probability at the receiver of the event after the message is received}}{\text{probability at the receiver of the event before the message is received}} \right]$$

$$= \log \left(\frac{p'}{p} \right) \quad \text{say} \tag{12}$$

as the fundamental definition of the information transmitted to a receiver by a message. In the "noiseless" case, *i.e.*, the case in which the probability after the message is received (p') is unity, Eq. (12) reduces to

$$\text{information received} = -\log p \tag{13}$$

With the definition of information given in Eqs. (12) and (13), each individual message transmits an amount of information to the receiver which depends upon the probability of the event which it describes; and in case the message is erroneous, the amount of information transmitted may actually be negative.

A fundamentally different definition of information is used by Shannon and others. Their definition of information is based upon entropy, and applies to average conditions in an ergodic sequence without specifying the information associated with individual messages. Thus in the noiseless case, the entropy of a long sample of n symbols duration of an ergodic sequence is determined in the form

$$H = H(x_1,x_2, \ . \ . \ . \ ,x_n) \tag{14}$$

and the rate R of transmission of information is taken to be

$$R = \frac{H}{T} \tag{15}$$

where T is the time duration of the n symbols.

According to the discussion given early in this appendix, the definition of information rate given by Eq. (15) may also be written as

$$R = \frac{\log M}{T} \tag{16}$$

where M is the number of different possible typical ergodic sequences of duration T. Therefore R, and more specifically the product RT, is a measure of available choice.

For a long ergodic sequence, R as given by Eqs. (15) and (16) is also equal to the average rate at which information is received, where information is defined as in Eq. (13). Thus for the discussion of average conditions, the two points of views concerning information rates are equivalent. As far as the design of systems or equipment is concerned, there is no practical difference between the two points of view, since systems and equipment are designed for average conditions. When, however, it comes to details, the point of view of Eq. (13) considers the information rate to fluctuate in time, being low for commonplace messages of high initial probability and being high for the unusual messages of low initial probability. The point of view of Eqs. (15) and (16) ignores these fluctuations.

APPENDIX V

The Quadrature Function and Hilbert Transforms

In Sec. 2.3, we introduced the quadrature function

$$G_1(t) = \frac{1}{\pi}\int_0^\infty b(\omega)\cos\omega t\,d\omega - \frac{1}{\pi}\int_0^\infty a(\omega)\sin\omega t\,d\omega \tag{1}$$

corresponding to the function

$$G(t) = \frac{1}{\pi}\int_0^\infty a(\omega)\cos\omega t\,d\omega + \frac{1}{\pi}\int_0^\infty b(\omega)\sin\omega t\,d\omega \tag{2}$$

We now wish, first of all, to show that a function $G(t)$ and its quadrature function $G_1(t)$ are what the mathematicians call Hilbert transforms.[1,2]

By definition, if

$$u(t) = \frac{1}{\pi}\int_{-\infty}^{+\infty}\frac{u_1(\tau)}{t-\tau}\,d\tau \tag{3}$$

and

$$u_1(t) = \frac{-1}{\pi}\int_{-\infty}^{+\infty}\frac{u(\tau)}{t-\tau}\,d\tau \tag{4}$$

then $u(t)$ and $u_1(t)$ are called Hilbert transforms. The principal value is used for the improper integral. The transformations (3) and (4) are inverse to each other, since the application of one of the transformations after the other brings back the original function.

Next, introduce the Fourier transforms of $u(t)$ and $u_1(t)$, namely,

$$\Omega(\omega) = \int_{-\infty}^{+\infty} u(t)\epsilon^{-j\omega t}dt \tag{5}$$

and

$$\Omega_1(\omega) = \int_{-\infty}^{+\infty} u_1(t)\epsilon^{-j\omega t}dt \tag{6}$$

If it can be shown that

$$\Omega_1(\omega) = \begin{cases} +j\Omega(\omega) & \text{if } \omega > 0 \\ -j\Omega(\omega) & \text{if } \omega < 0 \end{cases} \tag{7}$$

[1] For a discussion of Hilbert transforms, see Guillemin, *The Mathematics of Circuit Analysis*, pp. 338–349; and E. C. Titchmarsh, *Introduction to the Theory of Fourier Integrals*, Oxford, New York (1937), pp. 119 *ff.*

[2] C. R. Cahn is responsible for the proof of this equivalence which is given here.

then the relation of $u(t)$ and $u_1(t)$ will be the same as that of $G(t)$ and $G_1(t)$ in Eqs. (1) and (2). First

$$\Omega_1(\omega) = \int_{-\infty}^{+\infty} u_1(t)\epsilon^{-j\omega t}dt = -\frac{1}{\pi}\int_{-\infty}^{+\infty}\left[\int_{-\infty}^{+\infty}\frac{u(\tau)}{t-\tau}\,d\tau\right]e^{-j\omega t}dt$$

$$= \lim_{\delta\to 0}\left\{-\frac{1}{\pi}\left[\int_{-\infty}^{+\infty}\int_{\tau=-\infty}^{\tau=t-\delta}\frac{u(\tau)}{t-\tau}\,\epsilon^{-j\omega t}\,d\tau\,dt + \int_{-\infty}^{+\infty}\int_{\tau=t+\delta}^{\tau=+\infty}\frac{u(\tau)}{t-\tau}\,\epsilon^{-j\omega t}\,d\tau\,dt\right]\right\} \quad (8)$$

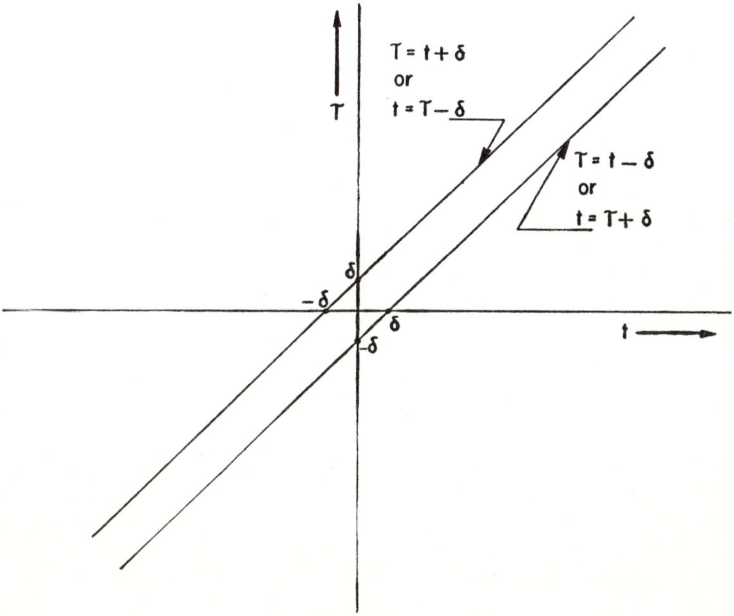

Fig. A-2.

Next reverse the order of integration. The corresponding changes in limits of integration are obtained by reference to Fig. A-2. Thus Eq. (8) becomes

$$\Omega_1(\omega) = \lim_{\delta\to 0}\left\{-\frac{1}{\pi}\int_{\tau=-\infty}^{\tau=+\infty} u(\tau)\left[\int_{t=\tau+\delta}^{t=+\infty}\frac{\epsilon^{-j\omega t}}{t-\tau}\,dt + \int_{t=-\infty}^{t=\tau-\delta}\frac{\epsilon^{-j\omega t}}{t-\tau}\,dt\right]d\tau\right\} \quad (9)$$

$$= \lim_{\delta\to 0}\left\{-\frac{1}{\pi}\int_{\tau=-\infty}^{\tau=+\infty} u(\tau)\left[\int_{t=\tau+\delta}^{t=+\infty}\frac{\epsilon^{-j\omega(t-\tau)}}{t-\tau}\,dt + \int_{t=-\infty}^{t=\tau-\delta}\frac{\epsilon^{-j\omega(t-\tau)}}{t-\tau}\,dt\right]\epsilon^{-j\omega\tau}d\tau\right\}$$

Now

$$\lim_{\delta\to 0}\int_{t=-\infty}^{t=\tau-\delta}\frac{\epsilon^{-j\omega(t-\tau)}}{t-\tau}\,dt = \lim_{\delta\to 0}\int_{v=-\omega(t-\tau)=+\omega\infty}^{v=-\omega(t-\tau)=\omega\delta}\frac{\epsilon^{jv}}{v}\,dv$$
$$= \lim_{\delta\to 0} -\int_{\omega\delta}^{+\omega\infty}\frac{\epsilon^{jv}}{v}\,dv \quad (10)$$

and

$$\lim_{\delta\to 0}\int_{t=\tau+\delta}^{t=+\infty}\frac{\epsilon^{-j\omega(t-\tau)}}{t-\tau}\,dt = \lim_{\delta\to 0}\int_{v=\omega(t-\tau)=\omega\delta}^{v=\omega(t-\tau)=+\omega\infty}\frac{\epsilon^{-jv}}{v}\,dv$$
$$= \lim_{\delta\to 0}\int_{\omega\delta}^{+\omega\infty}\frac{\epsilon^{-jv}}{v}\,dv \quad (11)$$

Therefore, Eq. (9) becomes

$$\Omega_1(\omega) = \lim_{\delta\to 0}\left\{-\frac{1}{\pi}\int_{\tau=-\infty}^{+\infty} u(\tau)\epsilon^{-j\omega\tau}\left[\int_{\omega\delta}^{+\omega\infty}\frac{\epsilon^{-jv}-\epsilon^{jv}}{v}\,dv\right]d\tau\right\}$$
$$= \frac{2j}{\pi}\int_{\tau=-\infty}^{\tau=+\infty} u(\tau)\epsilon^{-j\omega\tau}\left[\lim_{\delta\to 0}\int_{\omega\delta}^{+\omega\infty}\frac{\sin v}{v}\,dv\right]d\tau$$
$$= \begin{cases} j\Omega(\omega) & \text{if } \omega > 0 \\ -j\Omega(\omega) & \text{if } \omega < 0 \end{cases} \quad (12)$$

since

$$\int_0^{+\infty}\frac{\sin x}{x}\,dx = \frac{\pi}{2} \quad (13)$$

and

$$\int_0^{-\infty}\frac{\sin x}{x}\,dx = -\frac{\pi}{2} \quad (14)$$

If now we let $u(t)$ in Eq. (4) be $G(t)$ of Eq. (2), then it follows from Eqs. (1), (2), (5), (6), and (12) that $u_1(t)$ in Eq. (4) will be identical with $G_1(t)$ in Eq. (1). The quadrature function $G_1(t)$ is thus the Hilbert transform of $G(t)$.

We next wish to give some examples of important functions and their quadrature functions.

(a) First of all, we have

$$G(t) = A\sin\omega t \quad (15)$$

Then, by Eq. (4), the quadrature function is

$$G_1(t) = -\frac{1}{\pi}\int_{-\infty}^{+\infty}\frac{A\sin\omega\tau}{t-\tau}\,d\tau = \frac{-A}{\pi}\int_{-\infty}^{+\infty}\frac{-\sin[\omega(t-\tau)-\omega t]}{t-\tau}\,d\tau$$
$$= +\frac{A}{\pi}\int_{-\infty}^{+\infty}\frac{\sin[\omega(t-\tau)]\cos\omega t}{t-\tau}\,d(t-\tau)$$
$$-\frac{A}{\pi}\int_{-\infty}^{+\infty}\frac{\cos[\omega(t-\tau)]\sin\omega t}{t-\tau}\,d(t-\tau)$$
$$= \frac{A}{\pi}\cos\omega t\int_{-\infty}^{+\infty}\frac{\sin x}{x}\,dx - \frac{A}{\pi}\sin\omega t\int_{-\infty}^{+\infty}\frac{\cos x}{x}\,dx \quad (16)$$

Now

$$\int_{-\infty}^{+\infty}\frac{\sin x}{x}\,dx = \pi \quad (17)$$

and, as far as the principal part is concerned,

$$\int_{-\infty}^{+\infty} \frac{\cos x}{x}\, dx = 0 \tag{18}$$

Thus Eq. (16) reduces to

$$G_1(t) = A \cos \omega t \tag{19}$$

as was to be expected.

(b) Next consider the sampling function

$$G(t) = \frac{\sin \omega_0 t}{\omega_0 t} \tag{20}$$

In this case

$$\begin{aligned} a(\omega) &= \int_{-\infty}^{+\infty} \frac{\sin \omega_0 t \cos \omega t}{\omega_0 t}\, dt \\ &= \frac{1}{2}\int_{-\infty}^{+\infty} \frac{\sin (\omega + \omega_0)t}{\omega_0 t}\, dt - \frac{1}{2}\int_{-\infty}^{+\infty} \frac{\sin (\omega - \omega_0)t}{\omega_0 t}\, dt \\ &= \frac{1}{2\omega_0}\int_{-(\omega+\omega_0)\infty}^{+(\omega+\omega_0)\infty} \frac{\sin x}{x}\, dx - \frac{1}{2\omega_0}\int_{-(\omega-\omega_0)\infty}^{+(\omega-\omega_0)\infty} \frac{\sin x}{x}\, dx \end{aligned} \tag{21}$$

Now

$$\int_{-(\omega+\omega_0)\infty}^{+(\omega+\omega_0)\infty} \frac{\sin x}{x}\, dx = \begin{cases} \displaystyle\int_{-\infty}^{+\infty} \frac{\sin x}{x}\, dx = \pi & \text{if } \omega > -\omega_0 \\ \displaystyle\int_{+\infty}^{-\infty} \frac{\sin x}{x}\, dx = -\pi & \text{if } \omega < -\omega_0 \end{cases} \tag{22}$$

while

$$\int_{-(\omega-\omega_0)\infty}^{+(\omega-\omega_0)\infty} \frac{\sin x}{x}\, dx = \begin{cases} \displaystyle\int_{-\infty}^{+\infty} \frac{\sin x}{x}\, dx = \pi & \text{if } \omega > +\omega_0 \\ \displaystyle\int_{+\infty}^{-\infty} \frac{\sin x}{x}\, dx = -\pi & \text{if } \omega < +\omega_0 \end{cases} \tag{23}$$

Therefore

$$a(\omega) = \begin{cases} \dfrac{\pi}{\omega_0} & \text{if } (-\omega_0 < \omega < \omega_0) \\ 0 & \text{elsewhere} \end{cases} \tag{24}$$

Furthermore

$$b(\omega) = 0 \tag{25}$$

since $G(t)$ is an even function in Eq. (20). Consequently, by Eq. (1)

$$\begin{aligned} G_1(t) &= -\frac{1}{\pi}\int_0^{\omega_0} \frac{\pi}{\omega_0} \sin \omega t\, d\omega \\ &= +\frac{1}{\omega_0 t}(\cos \omega_0 t - 1) = -\frac{1}{\left(\frac{\omega_0 t}{2}\right)} \sin^2\left(\frac{\omega_0}{2} t\right) \end{aligned} \tag{26}$$

(c) Next let $G(t)$ be the rectangular pulse

$$G(t) = \begin{cases} 1 & -\frac{T}{2} < t < +\frac{T}{2} \\ 0 & \text{elsewhere} \end{cases} \tag{27}$$

Then

$$a(\omega) = \int_{-T/2}^{+T/2} \cos \omega t \, dt = \frac{1}{\omega}\left[\sin \frac{\omega T}{2} - \sin\left(-\frac{\omega T}{2}\right)\right] = \frac{2}{\omega} \sin \frac{\omega T}{2} \tag{28}$$

Also

$$b(\omega) = 0 \tag{29}$$

since $G(t)$ in Eq. (27) is an even function of t. Thus, by Eq. (1)

$$\begin{aligned} G_1(t) &= -\frac{1}{\pi}\int_0^\infty \frac{2}{\omega} \sin\left(\frac{\omega T}{2}\right) \sin \omega t \, d\omega \\ &= \frac{1}{\pi}\int_0^\infty \frac{\cos\left[\omega\left(t+\frac{T}{2}\right)\right] - \cos\left[\omega\left(t-\frac{T}{2}\right)\right]}{\omega} d\omega \\ &= \frac{1}{\pi}\int_0^\infty \frac{\cos\left[\omega\left(t+\frac{T}{2}\right)\right]}{\omega} d\omega - \frac{1}{\pi}\int_0^\infty \frac{\cos\left[\omega\left(t-\frac{T}{2}\right)\right]}{\omega} d\omega \end{aligned} \tag{30}$$

Equation (30) is an indeterminate form of the type $\infty - \infty$. In order to evaluate it, let us write

$$\begin{aligned} G_1(t) &= \lim_{x\to 0} \left\{\frac{1}{\pi}\int_x^\infty \frac{\cos\left[\omega\left(t+\frac{T}{2}\right)\right]}{\omega} d\omega - \frac{1}{\pi}\int_x^\infty \frac{\cos\left[\omega\left(t-\frac{T}{2}\right)\right]}{\omega} d\omega\right\} \\ &= \lim_{x\to 0} \left\{\frac{1}{\pi}\int_{x\left(t+\frac{T}{2}\right)}^{\infty\left(t+\frac{T}{2}\right)} \frac{\cos y}{y} dy - \frac{1}{\pi}\int_{x\left(t-\frac{T}{2}\right)}^{\infty\left(t-\frac{T}{2}\right)} \frac{\cos y}{y} dy\right\} \end{aligned} \tag{31}$$

We next introduce the cosine integral function defined by the equation

$$Ci(y) = -\int_y^\infty \frac{\cos u}{u} du \tag{32}$$

for positive values of y. The series expansion of $Ci(y)$ is well known to be

$$Ci(y) = \gamma + \log_e (y) - \frac{y^2}{2.2!} + \frac{y^4}{4.4!} \cdots \tag{33}$$

where

$$\gamma = 0.5772 \ldots \tag{34}$$

is Euler's constant. We furthermore note that if A is a positive quantity

$$\int_{-A}^{-\infty} \frac{\cos u}{u}\,du = \int_{A}^{+\infty} \frac{\cos u}{u}\,du \tag{35}$$

Consequently, Eq. (31) reduces to

$$G_1(t) = \lim_{x\to 0} \left\{ \frac{1}{\pi} Ci \left| x\left(t - \frac{T}{2}\right) \right| - \frac{1}{\pi} Ci \left| x\left(t + \frac{T}{2}\right) \right| \right\} \tag{36}$$

where the vertical bars stand for absolute-value signs. With the aid of Eq. (33), after noting which terms cancel or vanish as $x \to 0$, we have

$$\begin{aligned} G_1(t) &= \lim_{x\to 0} \left\{ \frac{1}{\pi} \log_\epsilon \left[\left| x\left(t - \frac{T}{2}\right) \right| \right] - \frac{1}{\pi} \log_\epsilon \left[\left| x\left(t + \frac{T}{2}\right) \right| \right] \right\} \\ &= \frac{1}{\pi} \log_\epsilon \left| \frac{t - \frac{T}{2}}{t + \frac{T}{2}} \right| \end{aligned} \tag{37}$$

(d) As an alternative and simpler method of solving for $G_1(t)$ in example (c), let us find $G_1(t)$ by using Eq. (4). Thus

$$G_1(t) = -\frac{1}{\pi} \int_{-T/2}^{+T/2} \frac{d\tau}{t - \tau} \tag{38}$$

Three cases must be considered, since the argument of a logarithm must be positive.

If t is greater than $T/2$

$$G_1(t) = +\frac{1}{\pi} \log_\epsilon (t - \tau) \Big]_{-T/2}^{T/2} = +\frac{1}{\pi} \log_\epsilon \frac{t - \frac{T}{2}}{t + \frac{T}{2}} \tag{39}$$

If t is in the range $-\frac{T}{2} < t < \frac{T}{2}$

$$\begin{aligned} G_1(t) &= \lim_{\delta\to 0} -\frac{1}{\pi} \left[\int_{-T/2}^{t-\delta} \frac{d\tau}{t - \tau} - \int_{t+\delta}^{T/2} \frac{d\tau}{\tau - t} \right] \\ &= \lim_{\delta\to 0} \frac{1}{\pi} \left[\log_\epsilon (t - \tau) \Big|_{-T/2}^{t-\delta} + \log_\epsilon (\tau - t) \Big|_{t+\delta}^{T/2} \right] \\ &= \lim_{\delta\to 0} \frac{1}{\pi} \left[\log_\epsilon \delta - \log_\epsilon \left(t + \frac{T}{2}\right) + \log_\epsilon \left(\frac{T}{2} - t\right) - \log_\epsilon \delta \right] \\ &= \frac{1}{\pi} \log_\epsilon \left(\frac{\frac{T}{2} - t}{t + \frac{T}{2}} \right) \end{aligned} \tag{40}$$

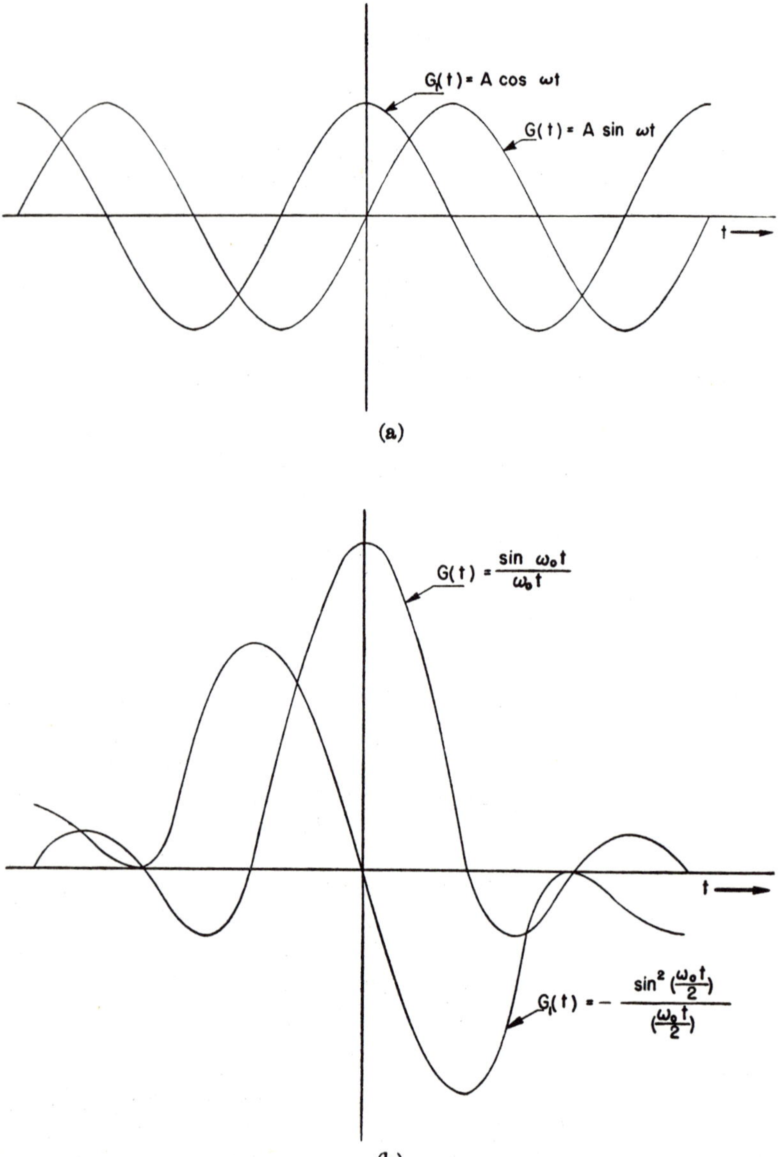

Fig. A-3. (a) $A \sin \omega t$ and its Hilbert transform, $A \cos \omega t$. (b) The sampling function $(\sin \omega_0 t/\omega_0 t)$ and its Hilbert transform, $(\omega_0 t/2) \sin^2 (\omega_0 t/2)$.

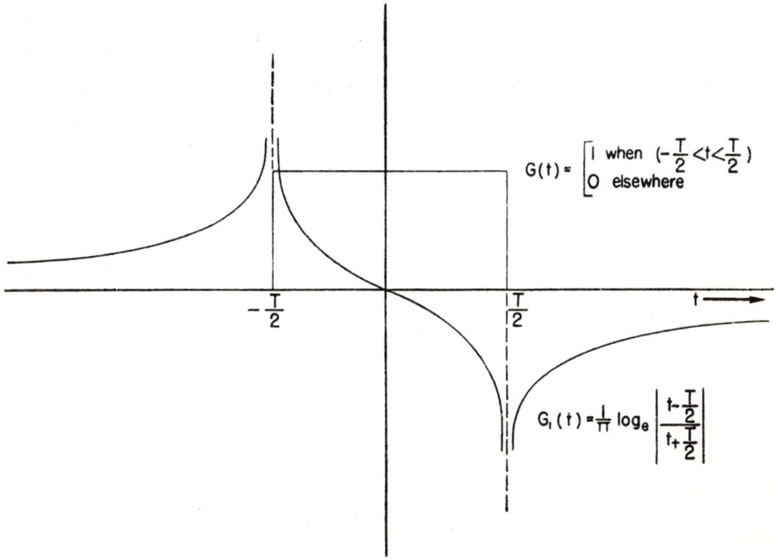

Fig. A-3. (c) The rectangular pulse and its Hilbert transform.

If $t < -T/2$

$$G_1(t) = \frac{1}{\pi}\int_{-T/2}^{T/2}\frac{d\tau}{\tau - t} = \frac{1}{\pi}\log_\epsilon(\tau - t)\Big]_{-T/2}^{T/2}$$

$$= \frac{1}{\pi}\log_\epsilon\left(\frac{\dfrac{T}{2} - t}{-\dfrac{T}{2} - t}\right) = \frac{1}{\pi}\log_\epsilon\left(\frac{t - \dfrac{T}{2}}{t + \dfrac{T}{2}}\right) \tag{41}$$

Consequently, we see that the general expression

$$G_1(t) = \frac{1}{\pi}\log_\epsilon\left|\frac{t - \dfrac{T}{2}}{t + \dfrac{T}{2}}\right| \tag{42}$$

is always true, in agreement with Eq. (37).

In Fig. A-3 are shown graphs of $G(t)$ and $G_1(t)$ for examples (a), (b), and (c). On page 94 of *Frequency Analysis, Modulation, and Noise* is a figure showing a set of quadrature functions which are important in studying single-sideband transmission in television. In that figure, the $Q(x)$ functions are the quadrature functions of the $P(x)$ functions.

PROBLEMS

1. (a) If $F_1(t)$ is the Hilbert transform of $F(t)$ and $G_1(t)$ is the Hilbert transform of $G(t)$, show that the Hilbert transform of $[F(t) + G(t)]$ is $F_1(t) + G_1(t)$.

(b) If $G_1(t)$ is the Hilbert transform of $G(t)$, show that $G_1(t - T)$ is the Hilbert transform of $G(t - T)$.

2. If $G(t)$ is random noise, show that its Hilbert transform $G_1(t)$ is also random noise.

3. Find the Hilbert transform of

$$G(t) = \frac{\sin (\omega_0 t + \theta)}{\omega_0 t + \theta}$$

APPENDIX VI

Proof That the Transformation between Time and Frequency Sampling Values Can Be Made an Orthogonal Measure Preserving Transformation[1]

Equation (58) of Chap. II gives us the equations

$$a_0 = \frac{1}{2W} \sum_{n=-TW}^{+TW} G\left(\frac{n}{2W}\right) \tag{1}$$

$$a\left(\frac{2\pi m}{T}\right) = \frac{1}{2W} \sum_{n=-TW}^{+TW} G\left(\frac{n}{2W}\right) \cos\frac{\pi mn}{WT} \tag{2}$$

$$b\left(\frac{2\pi m}{T}\right) = \frac{1}{2W} \sum_{n=-TW}^{+TW} G\left(\frac{n}{2W}\right) \sin\frac{\pi mn}{WT} \tag{3}$$

as a transformation between time and frequency sampling values.[2]

In order for the transformation to be orthogonal, it is necessary that it shall be measure preserving. Reference to Eq. (52) of Chap. II shows that the transformation can be made measure preserving by replacing a_0, a, and b with the new variables a_0', a', and b' defined by the equations

$$a_0' = \sqrt{\frac{2W}{T}}\, a_0 \tag{4}$$

$$a' = 2\sqrt{\frac{W}{T}}\, a \tag{5}$$

$$b' = 2\sqrt{\frac{W}{T}}\, b \tag{6}$$

Equation (52) of Chap. II can now be rewritten as

$$a_0'^2 + \sum_{n=1}^{TW} (a_n'^2 + b_n'^2) = \sum_{n=-TW}^{+TW} \left[G\left(\frac{n}{2W}\right)\right]^2 \tag{7}$$

[1] The proof given here is due to C. R. Cahn.

[2] The inverse of this transformation is Eq. (59) of Chap. II.

and Eqs. (1), (2), and (3) of this appendix become

$$a'_0 = \sqrt{\frac{1}{2TW}} \sum_{n=-TW}^{+TW} G\left(\frac{n}{2W}\right) \tag{8}$$

$$a'\left(\frac{2\pi m}{T}\right) = \sqrt{\frac{1}{TW}} \sum_{n=-TW}^{+TW} G\left(\frac{n}{2W}\right) \cos\frac{\pi mn}{WT} \tag{9}$$

and

$$b'\left(\frac{2\pi m}{T}\right) = \sqrt{\frac{1}{TW}} \sum_{n=-TW}^{+TW} G\left(\frac{n}{2W}\right) \sin\frac{\pi mn}{WT} \tag{10}$$

If a linear transformation is written in the form

$$\left.\begin{aligned} y_1 &= O_{11}x_1 + O_{12}x_2 + \cdots + O_{1n}x_n \\ y_2 &= O_{21}x_1 + O_{22}x_2 + \cdots + O_{2n}x_n \\ &\cdots\cdots\cdots\cdots\cdots\cdots \\ &\cdots\cdots\cdots\cdots\cdots\cdots \\ &\cdots\cdots\cdots\cdots\cdots\cdots \\ y_n &= O_{n1}x_1 + O_{n2}x_2 + \cdots + O_{nn}x_n \end{aligned}\right\} \tag{11}$$

the conditions for orthogonality are[1]

$$O_{r1}O_{s1} + O_{r2}O_{s2} + \cdots + O_{rn}O_{sn} = \begin{cases} 1 & \text{if } r = s \\ 0 & \text{if } r \neq s \end{cases} \tag{12a}$$

and

$$O_{1s}O_{1r} + O_{2s}O_{2r} + \cdots + O_{ns}O_{nr} = \begin{cases} 1 & \text{if } s = r \\ 0 & \text{if } s \neq r \end{cases} \tag{12b}$$

Actually, if either Eq. (12a) or (12b) is true, the other may be shown to follow from it. We will now show that Eqs. (8), (9), and (10) represent an orthogonal transformation. We first note that if $y_p = a'_0$, then

$$O_{p1} = O_{p2} = \cdots = O_{pn} = \frac{1}{\sqrt{2TW}} \tag{13}$$

and

$$O_{p1}^2 + O_{p2}^2 + \cdots + O_{pn}^2 = \frac{2TW+1}{2TW} = 1 \quad \text{(approx.)} \tag{14}$$

[1] See Guillemin *The Mathematics of Circuit Analysis*, pp. 81–85.

For the other y's

$$\cos\frac{\pi m_1 n}{WT}\cdot\cos\frac{\pi m_2 n}{WT} = \frac{1}{2}\left\{\cos\left[\frac{\pi n}{WT}(m_1 - m_2)\right] + \cos\left[\frac{\pi n}{WT}(m_1 + m_2)\right]\right\} \tag{15}$$

If $m_1 = m_2 = m$, this reduces to

$$\cos^2\left(\frac{\pi mn}{WT}\right) = \frac{1}{2}\left[1 + \cos\frac{2\pi mn}{WT}\right] \tag{16}$$

Also

$$\sin\frac{\pi m_1 n}{WT}\cdot\sin\frac{\pi m_2 n}{WT} = \frac{1}{2}\left\{\cos\left[\frac{\pi n}{WT}(m_1 - m_2)\right] - \cos\left[\frac{\pi n}{WT}(m_1 + m_2)\right]\right\} \tag{17}$$

If $m_1 = m_2 = m$, this reduces to

$$\sin^2\left(\frac{\pi mn}{WT}\right) = \frac{1}{2}\left[1 - \cos\frac{2\pi mn}{WT}\right] \tag{18}$$

Next if we consider Eqs. (8), (9), and (10) in the form of Eq. (11), and if we substitute into the left side of Equation (12a), we have as $TW \to \infty$

$$\frac{1}{TW}\sum_{n=-TW}^{+TW}\cos^2\left(\frac{\pi mn}{WT}\right) = \frac{1}{TW}\sum_{n=-TW}^{+TW}\frac{1}{2}\left[1 + \cos\frac{2\pi mn}{WT}\right] = 1 \quad \text{(approximately)} \tag{19}$$

since $\frac{1}{2TW} \to 0$ and the average value of $\cos\left(\frac{2\pi mn}{WT}\right)$ for all integral values of n approaches zero. At the same time

$$\frac{1}{TW}\sum_{n=-TW}^{+TW}\cos\frac{\pi m_1 n}{WT}\cos\frac{\pi m_2 n}{WT} = \frac{1}{2TW}\sum_{n=-TW}^{+TW}\left\{\cos\left[\frac{\pi n}{WT}(m_1 - m_2)\right] + \cos\left[\frac{\pi n}{WT}(m_1 + m_2)\right]\right\} = 0 \tag{20}$$

since the average values of

$$\cos\left[\frac{\pi n}{WT}(m_1 - m_2)\right] \quad \text{and of} \quad \cos\left[\frac{\pi n}{WT}(m_1 + m_2)\right]$$

for all integral values of n both approach zero. In a similar manner we could show that

$$\frac{1}{TW}\sum_{n=-TW}^{+TW} \sin^2\left(\frac{\pi mn}{WT}\right) \to 1 \tag{21}$$

$$\frac{1}{TW}\sum_{n=-TW}^{+TW} \sin\frac{\pi m_1 n}{WT} \sin\frac{\pi m_2 n}{WT} \to 0 \quad (\text{when } m_1 \neq m_2) \tag{22}$$

Furthermore, the average values of both $\cos\frac{\pi mn}{WT}$ and of $\sin\frac{\pi mn}{WT}$ for all integral values of n approach zero. This fact coupled with Eq. (14), shows that Eq. (12a) is satisfied for the case $y_p = a_0'$. Thus, in all cases, Eq. (12a) as a condition for orthogonality is satisfied. As already mentioned, Eq. (12b) will then also be satisfied. We have thus demonstrated that the transformation between the G values on the one hand and the a' and b' values on the other hand is an orthogonal transformation. This is what we set out to show.

APPENDIX VII

Derivation of the Formula for the Root-mean-square Displacement in the Case of the Random Walk in One Dimension[1]

The conditions of the problem are set up in Sec. 3.5 and will not be repeated here. Let D be the total displacement after N random displacements each of length d. Let M be the number of displacements in the positive direction. Then $(N - M)$ will be the number in the negative direction. Thus

$$D = [M - (N - M)]d = (2M - N)d \tag{1}$$

Suppose we let a particular sequence of N steps be represented by a sequence of N $(+)$ or $(-)$ signs, each $(+)$ sign standing for a step in the positive direction and each $(-)$ sign standing for a step in the negative direction. Thus a sequence might be

$$+ + - + + - - + - - + - + \ldots - + \tag{2}$$

Now each $(+)$ or $(-)$ step had a probability of $\frac{1}{2}$ before it was taken. Therefore all sequences of N steps have the probability

$$P_N = (\tfrac{1}{2})^N \tag{3}$$

and all such sequences are thus equally probable. Now, let us write

$$Q = \frac{D}{d} = 2M - N = \begin{bmatrix}\text{excess of steps in} \\ \text{positive direction}\end{bmatrix} \tag{4}$$

In order to have a positive excess of Q steps, out of a total number of N, it follows from Eq. (4) that

$$M = \frac{N + Q}{2} \tag{5}$$

steps must be taken in the positive direction, and

$$N - M = \frac{N - Q}{2} \tag{6}$$

steps must be taken in the negative direction.

[1] The procedure used here is that of S. Chandrasekhar, "Stochastic Problems in Physics and Astronomy," *Rev. Modern Physics*, vol. 15 January, 1943, p. 4.

The number of distinct sequences of N steps, of the type shown in Eq. (2), which will have such an excess of Q steps will therefore[1] be

$$\frac{N!}{\left(\frac{N+Q}{2}\right)!\left(\frac{N-Q}{2}\right)!} = C^N_{(N+Q)/2} \tag{7}$$

where $C^N_{(N+Q)/2}$ is a binomial coefficient using the ordinary notation.

From Eqs. (3) and (7) it follows that the probability of the displacement D in N steps is

$$P(N,D) = \frac{N!}{\left(\frac{N+Q}{2}\right)!\left(\frac{N-Q}{2}\right)!}\left(\frac{1}{2}\right)^N \tag{8}$$

From Eq. (8) it is apparent that $P(N,D)$ has a Bernoulli distribution.

If the Bernoulli distribution is written as

$$\frac{m!}{n!(m-n)!}\,p^n(1-p)^{m-n} \tag{9}$$

we know that the ensemble average value of n is[2]

$$\bar{n} = mp \tag{10}$$

and the mean-square deviation of n is[3]

$$\overline{(n-\bar{n})^2} = \bar{n}(1-p) \tag{11}$$

To put Eq. (9) in the form of Eq. (8) we write

$$m = N \tag{12}$$

$$n = \left(\frac{N+Q}{2}\right) \tag{13}$$

$$p = \tfrac{1}{2} \tag{14}$$

It then follows from Eq. (10) that

$$\bar{n} = \frac{\overline{N+Q}}{2} = \frac{N}{2} + \frac{\bar{Q}}{2} = \frac{N}{2} \tag{15}$$

Thus

$$\bar{Q} = 0 \tag{16}$$

The ensemble average of Q is therefore zero.

[1] See *F.A.,M., and N.*, p. 287, Eq. (4).
[2] *F.A.,M., and N.*, p. 293.
[3] *F.A.,M., and N.*, p. 306.

From Eq. (11) we have

$$\overline{\left(\frac{N+Q}{2}-\frac{N}{2}\right)^2} = \overline{\left(\frac{Q}{2}\right)^2} = \left(\frac{N}{2}\right)\left(\frac{1}{2}\right) \tag{17}$$

Thus

$$\overline{Q^2} = N \tag{18}$$

and finally

$$\overline{D^2} = \overline{(Qd)^2} = d^2\overline{Q^2} = d^2N \tag{19}$$

Equation (18), or its equivalent Eq. (19), is the desired random-walk formula.

APPENDIX VIII

Proof of the Inequality[1] $H(x,y) \leq H(x) + H(y)$ and Some Related Formulas and Properties of Entropy[2]

In a system having two variables x and y (*i.e.*, 2 degrees of freedom) each limited to a set of discrete values, suppose that the values of the variables are governed by chance. Let us consider the entropies

$$H(x,y) = -\sum_{i,j} P_{ij} \log P_{ij} \tag{1}$$

$$H(x) = -\sum_{i} P_i \log P_i \tag{2}$$

$$H(y) = -\sum_{j} P_j \log P_j \tag{3}$$

Here[3]

$$P_{ij} \equiv \begin{bmatrix}\text{probability of the simultaneous occur-} \\ \text{rence of value } i \text{ for } x \text{ and value } j \text{ for } y\end{bmatrix} \tag{4}$$

Then

$$\sum_{ij} P_{ij} = 1 \tag{5}$$

where the summation is taken over all possible values i of x and j of y.

Also

$$P_i \equiv \begin{bmatrix}\text{probability of occurrence of the value} \\ \text{of } i \text{ for } x\text{, regardless of the value of } y\end{bmatrix} \tag{6}$$

and

$$P_j \equiv \begin{bmatrix}\text{probability of occurrence of the value} \\ \text{of } j \text{ for } y\text{, regardless of the value of } x\end{bmatrix} \tag{7}$$

Then

$$\sum_{i} P_i = 1 \tag{8}$$

and

$$\sum_{j} P_j = 1 \tag{9}$$

[1] The proof given here is due to N. G. Parke, III, and E. W. Samson, "Distance and Equivalence in Sequence Space," a report of the Air Force Cambridge Research Laboratories.

[2] $H(x)$ is a simplified way of expressing "the entropy of the x distribution of probabilities." It is not a specific function of x such as $\sin x$ or x^3. Analogous remarks apply to $H(y)$ and $H(x,y)$.

[3] Note that P_{ij} differs from p_{ij} defined on p. 48 of Chap. I. The latter was the probability of the occurrence of the value j for y, if it is certain that i is the value of x.

It must then follow that

$$\sum_i P_{ij} = P_j \tag{10}$$

and

$$\sum_j P_{ij} = P_i \tag{11}$$

From Eqs. (1), (2), and (3), we have

$$H(x,y) - H(x) - H(y) = -\sum_{i,j} P_{ij} \log P_{ij} + \sum_i P_i \log P_i + \sum_j P_j \log P_j \tag{12}$$

In case the values of x and y are independent of one another, then probability theory tells us that

$$P_{ij} = P_i P_j \tag{13}$$

When this occurs, the right side of Eq. (12) vanishes so that

$$H(x,y) = H(x) + H(y) \tag{14}$$

which says that in this case the entropies of the individual variables are additive.

For the general case when x and y are not necessarily independent, let us next define the quantity

$$h_{ij} = P_{ij} - P_i P_j \tag{15}$$

Thus h_{ij} will vanish when x and y are independent. Furthermore, in any case,

$$\begin{aligned}\sum_{ij} h_{ij} &= \sum_{ij} P_{ij} - \sum_{ij} P_i P_j \\ &= \sum_{ij} P_{ij} - \sum_i P_i \sum_j P_j = 1 - 1 = 0\end{aligned} \tag{16}$$

We now recall the mean-value theorem from the differential calculus, namely,

$$f(x + h) - f(x) = hf'(x + \theta h) \tag{17}$$[1]

where

$$0 \leq \theta \leq 1 \tag{18}$$

If $f(x) = \log x$, we have

$$\log (x + h) = \log x + \frac{h}{x + \theta h} \tag{19}$$

In Eq. (19), let

$$x = P_i P_j \tag{20}$$

and

$$h = h_{ij} \tag{21}$$

[1] f' here stands for the derivative of f.

Then from Eqs. (15) and (19), we have

$$\log P_{ij} = \log P_i + \log P_j + \frac{h_{ij}}{P_iP_j + \theta h_{ij}} \tag{22}$$

Substituting into Eq. (12) and using Eqs. (10) and (11) and (15) we then have

$$\begin{aligned} H(x,y) - H(x) - H(y) &= -\sum_{ij} P_{ij}\left[\log P_i + \log P_j + \frac{h_{ij}}{P_iP_j + \theta h_{ij}}\right] \\ &\quad + \sum_i P_i \log P_i + \sum_j P_j \log P_j \\ = -\sum_i P_i \log P_i - \sum_j P_j \log P_j &- \sum_{ij} \frac{h_{ij}(P_iP_j + h_{ij})}{P_iP_j + \theta h_{ij}} \\ &\quad + \sum_i P_i \log P_i + \sum_j P_j \log P_j \\ = -\sum_{ij} \frac{h_{ij}(P_iP_j + h_{ij})}{P_iP_j + \theta h_{ij}} &= -\sum_{ij} h_{ij}\left[1 + \frac{(1-\theta)h_{ij}}{P_iP_j + \theta h_{ij}}\right] \end{aligned} \tag{23}$$

The last step in Eq. (23) is obtained by algebraic division. If now, we use Eq. (16), Eq. (23) reduces to

$$H(x) + H(y) - H(x,y) = +\sum_{ij} \frac{(1-\theta)h_{ij}^2}{P_iP_j + \theta h_{ij}} \tag{24}$$

The right side of Eq. (24) cannot be negative, since the numerators and denominators of all terms are positive. This is assured for the numerator by Eq. (18). The denominator is obviously positive if h_{ij} is positive. If h_{ij} is negative, it follows from Eqs. (18) and (15) that

$$P_iP_j + \theta h_{ij} \geq P_iP_j + h_{ij} \geq 0 \tag{25}$$

so that the denominator is still positive. Since the right side of Eq. (24) cannot be negative, we have the inequality

$$H(x,y) \leq H(x) + H(y) \tag{26}$$

In case the ranges of x and y are continuous, instead of being discrete, the analogue of Eq. (23) becomes

$$\begin{aligned} H(x,y) - H(x) - H(y) &= \\ -\iint P(x,y)&\left[\log P(x) + \log P(y) + \frac{h_{ij}}{P(x)P(y) + \theta h_{ij}}\right] dx\,dy \\ + \int P(x) \log P(x) dx &+ \int P(y) \log P(y) dy \\ &= -\iint \frac{(1-\theta)h_{ij}^2}{P(x)P(y) + \theta h_{ij}}\, dx\,dy \leq 0 \end{aligned} \tag{27}$$

We have thus established the important inequality, Eq. (26), for continuous as well as discrete cases.

Next, let

$$P_i(j) \equiv \begin{bmatrix} \text{probability that } y \text{ has the value } j, \\ \text{if it is known that } x \text{ has the value } i \end{bmatrix} \tag{28}$$

and

$$P_j(i) \equiv \begin{bmatrix} \text{probability that } x \text{ has the value } i, \\ \text{if it is known that } y \text{ has the value } j \end{bmatrix} \tag{29}$$

Then

$$P_i(j) = \frac{P_{ij}}{P_i} \qquad (30)^1$$

and

$$P_j(i) = \frac{P_{ij}}{P_j} \tag{31}$$

We now define the conditional entropies $H_x(y)$ and $H_y(x)$ as

$$H_x(y) = -\sum_{i,j} P_{ij} \log P_i(j) = -\sum_i P_i \sum_j P_i(j) \log P_i(j) \tag{32}$$

and

$$H_y(x) = -\sum_{i,j} P_{ij} \log P_j(i) = -\sum_j P_j \sum_i P_j(i) \log P_j(i) \tag{33}$$

$H_x(y)$ is thus the entropy of the probability distribution of y when x is known, averaged over all values of x. As such it is a measure of the choice or uncertainty in y when x is known. We may also say that $H_x(y)$ is the average information in y (in the noiseless case) when x is known, the average being taken over all values of x and y. $H_y(x)$ has an analogous significance.

From Eqs. (30) and (11)

$$\begin{aligned} H_x(y) &= -\sum_{i,j} P_{ij} \log P_{ij} + \sum_{i,j} P_{ij} \log P_i \\ &= -\sum_{i,j} P_{ij} \log P_{ij} + \sum_i P_i \log P_i = H(x,y) - H(x) \end{aligned} \tag{34}$$

Similarly

$$H_y(x) = H(x,y) - H(y) \tag{35}$$

It follows from Eqs. (26), (34), and (35) that

$$H(x) \geq H_y(x) \tag{36}$$

and

$$H(y) \geq H_x(y) \tag{37}$$

By using integrals instead of summations, it is readily shown that the foregoing results hold for continuous as well as discrete distributions.

In order to illustrate the significance of intersymbol constraints and conditional entropy, let us consider the following example. Suppose we

[1] $P_i(j)$ is equivalent to p_{ij} used in Sec. 1.9 of Chap. I.

have a system having the stochastic variables x and y, and let x have the three possible values A, B, and C, while y has the two possible values a and b. Let

$$P_{\mathrm{A}} = 0.4 \tag{38}$$

$$P_{\mathrm{B}} = 0.4 \tag{39}$$

$$P_{\mathrm{C}} = 0.2 \tag{40}$$

and

$$P_{\mathrm{a}} = 0.6 \tag{41}$$

$$P_{\mathrm{b}} = 0.4 \tag{42}$$

Table I represents three illustrative possibilities of intersymbol constraints. In Case I, there are no intersymbol constraints, so that

$$P_{\mathrm{Aa}} = P_{\mathrm{A}} \cdot P_{\mathrm{a}} \tag{43}$$

$$P_{\mathrm{Ab}} = P_{\mathrm{A}} \cdot P_{\mathrm{b}} \tag{44}$$

and so on. In this case

$$H(x,y) = H(x) + H(y) \tag{45}$$

$$H_x(y) = H(y) \tag{46}$$

and

$$H_y(x) = H(x) \tag{47}$$

In Case III, the value of x completely determines the value of y, so that

$$H(x,y) = H(x) \tag{48}$$

and

$$H_x(y) = 0 \tag{49}$$

The extent of the constraints in Case II is intermediate between those of Case I and Case III, and the entropy $H(x,y)$ also has an intermediate value in this case. All cases illustrate the general formulas such as Eqs. (26), (30), (31), (34), (35), (36), and (37).

The actual calculation of the probabilities and entropies is carried out according to the formulas

$$P_{\mathrm{a}}(\mathrm{A}) = \frac{P_{\mathrm{Aa}}}{P_{\mathrm{a}}} \tag{50}$$

$$P_{\mathrm{A}}(\mathrm{a}) = \frac{P_{\mathrm{Aa}}}{P_{\mathrm{A}}} \tag{51}$$

$$P_{\mathrm{a}}(\mathrm{B}) = \frac{P_{\mathrm{Ba}}}{P_{\mathrm{a}}} \tag{52}$$

$$P_{\mathrm{A}}(\mathrm{b}) = \frac{P_{\mathrm{Ab}}}{P_{\mathrm{A}}} \tag{53}$$

Table I

	Case I (unconstrained)	Case II (partially constrained)	Case III (y completely constrained)
P_{Aa}	0.24	0.3	0.4
P_{Ab}	0.16	0.1	0.0
P_{Ba}	0.24	0.2	0.0
P_{Bb}	0.16	0.2	0.4
P_{Ca}	0.12	0.1	0.2
P_{Cb}	0.08	0.1	0.0
$P_{\mathrm{a}}(\mathrm{A})$	0.4	0.5	0.667
$P_{\mathrm{a}}(\mathrm{B})$	0.4	0.333	0.0
$P_{\mathrm{a}}(\mathrm{C})$	0.2	0.167	0.333
$P_{\mathrm{b}}(\mathrm{A})$	0.4	0.25	0.0
$P_{\mathrm{b}}(\mathrm{B})$	0.4	0.50	1.0
$P_{\mathrm{b}}(\mathrm{C})$	0.2	0.25	0.0
$P_{\mathrm{A}}(\mathrm{a})$	0.6	0.75	1.0
$P_{\mathrm{A}}(\mathrm{b})$	0.4	0.25	0.0
$P_{\mathrm{B}}(\mathrm{a})$	0.6	0.50	0.0
$P_{\mathrm{B}}(\mathrm{b})$	0.4	0.50	1.0
$P_{\mathrm{C}}(\mathrm{a})$	0.6	0.50	1.0
$P_{\mathrm{C}}(\mathrm{b})$	0.4	0.50	0.0
$H(x)$	1.52 binits	1.52 binits	1.52 binits
$H(y)$	0.97 binit	0.97 binit	0.97 binit
$H(x,y)$	2.49 binits	2.45 binits	1.52 binits
$H_x(y)$	0.97 binit	0.93 binit	0.0 binit
$H_y(x)$	1.52 binits	1.48 binits	0.55 binit

and so on, for the probabilities, with

$$H(x) = -[P_{\mathrm{A}} \log P_{\mathrm{A}} + P_{\mathrm{B}} \log P_{\mathrm{B}} + P_{\mathrm{C}} \log P_{\mathrm{C}}] \tag{54}$$

$$H(y) = -[P_{\mathrm{a}} \log P_{\mathrm{a}} + P_{\mathrm{b}} \log P_{\mathrm{b}}] \tag{55}$$

$$\begin{aligned} H_x(y) = & - P_{\mathrm{A}}[P_{\mathrm{A}}(\mathrm{a}) \log P_{\mathrm{A}}(\mathrm{a}) + P_{\mathrm{A}}(\mathrm{b}) \log P_{\mathrm{A}}(\mathrm{b})] \\ & - P_{\mathrm{B}}[P_{\mathrm{B}}(\mathrm{a}) \log P_{\mathrm{B}}(\mathrm{a}) + P_{\mathrm{B}}(\mathrm{b}) \log P_{\mathrm{B}}(\mathrm{b})] \\ & - P_{\mathrm{C}}[P_{\mathrm{C}}(\mathrm{a}) \log P_{\mathrm{C}}(\mathrm{a}) + P_{\mathrm{C}}(\mathrm{b}) \log P_{\mathrm{C}}(\mathrm{b})] \end{aligned} \tag{56}$$

$$\begin{aligned} H_y(x) = & -P_{\mathrm{a}}[P_{\mathrm{a}}(\mathrm{A}) \log P_{\mathrm{a}}(\mathrm{A}) + P_{\mathrm{a}}(\mathrm{B}) \log P_{\mathrm{a}}(\mathrm{B}) + P_{\mathrm{a}}(\mathrm{C}) \log P_{\mathrm{a}}(\mathrm{C})] \\ & - P_{\mathrm{b}}[P_{\mathrm{b}}(\mathrm{A}) \log P_{\mathrm{b}}(\mathrm{A}) + P_{\mathrm{b}}(\mathrm{B}) \log P_{\mathrm{b}}(\mathrm{B}) + P_{\mathrm{b}}(\mathrm{C}) \log P_{\mathrm{b}}(\mathrm{C})] \end{aligned} \tag{57}$$

$$H(x,y) = H(x) + H_x(y) = H(y) + H_y(x) \tag{58}$$

for the entropies. The reader will note that it is only in Case II that it is necessary to substitute in Eq. (56) or (57) to find $H_x(y)$, $H_y(x)$, or $H(x,y)$. In Case I, these are found from $H(x)$ and $H(y)$ with the aid of Eqs. (45)

through (47), while in Case III Eqs. (48), (49), and (58) are used for this purpose.

PROBLEMS

1. If we have a discrete distribution of probabilities $p_1, p_2, \ldots, p_n$ with

$$\sum_i p_i = 1 \tag{A}$$

show that for a fixed value of n

$$H = -\sum_i p_i \log p_i \tag{B}$$

is a maximum if

$$p_1 = p_2 = \ldots = p_n \tag{C}$$

[The corresponding problem for the continuous case is solved is Sec. 4.3(a).]

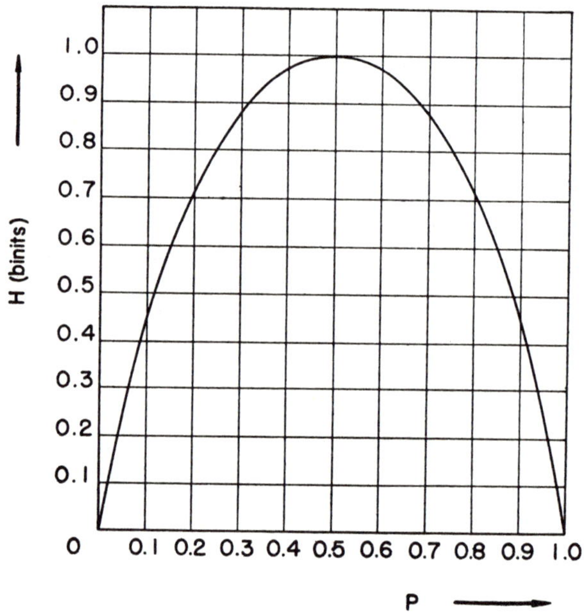

Fig. A-A. Entropy in the case of two possibilities with probabilities P and (1-P). (*After C. E. Shannon.*)

If there are just two possibilities

$$p_1 = P \quad \text{say} \tag{D}$$

and

$$p_2 = 1 - P \tag{E}$$

then

$$H = -P \log P - (1 - P) \log (1 - P) \tag{F}$$

Equation (F) is plotted in Fig. A-A and illustrates Eq. (C) for the case when $n = 2$.

2. In an ergodic sequence consisting only of the symbols A, B, and C, let

$$P_A = 0.4 \quad P_B = 0.4 \quad P_C = 0.2$$

Suppose that the entire intersymbol influence of the sequence is expressed by the digram probabilities shown as Cases I, II, and III in the following table.

	Case I	Case II	Case III
P_{AA}	0.16	0.4	0.2
P_{AB}	0.16	0.0	0.1
P_{AC}	0.08	0.0	0.1
P_{BA}	0.16	0.0	0.1
P_{BB}	0.16	0.4	0.2
P_{BC}	0.08	0.0	0.1
P_{CA}	0.08	0.0	0.1
P_{CB}	0.08	0.0	0.1
P_{CC}	0.04	0.2	0.0

Find the entropy per symbol of the sequence in each case.

Ans. Case I. 1.52 binits/symbol.
Case II. zero binits/symbol.
Case III. 1.40 binits/symbol.

APPENDIX IX

The Output of a Triangular Filter When the Input Is a Sampling Function

Suppose we have a transmission system whose gain is unity at zero frequency and then declines linearly to a gain of zero at frequency W (see Fig. A-4). Suppose further that there is no phase shift. Let a sampling function

$$G_1(t) = \frac{\sin 2\pi Wt}{2\pi Wt} \tag{1}$$

be the input signal to this transmission system. We wish to determine the output.

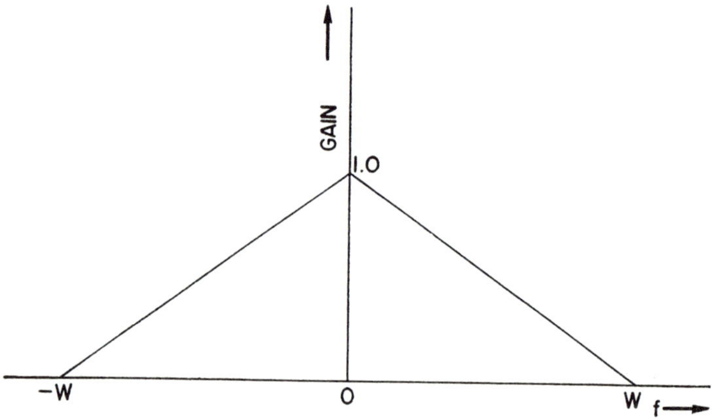

Fig. A-4.

The transmission characteristic of the system may be written

$$Y(\omega) = \begin{cases} 1 - \dfrac{|\omega|}{2\pi W} & \text{when } (-2\pi W \le \omega \le 2\pi W) \\ 0 & \text{otherwise} \end{cases} \tag{2}$$

The spectrum of $G_1(t)$ is well known[1] to be

$$\Omega_1(\omega) = \begin{cases} \dfrac{1}{2W} & \text{when } -2\pi W \le \omega \le 2\pi W \\ 0 & \text{otherwise} \end{cases} \tag{3}$$

[1] See *F.A.,M., and N.*, p. 133, footnote 1.

The spectrum of the output is therefore

$$\Omega_2(\omega) = \begin{cases} \left(1 - \dfrac{|\omega|}{2\pi W}\right)\dfrac{1}{2W} & \text{when } -2\pi W \leq \omega \leq 2\pi W \\ 0 & \text{otherwise} \end{cases} \tag{4}$$

We note that $\Omega_2(\omega)$ is an even function of ω. Under these circumstances $G_2(t)$ may be written very simply as

$$\begin{aligned} G_2(t) &= \frac{1}{\pi}\int_0^{2\pi W} \frac{1}{2W}\left(1 - \frac{\omega}{2\pi W}\right) \cos \omega t \, d\omega \\ &= \frac{1}{2\pi W}\int_0^{2\pi W} \cos \omega t \, d\omega - \frac{1}{4\pi^2 W^2}\int_0^{2\pi W} \omega \cos \omega t \, d\omega \\ &= \frac{\sin 2\pi W t}{2\pi W t} - \frac{1}{4\pi^2 W^2}\int_0^{2\pi W} \omega \cos \omega t \, d\omega \end{aligned} \tag{5}$$

Remembering that

$$\int x \cos x \, dx = \cos x + x \sin x \tag{6}$$

the last term in Eq. (5) is

$$\begin{aligned} -\frac{1}{4\pi^2 W^2}\int_0^{2\pi W} \omega \cos \omega t \, d\omega &= -\frac{1}{4\pi^2 W^2 t^2}\int_0^{2\pi W t} \omega t \cos \omega t \, d(\omega t) \\ &= \frac{-1}{4\pi^2 W^2 t^2}[\cos 2\pi W t + 2\pi W t \sin 2\pi W t - 1] \\ &= \frac{-1}{4\pi^2 W^2 t^2}[2\pi W t \sin (2\pi W t) - 2 \sin^2 (\pi W t)] \\ &= -\frac{\sin 2\pi W t}{2\pi W t} + \frac{\sin^2 (\pi W t)}{2(\pi W t)^2} \end{aligned} \tag{7}$$

Substituting Eq. (7) in Eq. (5), we obtain finally for the output signal

$$G_2(t) = \frac{\sin^2 (\pi W t)}{2(\pi W t)^2} \tag{8}$$

APPENDIX X

Some Further Notes on Physically Realizable Filters

For the reader who is acquainted with functions of a complex variable, a few additional notes to Sec. 8.2 on physically realizable filters may be of interest.

We will first show that the conditions imposed on $B(t)$ are derivable from the condition that no singularities of $Y(\omega)$ lie in the half plane Im $(\omega) < 0$. Thus consider Fig. A-5. The foregoing condition imposed on $Y(\omega)$ ensures

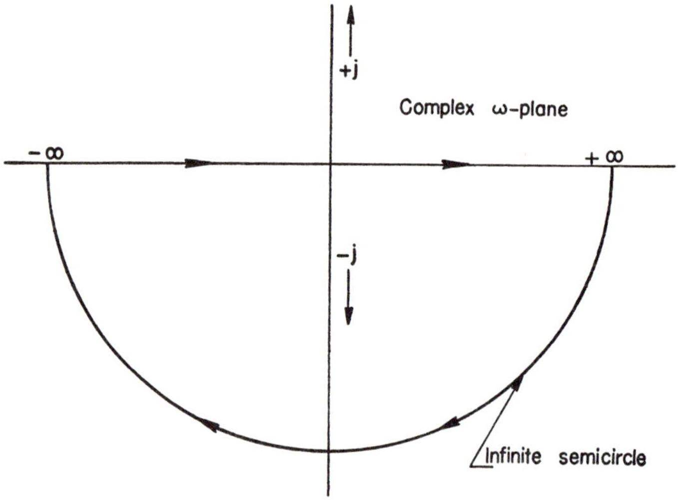

Fig. A-5.

the fact that, if t is negative, the function $\epsilon^{j\omega t}Y(\omega)$ has no singularity within the contour in Fig. A-5. Therefore the contour integral

$$\frac{1}{2\pi}\int \epsilon^{j\omega t}Y(\omega)\,d\omega$$

around the closed contour vanishes. Furthermore, since $\epsilon^{j\omega t}$ vanishes exponentially along the infinite semicircle except at the end points and $Y(\omega)$ vanishes as $|\omega| \to \infty$, it may be shown (Jordan's lemma) that the integral along the infinite semicircle vanishes separately. Therefore

$$B(t) = \frac{1}{2\pi} \int_{-\infty}^{+\infty} \epsilon^{j\omega t} Y(\omega) d\omega = 0 \quad \text{if } t < 0 \tag{1}$$

If t is positive, we integrate $1/2\pi \int \epsilon^{j\omega t} Y(\omega) d\omega$ along the closed contour shown in Fig. A-6. In this case, the integral again vanishes along the infinite

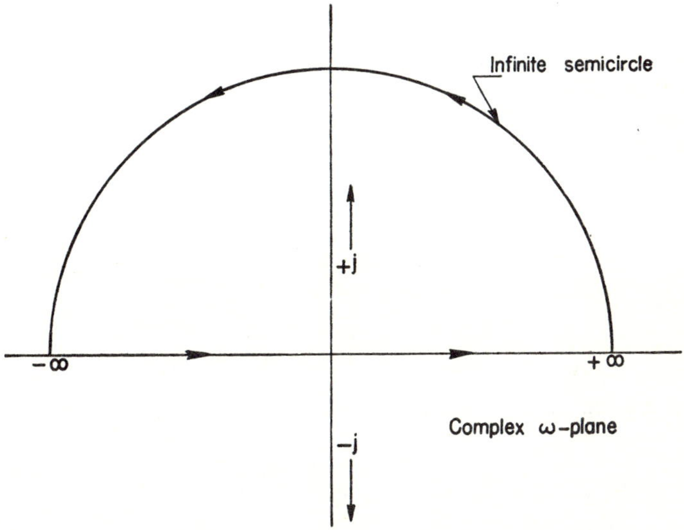

Fig. A-6.

semicircle. Therefore

$$\frac{1}{2\pi} \int_{-\infty}^{+\infty} \epsilon^{j\omega t} Y(\omega) d\omega = j \text{ times the sum of the residues of } \epsilon^{j\omega t} Y(\omega) \text{ in the upper half plane if } t > 0 \tag{2}$$

In case $Y(\omega)$ is expressible as the ratio of two polynomials

$$Y(\omega) = k \frac{(\omega - a_1)(\omega - a_2) \ . \ . \ . \ (\omega - a_n)}{(\omega - b_1)(\omega - b_2) \ . \ . \ . \ (\omega - b_n)} \tag{3}$$

which will be the case for a lumped network, then the residue at any singular point b_p will be[1]

$$k \frac{(b_p - a_1)(b_p - a_2) \ . \ . \ . \ (b_p - a_n)}{(b_p - b_1)(b_p - b_2) \ . \ . \ . \ (b_p - b_m)} \epsilon^{jb_pt} \tag{4}$$

[1] See Goldman, *Transformation Calculus and Electrical Transients*, p. 167.

where the denominator includes all factors of the form $(b_p - b_q)$ for $q = 1, \ldots m$ excepting $(b_p - b_p)$. Consequently

$$B(t) = \frac{1}{2\pi}\int_{-\infty}^{+\infty} \epsilon^{j\omega t} Y(\omega)d\omega$$
$$= j\sum_{p=1}^{m} k\frac{(b_p - a_1)(b_p - a_2)\ldots(b_p - a_n)}{(b_p - b_1)(b_p - b_2)\ldots(b_p - b_m)}(b_p - b_p)\epsilon^{jb_pt} \quad (5)$$

The $(b_p - b_p)$ multipliers cancel $(b_p - b_p)$ factors in the denominator.

Since the imaginary part of every b_p is positive, the real part of the exponent of every ϵ^{jb_pt} in Eq. (5) will be negative. $B(t)$ will therefore consist of a sum of terms, each of which decreases exponentially with time and vanishes at infinity.

By way of illustration, we will use Eq. (5) to calculate $B(t)$ for the filter in Fig. 8-5(a) of Chap. VIII. For this filter

$$Y(\omega) = \frac{R}{2R_0 + R}\,\frac{\omega - \dfrac{j}{RC}}{\omega - \dfrac{j}{(2R_0 + R)C}} \quad (6)$$

Susbtituting in Eq. (5), we obtain for the impulsive response of the filter

$$B(t) = \frac{jR}{(2R_0 + R)}\,\frac{\left[\dfrac{j}{(2R_0 + R)C} - \dfrac{j}{RC}\right]}{1}\epsilon^{j[j/(2R_0+R)C]t}$$
$$= \frac{2R_0}{(2R_0 + R)^2C}\epsilon^{-t/(2R_0+R)C} \quad \text{for } t > 0 \quad (7)$$

APPENDIX XI

Adjustment of the Weighting Coefficients for Minimum Mean-Square Error in a Linear Prediction

Suppose we have a number of stochastic variables $x_1, x_2, \ldots, x_n$ which are statistically independent,[1] or at least the ensemble average of the cross product of any two, $\widetilde{x_m x_n}$, vanishes. Let y be another stochastic variable, which is correlated with $x_1, \ldots, x_n$. Now, suppose we wish to obtain an optimum prediction of y, in the least square-error sense, by means of a linear operation on the x's. In other words, suppose we write

$$y_0 = \sum_{i=1}^{n} a_i x_i \tag{1}$$

where y_0 is the predicted value of y. Our problem is to find the set of a_i's which will make the ensemble average of $(y - y_0)^2$ a minimum.

For the ensemble average of the mean-square error we may write

$$\begin{aligned} \widetilde{(y - y_0)^2} &= \widetilde{[y - \sum_{i=1}^{n} a_i x_i]^2} \\ &= \widetilde{y^2} - 2 \sum_{i=1}^{n} a_i \widetilde{x_i y} + \sum_{i=1}^{n} \sum_{j=1}^{n} a_i a_j \widetilde{x_i x_j} \end{aligned} \tag{2}$$

By hypothesis,

$$\widetilde{x_i x_j} = 0 \quad \text{if } i \neq j \tag{3}$$

Therefore Eq. (2) reduces to

$$\widetilde{(y - y_0)^2} = \widetilde{y^2} - 2 \sum_{i=1}^{n} \widetilde{a_i x_i y} + \sum_{i=1}^{n} a_i^2 \widetilde{x_i^2} \tag{4}$$

In order to find the set of a_i's which will make $\widetilde{(y - y_0)^2}$ a minimum, we set the partial derivatives equal to zero in the usual manner. Thus for every a_i

$$\frac{\partial}{\partial a_i} \widetilde{[(y - y_0)^2]} = -2\widetilde{x_i y} + 2a_i \widetilde{x_i^2} = 0 \tag{5}$$

Therefore

$$a_i = \frac{\widetilde{x_i y}}{\widetilde{x_i^2}} \tag{6}$$

[1] In other words, there is no intersymbol influence between the variables.

The ensemble averages $\widetilde{x_i y}$ and $\widetilde{x_i^2}$ are a statistical property of the process and may be presumed known. Therefore the n equations of the form (6) for the n different values of i give a complete solution to the problem and allow y to be predicted from a knowledge of the x_i's with the least mean-square error.

We note in particular that each of the n minimizing equations of the form (5) (or 6) involves only one of the a_i's. This shows that minimizing $(y - y_0)^2$ on all the a_i's is accomplished by minimizing on each individual a_i separately, so that the value of any one a_i does not depend on the values of the others. In particular, if some but not all of the a_i's are available for adjustment, the same values will be assigned to the adjustable a_i's for minimum mean-square error of prediction as though all the a_i's were available for adjustment. This information will be found valuable in designing the optimum predicting filter in the presence of noise in Chap. VIII.

APPENDIX XII

Complete Solutions of Some of the Problems

PROBLEM 2, p. 7:

Question: In a certain community, 25% of all girls are blondes and 75% of all blondes have blue eyes. Also 50% of all girls in the community have blue eyes. If you know that a girl has blue eyes, how much additional information do you get by being informed that she is blonde?

Solution: Let

$p_1 = p$ (blonde) $= 0.25 =$ probability that a girl is blonde in the absence of knowledge of the color of her eyes

$p_2 = p_{\text{blonde}}$ (blue eyes) $= 0.75 =$ probability that a blonde has blue eyes

$p_3 = p$ (blue eyes) $= 0.50 =$ probability that a girl has blue eyes in the absence of knowledge of the color of her hair

$p_x = p_{\text{blue eyes}}$ (blonde) $= ? =$ probability that a blue-eyed girl is blonde

$p_4 = p$ (blonde, blue eyes) $=$ probability that a girl is blonde and has blue eyes

Then

$$p_4 = p_1 \cdot p_2 = p_3 p_x \tag{1}$$

or

$$p_x = \frac{p_1 \cdot p_2}{p_3} \tag{2}$$

If you know that a girl has blue eyes, the additional information you get by being informed that she is blonde is

$$\begin{aligned} \log \frac{1}{p_x} &= \log \left(\frac{p_3}{p_1 \cdot p_2} \right) \\ &= -\log p_1 - \log p_2 + \log p_3 = \log 4 + \log \tfrac{4}{3} - \log 2 \qquad (3) \\ &= 1.42 \text{ binits} \end{aligned}$$

PROBLEM 1, p. 38:

Question: Find the code capacities of the following code channels

	Symbols in alphabet	Symbol durations	Other constraints
(a)	dot	2	None
	dash	4	
	letter space	3	
	word space	6	
(b)	A	1	No letter may follow itself
	B	1	
	C	1	
(c)	A	1	None
	B	2	
	C	3	
	D	4	
(d)	A	1	No letter may occur twice in any group of 4, *i.e.*, ACBA is not allowed
	B	1	
	C	1	
	D	1	
(e)	A	1	The A's always occur in pairs
	B	1	
	C	1	
(f)	X	1	There is a space after every three consecutive letters
	Y	1	
	Z	1	
	Space	1	

Solution: Part (a). Since there are no constraints in this case, there is only one state and there are four symbols. Using the notation of Appendix I, we write

$$b_{11}^{(\text{dot})} = 2, \quad b_{11}^{(\text{dash})} = 4, \quad b_{11}^{(\text{letter space})} = 3, \quad b_{11}^{(\text{word space})} = 6$$

Thus, Eq. (20) of Appendix I becomes

$$W^{-6} + W^{-4} + W^{-3} + W^{-2} - 1 = 0 \tag{1}$$

Solving Eq. (1) graphically or by algebraic means, we obtain

$$W = 1.51 \tag{2}$$

$$C = \log W = 0.6 \text{ binit/unit of time} \tag{3}$$

Part (b). In this case we have three states, determined by the last letter which has so far occurred in the sequence. In

State No. 1, the last letter was A and it can be followed by B or C
State No. 2, the last letter was B and it can be followed by A or C
State No. 3, the last letter was C and it can be followed by A or B

Again using the notation of Appendix I, we have

$$\begin{array}{lll} b_{11} \text{ (none)} & b_{21}^{(A)} = 1 & b_{31}^{(A)} = 1 \\ b_{12}^{(B)} = 1 & b_{22} \text{ (none)} & b_{32}^{(B)} = 1 \\ b_{13}^{(C)} = 1 & b_{23}^{(C)} = 1 & b_{33} \text{ (none)} \end{array}$$

Equation (20) of Appendix I now becomes

$$\begin{vmatrix} -1 & W^{-1} & W^{-1} \\ W^{-1} & -1 & W^{-1} \\ W^{-1} & W^{-1} & -1 \end{vmatrix} = 0 \tag{4}$$

Thus

$$W = 2 \tag{5}$$

$$C = \log W = 1 \text{ binit/unit of time} \tag{6}$$

Part (c). In this case, there is only one state and there are four symbols. Here

$$b_{11}^{(A)} = 1, \quad b_{11}^{(B)} = 2, \quad b_{11}^{(C)} = 3, \quad b_{11}^{(D)} = 4$$

Equation (20) of Appendix I here becomes

$$W^{-4} + W^{-3} + W^{-2} + W^{-1} - 1 = 0 \tag{7}$$

Solving Eq. (7) graphically or by algebraic means we obtain

$$W = 1.92 \tag{8}$$

$$C = \log W = 0.94 \text{ binit/unit of time} \tag{9}$$

Part (d). This problem can be solved by using the formulas of Appendix I, just as preceding parts have been solved. However, it can also be solved directly by inspection. Thus wherever we are in one of these sequences, the future of the sequence is completely determined by the past. There are just 24 different possible permutations of the letters A, B, C, and D taken three at a time. There are consequently 24 different possible sequences of length T, regardless of the magnitude of T. Thus

$$N(T) = AW^T = 24(1)^T \tag{10}$$

Thus

$$W = 1 \tag{11}$$

and

$$C = \log W = 0 \tag{12}$$

Part (e). This is probably solved most easily by considering that the channel has the three symbols

AA
B
C

There is then only one state, and we have

$$W^{-2} + W^{-1} + W^{-1} - 1 = 0 \tag{13}$$

Thus

$$W = 2.414 \tag{14}$$

and

$$C = \log W = 1.27 \tag{15}$$

Part (f). In this case, there are four states depending upon the distance from the last space. These may be specified as

State No. 1 after a space
State No. 2 after a space plus one letter
State No. 3 after a space plus two letters
State No. 4 after a space plus three letters

Using the notation of Appendix I, we have

$$\begin{array}{llll} b_{12}^{(X)} = 1 & b_{23}^{(X)} = 1 & b_{34}^{(X)} = 1 & b_{41}^{(\text{space})} = 1 \\ b_{12}^{(Y)} = 1 & b_{23}^{(Y)} = 1 & b_{34}^{(Y)} = 1 & \\ b_{12}^{(Z)} = 1 & b_{23}^{(Z)} = 1 & b_{34}^{(Z)} = 1 & \end{array}$$

No other b's exist in this case. Equation (20) of Appendix I now becomes

$$\begin{vmatrix} -1 & 0 & 0 & W^{-1} \\ 3W^{-1} & -1 & 0 & 0 \\ 0 & 3W^{-1} & -1 & 0 \\ 0 & 0 & 3W^{-1} & -1 \end{vmatrix} = 1 - 27W^{-4} = 0 \tag{16}$$

Thus

$$W = (27)^{\frac{1}{4}} \tag{17}$$

and

$$C = \log W = \tfrac{1}{4} \log 27 \tag{18}$$

Problem 1, p. 53

Question: An alphabet consists of 8 consonants and 8 vowels. Suppose that all letters of the alphabet are equally probable and that there is no intersymbol influence. If consonants are always understood correctly, but vowels are understood correctly only half the time, being mistaken for other vowels the other half of the time, all vowels being involved in errors the same percentage of the time, what is the average rate of information transmission per symbol?

Solution: According to the conditions of the problem, 50% of the letters received will be correct consonants, 25% will be correct vowels, and 25% incorrect vowels. Let us first calculate the amount of information received in each of these three cases.

Case I: Correct consonants. Here

$$p_A = \text{probability before reception} = \tfrac{1}{16}$$
$$p_B = \text{probability after reception} = 1$$

Thus

$$\text{information/letter} = \log\left(\frac{1}{\frac{1}{16}}\right) = 4 \text{ binits/letter}$$

Case II: Correct vowels. Here

$$p_A = \text{probability before reception} = \tfrac{1}{16}$$
$$p_B = \text{probability after reception} = \tfrac{1}{2}$$

Thus

$$\text{information/letter} = \log\frac{\frac{1}{2}}{\frac{1}{16}} = 3 \text{ binits/letter}$$

Case III: Incorrect Vowels. Here

$$p_A = \text{probability before reception} = \tfrac{1}{16}$$

To find the probability after reception, let us refer to Fig. A-7. To be specific, suppose that an E is sent and a U is received. We wish to find the

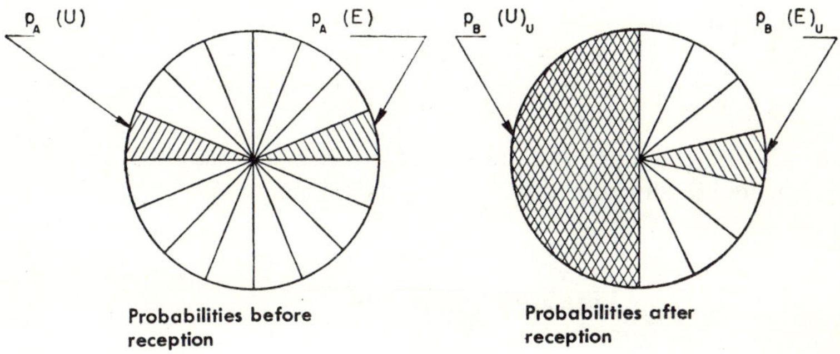

Fig. A-7.

probability that an E was sent as a consequence of receiving a U. Before reception, the probability that an E was sent was $\frac{1}{16}$ since all 16 letters are equally probable. After receiving a U, we know that a vowel was sent, and the probability that it was a U is $\frac{1}{2}$. The other $\frac{1}{2}$ of probability is divided equally among the other 7 vowels, one of which is E. Therefore the probability that E was sent is $\frac{1}{7} \times \frac{1}{2} = \frac{1}{14}$.

Therefore

$$p_B = p_B(\mathrm{E})_{\mathrm{U}} = \text{probability after reception} = \tfrac{1}{14}$$

Thus

$$\begin{aligned}\text{information/letter} &= \log\left(\frac{\frac{1}{14}}{\frac{1}{16}}\right)\\ &= \log 16 - \log 14 = 4 - 3.8 = 0.2 \text{ binit/letter}\end{aligned}$$

It is interesting to note that the information received with an incorrect vowel is positive in this case, since the reception of the incorrect vowel actually increases the probability of transmission of the correct vowel.

Finally, for the average information per symbol, we take 50% of case I and 25% each of cases II and III. Thus

$$\begin{aligned}\text{average information/symbol} &= 0.5 \times 4 + 0.25 \times 3 + 0.25 \times 0.2\\ &= 2.8 \text{ binits/symbol}\end{aligned}$$

PROBLEM 1, p. 83

Question: The Sampling Theorem For Periodic Functions (Due to C. R. Cahn). Making use of the Identities

$$\sum_{q=0}^{2N} \epsilon^{j\frac{2\pi nq}{2N+1}} = \sum_{q=0}^{2N} \epsilon^{j\frac{2\pi n}{T}\cdot\frac{qT}{(2N+1)}} \begin{cases} = 2N+1 & \text{if } n \text{ is an integral multiple of } (2N+1) \text{, including zero} \\ = 0 & \text{for other values of } n \end{cases}$$

and
$$\sum_{n=-N}^{+N} \epsilon^{j\frac{2\pi n}{T}x} = 1 + 2\sum_{n=1}^{N} \cos\frac{2\pi n}{T}x = \frac{\sin\left[\frac{(2N+1)\pi}{T}x\right]}{\sin\left(\frac{\pi}{T}x\right)}$$

show that if $f(t)$ is a periodic function of period T, and that if all the Fourier coefficients vanish above the Nth harmonic, then

$$f(t) = \sum_{q=0}^{2N} f\left(\frac{qT}{2N+1}\right) \frac{\sin\left\{(2N+1)\frac{\pi}{T}\left(t - \frac{qT}{2N+1}\right)\right\}}{(2N+1)\sin\left\{\frac{\pi}{T}\left(t - \frac{qT}{2N+1}\right)\right\}}$$

Therefore $f(t)$ has only $(2N + 1)$ sampling values and these have a spacing of $\left(\frac{T}{2N+1}\right)$ between them.

Solution: If a periodic function has no Fourier components above the Nth harmonic, it may be expressed as

$$f(t) = \sum_{n=-N}^{N} C_n \epsilon^{j\frac{2\pi n}{T}t} \tag{1}$$

At this point, we recall the identity

$$\sum_{q=0}^{2N} \epsilon^{j\frac{2\pi nq}{2N+1}} = \sum_{q=0}^{2N} \epsilon^{j\frac{2\pi n}{T}\left(\frac{qT}{2N+1}\right)} = \begin{cases} 2N+1 \text{ if } n = 0 \text{ or a multiple of } (2N+1) \\ 0 \text{ for other values of } n \end{cases} \tag{2}$$

Equation (2) says in effect that the sum of the $(2N + 1)$ vectors of the form

$$\epsilon^{j\frac{2\pi nq}{2N+1}}$$

adds up to zero, unless they are all in phase. This is illustrated in Fig. A-8.

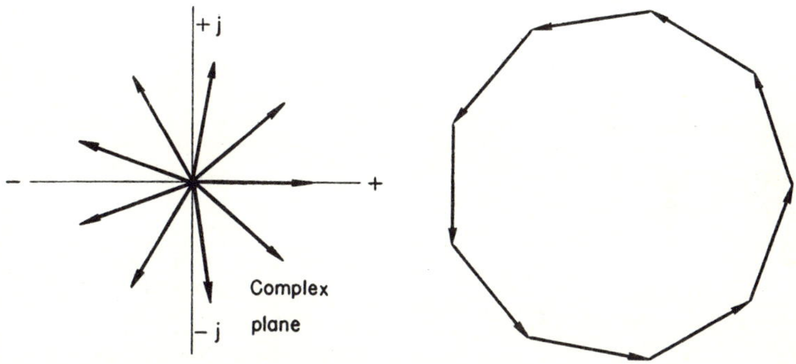

(a) The individual vectors of equation (2)

(b) Vectorial addition of vectors in (a)

Fig. A-8.

From Eq. (1), we have

$$f\left(\frac{qT}{2N+1}\right) = \sum_{n=-N}^{+N} C_n \epsilon^{j\frac{2\pi nq}{2N+1}} \tag{3}$$

Using Eq. (3), form the sum

$$\sum_{q=0}^{2N} f\left(\frac{qT}{2N+1}\right) \epsilon^{-j\frac{2\pi mq}{2N+1}} = \sum_{n=-N}^{N} C_n \sum_{q=0}^{2N} \epsilon^{j\frac{2\pi}{2N+1}q(n-m)} \tag{4}$$[1]

[1] Note that Eqs. (4) and (5) are analogous to the method used for finding the Fourier coefficients C_m in the case of an infinite Fourier series. There we take the expansion

$$f(t) = \sum_{n=-\infty}^{+\infty} C_n \epsilon^{j(2\pi nt/T)}$$

and multiply both sides by $\epsilon^{-j(2\pi mt/T)}$ and integrate from 0 to T. This gives

$$\int_0^T f(t)\epsilon^{-j(2\pi mt/T)}\,dt = C_m T$$

where $-N \leq m \leq N$
From Eqs. (2) and (4), we have

$$\sum_{q=0}^{2N} f\left(\frac{qT}{2N+1}\right) \epsilon^{-j\frac{2\pi mq}{2N+1}} = (2N+1)C_m \tag{5}$$

Substituting the values of C_m from Eq. (5) into Eq. (1), we have

$$\begin{aligned} f(t) &= \sum_{n=-N}^{N} \left\{\frac{1}{2N+1} \sum_{q=0}^{2N} f\left(\frac{qT}{2N+1}\right) \epsilon^{-j\frac{2\pi nq}{2N+1}}\right\} \epsilon^{j\frac{2\pi nt}{T}} \\ &= \sum_{q=0}^{2N} f\left(\frac{qT}{2N+1}\right) \cdot \frac{1}{2N+1} \sum_{n=-N}^{N} \epsilon^{j\frac{2\pi n}{T}\left(t-\frac{qT}{2N+1}\right)} \end{aligned} \tag{6}$$

Now, it may be shown[1] that

$$\sum_{n=-N}^{N} \epsilon^{j\frac{2\pi n}{T}x} = 1 + 2\sum_{n=1}^{N} \cos \frac{2\pi n}{T} x = \frac{\sin (2N+1)\frac{\pi}{T}x}{\sin \frac{\pi}{T}x} \tag{7}$$

Substituting Eq. (7) into Eq. (6), we have finally the sampling formula

$$f(t) = \sum_{q=0}^{2N} f\left(\frac{qT}{2N+1}\right) \frac{\sin\left\{(2N+1)\frac{\pi}{T}\left(t - \frac{qT}{2N+1}\right)\right\}}{(2N+1)\sin\left\{\frac{\pi}{T}\left(t - \frac{qT}{2N+1}\right)\right\}} \tag{8}$$

[1] Guillemin, *The Mathematics of Circuit Analysis.*

APPENDIX XIII

Logarithms to the Base 2 of Numbers from 1 to 10,000

1	0.00000	51	5.67242	101	6.65821	151	7.23840
2	1.00000	52	5.70044	102	6.67242	152	7.24793
3	1.58496	53	5.72792	103	6.68650	153	7.25739
4	2.00000	54	5.75489	104	6.70044	154	7.26679
5	2.32193	55	5.78136	105	6.71425	155	7.27613
6	2.58496	56	5.80735	106	6.72792	156	7.28540
7	2.80735	57	5.83289	107	6.74147	157	7.29462
8	3.00000	58	5.85798	108	6.75489	158	7.30378
9	3.16993	59	5.88264	109	6.76818	159	7.31288
10	3.32193	60	5.90689	110	6.78136	160	7.32192
11	3.45943	61	5.93074	111	6.79442	161	7.33092
12	3.58496	62	5.95420	112	6.80735	162	7.33985
13	3.70044	63	5.97728	113	6.82018	163	7.34873
14	3.80735	64	6.00000	114	6.83289	164	7.35755
15	3.90689	65	6.02237	115	6.84549	165	7.36632
16	4.00000	66	6.04439	116	6.85798	166	7.37504
17	4.08746	67	6.06609	117	6.87036	167	7.38370
18	4.16993	68	6.08746	118	6.88264	168	7.39232
19	4.24793	69	6.10852	119	6.89482	169	7.40088
20	4.32193	70	6.12928	120	6.90689	170	7.40939
21	4.39232	71	6.14975	121	6.91886	171	7.41785
22	4.45943	72	6.16992	122	6.93074	172	7.42626
23	4.52356	73	6.18982	123	6.94252	173	7.43463
24	4.58496	74	6.20945	124	6.95420	174	7.44294
25	4.64386	75	6.22882	125	6.96578	175	7.45121
26	4.70044	76	6.24793	126	6.97728	176	7.45943
27	4.75489	77	6.26679	127	6.98869	177	7.46760
28	4.80735	78	6.28540	128	7.00000	178	7.47573
29	4.85798	79	6.30378	129	7.01123	179	7.48382
30	4.90689	80	6.32193	130	7.02237	180	7.49185
31	4.95420	81	6.33985	131	7.03342	181	7.49985
32	5.00000	82	6.35755	132	7.04439	182	7.50779
33	5.04439	83	6.37504	133	7.05528	183	7.51570
34	5.08746	84	6.39232	134	7.06609	184	7.52356
35	5.12928	85	6.40939	135	7.07682	185	7.53138
36	5.16993	86	6.42626	136	7.08746	186	7.53916
37	5.20945	87	6.44294	137	7.09803	187	7.54690
38	5.24793	88	6.45943	138	7.10852	188	7.55459
39	5.28540	89	6.47573	139	7.11894	189	7.56224
40	5.32193	90	6.49185	140	7.12928	190	7.56986
41	5.35755	91	6.50779	141	7.13955	191	7.57743
42	5.39232	92	6.52356	142	7.14975	192	7.58496
43	5.42626	93	6.53916	143	7.15987	193	7.59246
44	5.45943	94	6.55459	144	7.16992	194	7.59991
45	5.49185	95	6.56986	145	7.17991	195	7.60733
46	5.52356	96	6.58496	146	7.18982	196	7.61471
47	5.55459	97	6.59991	147	7.19967	197	7.62205
48	5.58496	98	6.61471	148	7.20945	198	7.62936
49	5.61471	99	6.62936	149	7.21917	199	7.63662
50	5.64386	100	6.64386	150	7.22882	200	7.64386

201	7.65105	251	7.97154	301	8.23362	351	8.45533
202	7.65821	252	7.97728	302	8.23840	352	8.45943
203	7.66534	253	7.98299	303	8.24317	353	8.46352
204	7.67242	254	7.98869	304	8.24793	354	8.46760
205	7.67948	255	7.99435	305	8.25267	355	8.47167
206	7.68650	256	8.00000	306	8.25739	356	8.47573
207	7.69349	257	8.00562	307	8.26209	357	8.47978
208	7.70044	258	8.01123	308	8.26679	358	8.48382
209	7.70736	259	8.01681	309	8.27146	359	8.48784
210	7.71424	260	8.02237	310	8.27613	360	8.49185
211	7.72110	261	8.02791	311	8.28077	361	8.49585
212	7.72792	262	8.03342	312	8.28540	362	8.49985
213	7.73471	263	8.03892	313	8.29002	363	8.50383
214	7.74147	264	8.04440	314	8.29462	364	8.50779
215	7.74819	265	8.04985	315	8.29921	365	8.51175
216	7.75489	266	8.05528	316	8.30378	366	8.51570
217	7.76155	267	8.06069	317	8.30834	367	8.51964
218	7.76818	268	8.06609	318	8.31288	368	8.52356
219	7.77479	269	8.07146	319	8.31741	369	8.52748
220	7.78136	270	8.07682	320	8.32193	370	8.53138
221	7.78790	271	8.08215	321	8.32643	371	8.53528
222	7.79442	272	8.08746	322	8.33092	372	8.53916
223	7.80090	273	8.09276	323	8.33539	373	8.54303
224	7.80735	274	8.09803	324	8.33985	374	8.54690
225	7.81378	275	8.10329	325	8.34429	375	8.55075
226	7.82018	276	8.10852	326	8.34873	376	8.55459
227	7.82655	277	8.11374	327	8.35315	377	8.55842
228	7.83289	278	8.11894	328	8.35755	378	8.56224
229	7.83920	279	8.12412	329	8.36194	379	8.56605
230	7.84549	280	8.12928	330	8.36632	380	8.56986
231	7.85175	281	8.13442	331	8.37069	381	8.57365
232	7.85798	282	8.13955	332	8.37504	382	8.57743
233	7.86419	283	8.14466	333	8.37938	383	8.58120
234	7.87036	284	8.14975	334	8.38370	384	8.58496
235	7.87652	285	8.15482	335	8.38802	385	8.58872
236	7.88264	286	8.15987	336	8.39232	386	8.59246
237	7.88874	287	8.16491	337	8.39660	387	8.59619
238	7.89482	288	8.16992	338	8.40088	388	8.59991
239	7.90087	289	8.17493	339	8.40514	389	8.60363
240	7.90689	290	8.17991	340	8.40939	390	8.60733
241	7.91289	291	8.18488	341	8.41363	391	8.61103
242	7.91886	292	8.18982	342	8.41785	392	8.61471
243	7.92481	293	8.19476	343	8.42206	393	8.61839
244	7.93074	294	8.19967	344	8.42626	394	8.62205
245	7.93664	295	8.20457	345	8.43045	395	8.62571
246	7.94251	296	8.20945	346	8.43440	396	8.62936
247	7.94837	297	8.21432	347	8.43879	397	8.63300
248	7.95420	298	8.21917	348	8.44294	398	8.63662
249	7.96000	299	8.22400	349	8.44708	399	8.64025
250	7.96578	300	8.22882	350	8.45121	400	8.64386

Logarithms to the Base 2 of Numbers from 1 to 10,000

401	8.64746	451	8.81698	501	8.96867	551	9.10591
402	8.65105	452	8.82018	502	8.97154	552	9.10852
403	8.65464	453	8.82337	503	8.97441	553	9.11113
404	8.65821	454	8.82655	504	8.97728	554	9.11375
405	8.66178	455	8.82972	505	8.98014	555	9.11634
406	8.66534	456	8.83289	506	8.98299	556	9.11894
407	8.66888	457	8.83605	507	8.98584	557	9.12153
408	8.67242	458	8.83920	508	8.98869	558	9.12412
409	8.67596	459	8.84235	509	8.99152	559	9.12670
410	8.67948	460	8.84549	510	8.99435	560	9.12928
411	8.68299	461	8.84862	511	8.99718	561	9.13186
412	8.68650	462	8.85175	512	9.00000	562	9.13443
413	8.69000	463	8.85487	513	9.00281	563	9.13699
414	8.69349	464	8.85798	514	9.00562	564	9.13955
415	8.69697	465	8.86109	515	9.00843	565	9.14211
416	8.70044	466	8.86419	516	9.01123	566	9.14466
417	8.70390	467	8.86728	517	9.01402	567	9.14720
418	8.70736	468	8.87036	518	9.01681	568	9.14975
419	8.71081	469	8.87344	519	9.01959	569	9.15228
420	8.71424	470	8.87652	520	9.02237	570	9.15482
421	8.71768	471	8.87958	521	9.02514	571	9.15735
422	8.72110	472	8.88264	522	9.02791	572	9.15987
423	8.72451	473	8.88570	523	9.03067	573	9.16239
424	8.72792	474	8.88874	524	9.03342	574	9.16491
425	8.73132	475	8.89178	525	9.03617	575	9.16742
426	8.73471	476	8.89482	526	9.03892	576	9.16992
427	8.73809	477	8.89784	527	9.04166	577	9.17243
428	8.74147	478	8.90087	528	9.04439	578	9.17493
429	8.74483	479	8.90388	529	9.04712	579	9.17742
430	8.74819	480	8.90689	530	9.04985	580	9.17991
431	8.75154	481	8.90989	531	9.05257	581	9.18239
432	8.75489	482	8.91289	532	9.05528	582	9.18488
433	8.75822	483	8.91588	533	9.05799	583	9.18735
434	8.76155	484	8.91886	534	9.06069	584	9.18982
435	8.76487	485	8.92184	535	9.06340	585	9.19229
436	8.76818	486	8.92481	536	9.06609	586	9.19475
437	8.77149	487	8.92778	537	9.06878	587	9.19722
438	8.77479	488	8.93074	538	9.07146	588	9.19967
439	8.77808	489	8.93369	539	9.07414	589	9.20212
440	8.78136	490	8.93664	540	9.07682	590	9.20457
441	8.78464	491	8.93958	541	9.07948	591	9.20701
442	8.78790	492	8.94251	542	9.08215	592	9.20945
443	8.79116	493	8.94544	543	9.08481	593	9.21189
444	8.79442	494	8.94837	544	9.08746	594	9.21432
445	8.79766	495	8.95128	545	9.09011	595	9.21675
446	8.80090	496	8.95420	546	9.09276	596	9.21917
447	8.80413	497	8.95710	547	9.09540	597	9.22159
448	8.80735	498	8.96000	548	9.09803	598	9.22400
449	8.81057	499	8.96290	549	9.10066	599	9.22641
450	8.81378	500	8.96578	550	9.10329	600	9.22882

Logarithms to the Base 2 of Numbers from 1 to 10,000

601	9.23122	651	9.34651	701	9.45327	751	9.55267
602	9.23362	652	9.34873	702	9.45533	752	9.55459
603	9.23601	653	9.35094	703	9.45738	753	9.55651
604	9.23840	654	9.35315	704	9.45943	754	9.55842
605	9.24079	655	9.35535	705	9.46148	755	9.56033
606	9.24317	656	9.35755	706	9.46353	756	9.56224
607	9.24555	657	9.35975	707	9.46556	757	9.56415
608	9.24793	658	9.36194	708	9.46760	758	9.56605
609	9.25030	659	9.36413	709	9.46964	759	9.56796
610	9.25267	660	9.36632	710	9.47167	760	9.56986
611	9.25503	661	9.36850	711	9.47371	761	9.57175
612	9.25739	662	9.37069	712	9.47573	762	9.57365
613	9.25974	663	9.37287	713	9.47776	763	9.57554
614	9.26209	664	9.37504	714	9.47978	764	9.57743
615	9.26444	665	9.37721	715	9.48180	765	9.57931
616	9.26679	666	9.37938	716	9.48382	766	9.58120
617	9.26913	667	9.38154	717	9.48583	767	9.58308
618	9.27146	668	9.38370	718	9.48784	768	9.58496
619	9.27380	669	9.38586	719	9.48985	769	9.58684
620	9.27613	670	9.38802	720	9.49185	770	9.58872
621	9.27845	671	9.39017	721	9.49385	771	9.59059
622	9.28077	672	9.39232	722	9.49585	772	9.59246
623	9.28309	673	9.39446	723	9.49785	773	9.59432
624	9.28540	674	9.39660	724	9.49985	774	9.59619
625	9.28771	675	9.39874	725	9.50184	775	9.59805
626	9.29002	676	9.40088	726	9.50383	776	9.59991
627	9.29232	677	9.40301	727	9.50581	777	9.60177
628	9.29462	678	9.40514	728	9.50779	778	9.60363
629	9.29692	679	9.40727	729	9.50978	779	9.60548
630	9.29921	680	9.40939	730	9.51175	780	9.60733
631	9.30149	681	9.41151	731	9.51373	781	9.60918
632	9.30378	682	9.41363	732	9.51570	782	9.61103
633	9.30606	683	9.41574	733	9.51767	783	9.61287
634	9.30834	684	9.41785	734	9.51964	784	9.61471
635	9.31061	685	9.41996	735	9.52160	785	9.61655
636	9.31288	686	9.42206	736	9.52356	786	9.61839
637	9.31515	687	9.42417	737	9.52552	787	9.62022
638	9.31741	688	9.42626	738	9.52748	788	9.62205
639	9.31967	689	9.42836	739	9.52943	789	9.62388
640	9.32193	690	9.43045	740	9.53138	790	9.62571
641	9.32418	691	9.43254	741	9.53333	791	9.62753
642	9.32643	692	9.43463	742	9.53528	792	9.62936
643	9.32867	693	9.43671	743	9.53722	793	9.63118
644	9.33092	694	9.43879	744	9.53916	794	9.63300
645	9.33316	695	9.44087	745	9.54110	795	9.63481
646	9.33539	696	9.44294	746	9.54303	796	9.63662
647	9.33762	697	9.44502	747	9.54497	797	9.63843
648	9.33985	698	9.44708	748	9.54690	798	9.64025
649	9.34208	699	9.44915	749	9.54882	799	9.64205
650	9.34429	700	9.45121	750	9.55075	800	9.64386

801	9.64566	851	9.73302	901	9.81538	951	9.89330
802	9.64746	852	9.73471	902	9.81698	952	9.89482
803	9.64926	853	9.73640	903	9.81858	953	9.89633
804	9.65105	854	9.73809	904	9.82018	954	9.89784
805	9.65285	855	9.73978	905	9.82178	955	9.89936
806	9.65464	856	9.74147	906	9.82337	956	9.90087
807	9.65643	857	9.74315	907	9.82496	957	9.90238
808	9.65821	858	9.74483	908	9.82655	958	9.90388
809	9.66000	859	9.74651	909	9.82814	959	9.90539
810	9.66178	860	9.74819	910	9.82972	960	9.90689
811	9.66356	861	9.74987	911	9.83131	961	9.90839
812	9.66534	862	9.75154	912	9.83289	962	9.90989
813	9.66711	863	9.75322	913	9.83447	963	9.91139
814	9.66888	864	9.75489	914	9.83605	964	9.91289
815	9.67066	865	9.75656	915	9.83763	965	9.91438
816	9.67242	866	9.75822	916	9.83920	966	9.91588
817	9.67419	867	9.75989	917	9.84078	967	9.91737
818	9.67596	868	9.76155	918	9.84235	968	9.91886
819	9.67772	869	9.76321	919	9.84392	969	9.92035
820	9.67948	870	9.76487	920	9.84549	970	9.92184
821	9.68124	871	9.76653	921	9.84706	971	9.92333
822	9.68299	872	9.76818	922	9.84862	972	9.92481
823	9.68475	873	9.76984	923	9.85019	973	9.92630
824	9.68650	874	9.77149	924	9.85175	974	9.92778
825	9.68825	875	9.77314	925	9.85331	975	9.92926
826	9.69000	876	9.77479	926	9.85487	976	9.93074
827	9.69174	877	9.77643	927	9.85643	977	9.93222
828	9.69349	878	9.77808	928	9.85798	978	9.93369
829	9.69523	879	9.77972	929	9.85954	979	9.93517
830	9.69697	880	9.78136	930	9.86109	980	9.93664
831	9.69870	881	9.78300	931	9.86264	981	9.93811
832	9.70044	882	9.78464	932	9.86419	982	9.93958
833	9.70217	883	9.78627	933	9.86573	983	9.94105
834	9.70390	884	9.78790	934	9.86728	984	9.94251
835	9.70563	885	9.78953	935	9.86882	985	9.94398
836	9.70736	886	9.79116	936	9.87036	986	9.94544
837	9.70908	887	9.79279	937	9.87191	987	9.94691
838	9.71081	888	9.79442	938	9.87344	988	9.94837
839	9.71253	889	9.79604	939	9.87498	989	9.94983
840	9.71424	890	9.79766	940	9.87652	990	9.95128
841	9.71596	891	9.79928	941	9.87805	991	9.95274
842	9.71768	892	9.80090	942	9.87958	992	9.95420
843	9.71939	893	9.80251	943	9.88111	993	9.95565
844	9.72110	894	9.80413	944	9.88264	994	9.95710
845	9.72281	895	9.80574	945	9.88417	995	9.95855
846	9.72451	896	9.80735	946	9.88570	996	9.96000
847	9.72622	897	9.80896	947	9.88722	997	9.96145
848	9.72792	898	9.81057	948	9.88874	998	9.96290
849	9.72962	899	9.81218	949	9.89026	999	9.96434
850	9.73132	900	9.81378	950	9.89178	1000	9.96578

Logarithms to the Base 2 of Numbers from 1 to 10,000

1010	9.98014	1510	10.56033	2010	10.97298	2510	11.29347
1020	9.99435	1520	10.56986	2020	10.98014	2520	11.29921
1030	10.00843	1530	10.57931	2030	10.98726	2530	11.30492
1040	10.02237	1540	10.58872	2040	10.99435	2540	11.31061
1050	10.03617	1550	10.59805	2050	11.00141	2550	11.31628
1060	10.04985	1560	10.60733	2060	11.00843	2560	11.32193
1070	10.06340	1570	10.61655	2070	11.01541	2570	11.32755
1080	10.07682	1580	10.62571	2080	11.02237	2580	11.33316
1090	10.09011	1590	10.63481	2090	11.02929	2590	11.33874
1100	10.10329	1600	10.64386	2100	11.03617	2600	11.34429
1110	10.11634	1610	10.65285	2110	11.04302	2610	11.34984
1120	10.12928	1620	10.66178	2120	11.04985	2620	11.35535
1130	10.14211	1630	10.67066	2130	11.05664	2630	11.36085
1140	10.15482	1640	10.67948	2140	11.06340	2640	11.36633
1150	10.16742	1650	10.68825	2150	11.07012	2650	11.37178
1160	10.17991	1660	10.69697	2160	11.07682	2660	11.37721
1170	10.19229	1670	10.70563	2170	11.08348	2670	11.38262
1180	10.20457	1680	10.71424	2180	11.09011	2680	11.38802
1190	10.21675	1690	10.72280	2190	11.09671	2690	11.39339
1200	10.22882	1700	10.73132	2200	11.10329	2700	11.39874
1210	10.24079	1710	10.73978	2210	11.10983	2710	11.40408
1220	10.25267	1720	10.74819	2220	11.11634	2720	11.40939
1230	10.26444	1730	10.75656	2230	11.12283	2730	11.41469
1240	10.27613	1740	10.76487	2240	11.12928	2740	11.41996
1250	10.28771	1750	10.77314	2250	11.13571	2750	11.42522
1260	10.29921	1760	10.78136	2260	11.14211	2760	11.43045
1270	10.31061	1770	10.78953	2270	11.14848	2770	11.43567
1280	10.32193	1780	10.79766	2280	11.15482	2780	11.44087
1290	10.33316	1790	10.80574	2290	11.16113	2790	11.44605
1300	10.34429	1800	10.81378	2300	11.16742	2800	11.45121
1310	10.35535	1810	10.82178	2310	11.17368	2810	11.45635
1320	10.36632	1820	10.82972	2320	11.17991	2820	11.46148
1330	10.37721	1830	10.83763	2330	11.18611	2830	11.46658
1340	10.38802	1840	10.84549	2340	11.19229	2840	11.47167
1350	10.39874	1850	10.85331	2350	11.19845	2850	11.47675
1360	10.40939	1860	10.86109	2360	11.20457	2860	11.48180
1370	10.41996	1870	10.86882	2370	11.21067	2870	11.48684
1380	10.43045	1880	10.87652	2380	11.21675	2880	11.49185
1390	10.44087	1890	10.88417	2390	11.22279	2890	11.49685
1400	10.45121	1900	10.89178	2400	11.22882	2900	11.50184
1410	10.46148	1910	10.89936	2410	11.23482	2910	11.50680
1420	10.47167	1920	10.90689	2420	11.24079	2920	11.51175
1430	10.48180	1930	10.91438	2430	11.24674	2930	11.51669
1440	10.49185	1940	10.92184	2440	11.25267	2940	11.52160
1450	10.50184	1950	10.92926	2450	11.25857	2950	11.52650
1460	10.51175	1960	10.93664	2460	11.26444	2960	11.53138
1470	10.52160	1970	10.94398	2470	11.27030	2970	11.53625
1480	10.53138	1980	10.95128	2480	11.27613	2980	11.54110
1490	10.54110	1990	10.95855	2490	11.28193	2990	11.54593
1500	10.55075	2000	10.96578	2500	11.28771	3000	11.55075

3010	11.55555	3510	11.77725	4010	11.96938	4510	12.13891
3020	11.56033	3520	11.78136	4020	11.97298	4520	12.14211
3030	11.56510	3530	11.78545	4030	11.97656	4530	12.14529
3040	11.56986	3540	11.78953	4040	11.98014	4540	12.14848
3050	11.57459	3550	11.79360	4050	11.98371	4550	12.15165
3060	11.57931	3560	11.79766	4060	11.98726	4560	12.15482
3070	11.58402	3570	11.80171	4070	11.99081	4570	12.15798
3080	11.58872	3580	11.80574	4080	11.99435	4580	12.16113
3090	11.59339	3590	11.80977	4090	11.99788	4590	12.16428
3100	11.59805	3600	11.81378	4100	12.00141	4600	12.16742
3110	11.60270	3610	11.81778	4110	12.00492	4610	12.17055
3120	11.60733	3620	11.82178	4120	12.00843	4620	12.17368
3130	11.61195	3630	11.82575	4130	12.01193	4630	12.17680
3140	11.61655	3640	11.82972	4140	12.01541	4640	12.17991
3150	11.62114	3650	11.83368	4150	12.01889	4650	12.18301
3160	11.62571	3660	11.83763	4160	12.02237	4660	12.18611
3170	11.63027	3670	11.84156	4170	12.02583	4670	12.18921
3180	11.63481	3680	11.84549	4180	12.02929	4680	12.19229
3190	11.63934	3690	11.84940	4190	12.03273	4690	12.19537
3200	11.64386	3700	11.85332	4200	12.03617	4700	12.19845
3210	11.64836	3710	11.85720	4210	12.03960	4710	12.20151
3220	11.65285	3720	11.86109	4220	12.04303	4720	12.20457
3230	11.65732	3730	11.86496	4230	12.04644	4730	12.20762
3240	11.66178	3740	11.86882	4240	12.04985	4740	12.21067
3250	11.66622	3750	11.87267	4250	12.05325	4750	12.21371
3260	11.67066	3760	11.87652	4260	12.05664	4760	12.21675
3270	11.67508	3770	11.88035	4270	12.06002	4770	12.21977
3280	11.67948	3780	11.88417	4280	12.06340	4780	12.22279
3290	11.68387	3790	11.88798	4290	12.06676	4790	12.22581
3300	11.68825	3800	11.89178	4300	12.07012	4800	12.22882
3310	11.69262	3810	11.89558	4310	12.07347	4810	12.23182
3320	11.69697	3820	11.89936	4320	12.07682	4820	12.23482
3330	11.70131	3830	11.90313	4330	12.08015	4830	12.23781
3340	11.70563	3840	11.90689	4340	12.08348	4840	12.24079
3350	11.70995	3850	11.91064	4350	12.08680	4850	12.24377
3360	11.71424	3860	11.91438	4360	12.09011	4860	12.24674
3370	11.71853	3870	11.91812	4370	12.09342	4870	12.24971
3380	11.72281	3880	11.92184	4380	12.09671	4880	12.25267
3390	11.72707	3890	11.92556	4390	12.10001	4890	12.25562
3400	11.73132	3900	11.92926	4400	12.10329	4900	12.25857
3410	11.73555	3910	11.93295	4410	12.10656	4910	12.26151
3420	11.73978	3920	11.93664	4420	12.10983	4920	12.26444
3430	11.74399	3930	11.94031	4430	12.11309	4930	12.26737
3440	11.74819	3940	11.94398	4440	12.11634	4940	12.27030
3450	11.75238	3950	11.94764	4450	12.11959	4950	12.27321
3460	11.75633	3960	11.95128	4460	12.12283	4960	12.27613
3470	11.76072	3970	11.95492	4470	12.12606	4970	12.27903
3480	11.76487	3980	11.95855	4480	12.12928	4980	12.28193
3490	11.76901	3990	11.96217	4490	12.13250	4990	12.28483
3500	11.77314	4000	11.96578	4500	12.13571	5000	12.28771

5010	12.29060	5510	12.42784	6010	12.55315	6510	12.66844
5020	12.29347	5520	12.43045	6020	12.55555	6520	12.67066
5030	12.29634	5530	12.43306	6030	12.55794	6530	12.67287
5040	12.29921	5540	12.43568	6040	12.56033	6540	12.67508
5050	12.30207	5550	12.43827	6050	12.56272	6550	12.67728
5060	12.30492	5560	12.44087	6060	12.56510	6560	12.67948
5070	12.30777	5570	12.44346	6070	12.56748	6570	12.68168
5080	12.31061	5580	12.44605	6080	12.56986	6580	12.68387
5090	12.31345	5590	12.44863	6090	12.57223	6590	12.68606
5100	12.31628	5600	12.45121	6100	12.57459	6600	12.68825
5110	12.31911	5610	12.45379	6110	12.57696	6610	12.69043
5120	12.32193	5620	12.45635	6120	12.57931	6620	12.69262
5130	12.32474	5630	12.45892	6130	12.58170	6630	12.69479
5140	12.32755	5640	12.46148	6140	12.58402	6640	12.69697
5150	12.33036	5650	12.46403	6150	12.58640	6650	12.69914
5160	12.33316	5660	12.46658	6160	12.58872	6660	12.70131
5170	12.33595	5670	12.46913	6170	12.59105	6670	12.70347
5180	12.33874	5680	12.47167	6180	12.59339	6680	12.70563
5190	12.34152	5690	12.47421	6190	12.59572	6690	12.70779
5200	12.34429	5700	12.47675	6200	12.59805	6700	12.70995
5210	12.34707	5710	12.47927	6210	12.60038	6710	12.71210
5220	12.34984	5720	12.48180	6220	12.60270	6720	12.71424
5230	12.35260	5730	12.48432	6230	12.60502	6730	12.71639
5240	12.35535	5740	12.48684	6240	12.60733	6740	12.71853
5250	12.35810	5750	12.48935	6250	12.60964	6750	12.72067
5260	12.36085	5760	12.49185	6260	12.61195	6760	12.72281
5270	12.36359	5770	12.49436	6270	12.61425	6770	12.72494
5280	12.36632	5780	12.49685	6280	12.61655	6780	12.72707
5290	12.36905	5790	12.49935	6290	12.61885	6790	12.72920
5300	12.37178	5800	12.50184	6300	12.62114	6800	12.73132
5310	12.37450	5810	12.50432	6310	12.62342	6810	12.73344
5320	12.37721	5820	12.50680	6320	12.62571	6820	12.73555
5330	12.37992	5830	12.50928	6330	12.62799	6830	12.73767
5340	12.38262	5840	12.51175	6340	12.63027	6840	12.73978
5350	12.38532	5850	12.51422	6350	12.63254	6850	12.74189
5360	12.38802	5860	12.51668	6360	12.63481	6860	12.74399
5370	12.39071	5870	12.51914	6370	12.63708	6870	12.74609
5380	12.39339	5880	12.52160	6380	12.63934	6880	12.74819
5390	12.39607	5890	12.52405	6390	12.64160	6890	12.75029
5400	12.39874	5900	12.52650	6400	12.64386	6900	12.75238
5410	12.40141	5910	12.52894	6410	12.64611	6910	12.75447
5420	12.40408	5920	12.53138	6420	12.64836	6920	12.75656
5430	12.40674	5930	12.53382	6430	12.65060	6930	12.75864
5440	12.40939	5940	12.53625	6440	12.65285	6940	12.76072
5450	12.41204	5950	12.53867	6450	12.65508	6950	12.76280
5460	12.41469	5960	12.54110	6460	12.65732	6960	12.76487
5470	12.41732	5970	12.54351	6470	12.65955	6970	12.76694
5480	12.41996	5980	12.54593	6480	12.66178	6980	12.76901
5490	12.42259	5990	12.54834	6490	12.66400	6990	12.77108
5500	12.42522	6000	12.55075	6500	12.66622	7000	12.77314

7010	12.77520	7510	12.87460	8010	12.96759	8510	13.05494
7020	12.77725	7520	12.87652	8020	12.96938	8520	13.05664
7030	12.77931	7530	12.87843	8030	12.97119	8530	13.05833
7040	12.78136	7540	12.88035	8040	12.97298	8540	13.06002
7050	12.78341	7550	12.88261	8050	12.97477	8550	13.06171
7060	12.78545	7560	12.88417	8060	12.97656	8560	13.06340
7070	12.78749	7570	12.88608	8070	12.97835	8570	13.06508
7080	12.78953	7580	12.88798	8080	12.98014	8580	13.06676
7090	12.79157	7590	12.88988	8090	12.98193	8590	13.06844
7100	12.79360	7600	12.89178	8100	12.98371	8600	13.07012
7110	12.79563	7610	12.89368	8110	12.98549	8610	13.07180
7120	12.79766	7620	12.89558	8120	12.98726	8620	13.07347
7130	12.79969	7630	12.89747	8130	12.98904	8630	13.07515
7140	12.80171	7640	12.89936	8140	12.99081	8640	13.07682
7150	12.80373	7650	12.90124	8150	12.99259	8650	13.07848
7160	12.80574	7660	12.90313	8160	12.99435	8660	13.08015
7170	12.80776	7670	12.90501	8170	12.99612	8670	13.08182
7180	12.80977	7680	12.90689	8180	12.99788	8680	13.08348
7190	12.81178	7690	12.90877	8190	12.99965	8690	13.08514
7200	12.81378	7700	12.91064	8200	13.00140	8700	13.08680
7210	12.81578	7710	12.91251	8210	13.00317	8710	13.08846
7220	12.81778	7720	12.91438	8220	13.00492	8720	13.09011
7230	12.81978	7730	12.91625	8230	13.00668	8730	13.09176
7240	12.82178	7740	12.91812	8240	13.00843	8740	13.09342
7250	12.82376	7750	12.91998	8250	13.01018	8750	13.09507
7260	12.82575	7760	12.92184	8260	13.01193	8760	13.09671
7270	12.82774	7770	12.92370	8270	13.01367	8770	13.09836
7280	12.82972	7780	12.92556	8280	13.01541	8780	13.10001
7290	12.83170	7790	12.92741	8290	13.01716	8790	13.10165
7300	12.83368	7800	12.92926	8300	13.01889	8800	13.10329
7310	12.83565	7810	12.93111	8310	13.02063	8810	13.10493
7320	12.83763	7820	12.93295	8320	13.02237	8820	13.10656
7330	12.83960	7830	12.93480	8330	13.02410	8830	13.10820
7340	12.84156	7840	12.93664	8340	13.02583	8840	13.10983
7350	12.84353	7850	12.93848	8350	13.02756	8850	13.11146
7360	12.84549	7860	12.94031	8360	13.02929	8860	13.11309
7370	12.84745	7870	12.94215	8370	13.03101	8870	13.11472
7380	12.84940	7880	12.94398	8380	13.03273	8880	13.11634
7390	12.85136	7890	12.94581	8390	13.03445	8890	13.11797
7400	12.85331	7900	12.94764	8400	13.03617	8900	13.11959
7410	12.85526	7910	12.94946	8410	13.03789	8910	13.12121
7420	12.85720	7920	12.95128	8420	13.03960	8920	13.12283
7430	12.85915	7930	12.95310	8430	13.04132	8930	13.12444
7440	12.86109	7940	12.95492	8440	13.04303	8940	13.12606
7450	12.86302	7950	12.95674	8450	13.04474	8950	13.12767
7460	12.86496	7960	12.95855	8460	13.04644	8960	13.12928
7470	12.86689	7970	12.96036	8470	13.04814	8970	13.13089
7480	12.86882	7980	12.96217	8480	13.04985	8980	13.13250
7490	12.87075	7990	12.96398	8490	13.05155	8990	13.13411
7500	12.87267	8000	12.96578	8500	13.05325	9000	13.13571

9010	13.13731	9260	13.17680	9510	13.21523	9760	13.25267
9020	13.13891	9270	13.17835	9520	13.21675	9770	13.25414
9030	13.14051	9280	13.17991	9530	13.21826	9780	13.25562
9040	13.14211	9290	13.18146	9540	13.21977	9790	13.25709
9050	13.14370	9300	13.18301	9550	13.22128	9800	13.25857
9060	13.14529	9310	13.18457	9560	13.22279	9810	13.26004
9070	13.14689	9320	13.18611	9570	13.22430	9820	13.26151
9080	13.14848	9330	13.18766	9580	13.22581	9830	13.26298
9090	13.15006	9340	13.18921	9590	13.22732	9840	13.26444
9100	13.15165	9350	13.19075	9600	13.22882	9850	13.26591
9110	13.15323	9360	13.19229	9610	13.23032	9860	13.26737
9120	13.15482	9370	13.19383	9620	13.23182	9870	13.26883
9130	13.15640	9380	13.19537	9630	13.23332	9880	13.27030
9140	13.15798	9390	13.19691	9640	13.23482	9890	13.27175
9150	13.15956	9400	13.19845	9650	13.23631	9900	13.27321
9160	13.16113	9410	13.19998	9660	13.23781	9910	13.27467
9170	13.16270	9420	13.20151	9670	13.23930	9920	13.27613
9180	13.16428	9430	13.20304	9680	13.24079	9930	13.27758
9190	13.16585	9440	13.20457	9690	13.24228	9940	13.27903
9200	13.16742	9450	13.20610	9700	13.24377	9950	13.28048
9210	13.16899	9460	13.20762	9710	13.24525	9960	13.28193
9220	13.17055	9470	13.20915	9720	13.24674	9970	13.28338
9230	13.17212	9480	13.21067	9730	13.24822	9980	13.28483
9240	13.17368	9490	13.21219	9740	13.24971	9990	13.28627
9250	13.17524	9500	13.21371	9750	13.25119	10000	13.28771

INDEX

(*f* indicates footnote; *p* indicates problem)

T

W

Y

SOME DOVER SCIENCE BOOKS

SOME DOVER SCIENCE BOOKS

WHAT IS SCIENCE?,
Norman Campbell

This excellent introduction explains scientific method, role of mathematics, types of scientific laws. Contents: 2 aspects of science, science & nature, laws of science, discovery of laws, explanation of laws, measurement & numerical laws, applications of science. 192pp. 5⅜ x 8. Paperbound $1.25

FADS AND FALLACIES IN THE NAME OF SCIENCE,
Martin Gardner

Examines various cults, quack systems, frauds, delusions which at various times have masqueraded as science. Accounts of hollow-earth fanatics like Symmes; Velikovsky and wandering planets; Hoerbiger; Bellamy and the theory of multiple moons; Charles Fort; dowsing, pseudoscientific methods for finding water, ores, oil. Sections on naturopathy, iridiagnosis, zone therapy, food fads, etc. Analytical accounts of Wilhelm Reich and orgone sex energy; L. Ron Hubbard and Dianetics; A. Korzybski and General Semantics; many others. Brought up to date to include Bridey Murphy, others. Not just a collection of anecdotes, but a fair, reasoned appraisal of eccentric theory. Formerly titled *In the Name of Science*. Preface. Index. x + 384pp. 5⅜ x 8.

Paperbound $1.85

PHYSICS, THE PIONEER SCIENCE,
L. W. Taylor

First thorough text to place all important physical phenomena in cultural-historical framework; remains best work of its kind. Exposition of physical laws, theories developed chronologically, with great historical, illustrative experiments diagrammed, described, worked out mathematically. Excellent physics text for self-study as well as class work. Vol. 1: Heat, Sound: motion, acceleration, gravitation, conservation of energy, heat engines, rotation, heat, mechanical energy, etc. 211 illus. 407pp. 5⅜ x 8. Vol. 2: Light, Electricity: images, lenses, prisms, magnetism, Ohm's law, dynamos, telegraph, quantum theory, decline of mechanical view of nature, etc. Bibliography. 13 table appendix. Index. 551 illus. 2 color plates. 508pp. 5⅜ x 8.

Vol. 1 Paperbound $2.25, Vol. 2 Paperbound $2.25,
The set $4.50

THE EVOLUTION OF SCIENTIFIC THOUGHT FROM NEWTON TO EINSTEIN,
A. d'Abro

Einstein's special and general theories of relativity, with their historical implications, are analyzed in non-technical terms. Excellent accounts of the contributions of Newton, Riemann, Weyl, Planck, Eddington, Maxwell, Lorentz and others are treated in terms of space and time, equations of electromagnetics, finiteness of the universe, methodology of science. 21 diagrams. 482pp. 5⅜ x 8.

Paperbound $2.50

CHANCE, LUCK AND STATISTICS: THE SCIENCE OF CHANCE,
Horace C. Levinson

Theory of probability and science of statistics in simple, non-technical language. Part I deals with theory of probability, covering odd superstitions in regard to "luck," the meaning of betting odds, the law of mathematical expectation, gambling, and applications in poker, roulette, lotteries, dice, bridge, and other games of chance. Part II discusses the misuse of statistics, the concept of statistical probabilities, normal and skew frequency distributions, and statistics applied to various fields—birth rates, stock speculation, insurance rates, advertising, etc. "Presented in an easy humorous style which I consider the best kind of expository writing," Prof. A. C. Cohen, Industry Quality Control. Enlarged revised edition. Formerly titled *The Science of Chance.* Preface and two new appendices by the author. Index. xiv + 365pp. 5⅜ x 8. Paperbound $2.00

BASIC ELECTRONICS,
prepared by the U.S. Navy Training Publications Center

A thorough and comprehensive manual on the fundamentals of electronics. Written clearly, it is equally useful for self-study or course work for those with a knowledge of the principles of basic electricity. Partial contents: Operating Principles of the Electron Tube; Introduction to Transistors; Power Supplies for Electronic Equipment; Tuned Circuits; Electron-Tube Amplifiers; Audio Power Amplifiers; Oscillators; Transmitters; Transmission Lines; Antennas and Propagation; Introduction to Computers; and related topics. Appendix. Index. Hundreds of illustrations and diagrams. vi + 471pp. 6½ x 9¼.
Paperbound $2.75

BASIC THEORY AND APPLICATION OF TRANSISTORS,
prepared by the U.S. Department of the Army

An introductory manual prepared for an army training program. One of the finest available surveys of theory and application of transistor design and operation. Minimal knowledge of physics and theory of electron tubes required. Suitable for textbook use, course supplement, or home study. Chapters: Introduction; fundamental theory of transistors; transistor amplifier fundamentals; parameters, equivalent circuits, and characteristic curves; bias stabilization; transistor analysis and comparison using characteristic curves and charts; audio amplifiers; tuned amplifiers; wide-band amplifiers; oscillators; pulse and switching circuits; modulation, mixing, and demodulation; and additional semiconductor devices. Unabridged, corrected edition. 240 schematic drawings, photographs, wiring diagrams, etc. 2 Appendices. Glossary. Index. 263pp. 6½ x 9¼. Paperbound $1.25

GUIDE TO THE LITERATURE OF MATHEMATICS AND PHYSICS,
N. G. Parke III

Over 5000 entries included under approximately 120 major subject headings of selected most important books, monographs, periodicals, articles in English, plus important works in German, French, Italian, Spanish, Russian (many recently available works). Covers every branch of physics, math, related engineering. Includes author, title, edition, publisher, place, date, number of volumes, number of pages. A 40-page introduction on the basic problems of research and study provides useful information on the organization and use of libraries, the psychology of learning, etc. This reference work will save you hours of time. 2nd revised edition. Indices of authors, subjects, 464pp. 5⅜ x 8.
Paperbound $2.75

THE RISE OF THE NEW PHYSICS (formerly THE DECLINE OF MECHANISM), *A. d'Abro*

This authoritative and comprehensive 2-volume exposition is unique in scientific publishing. Written for intelligent readers not familiar with higher mathematics, it is the only thorough explanation in non-technical language of modern mathematical-physical theory. Combining both history and exposition, it ranges from classical Newtonian concepts up through the electronic theories of Dirac and Heisenberg, the statistical mechanics of Fermi, and Einstein's relativity theories. "A must for anyone doing serious study in the physical sciences," *J. of Franklin Inst.* 97 illustrations. 991pp. 2 volumes.

Vol. 1 Paperbound \$2.25, Vol. 2 Paperbound \$2.25,
The set \$4.50

THE STRANGE STORY OF THE QUANTUM, AN ACCOUNT FOR THE GENERAL READER OF THE GROWTH OF IDEAS UNDERLYING OUR PRESENT ATOMIC KNOWLEDGE, *B. Hoffmann*

Presents lucidly and expertly, with barest amount of mathematics, the problems and theories which led to modern quantum physics. Dr. Hoffmann begins with the closing years of the 19th century, when certain trifling discrepancies were noticed, and with illuminating analogies and examples takes you through the brilliant concepts of Planck, Einstein, Pauli, de Broglie, Bohr, Schroedinger, Heisenberg, Dirac, Sommerfeld, Feynman, etc. This edition includes a new, long postscript carrying the story through 1958. "Of the books attempting an account of the history and contents of our modern atomic physics which have come to my attention, this is the best," H. Margenau, Yale University, in *American Journal of Physics.* 32 tables and line illustrations. Index. 275pp. 5⅜ x 8.

Paperbound \$1.75

GREAT IDEAS AND THEORIES OF MODERN COSMOLOGY, *Jagjit Singh*

The theories of Jeans, Eddington, Milne, Kant, Bondi, Gold, Newton, Einstein, Gamow, Hoyle, Dirac, Kuiper, Hubble, Weizsäcker and many others on such cosmological questions as the origin of the universe, space and time, planet formation, "continuous creation," the birth, life, and death of the stars, the origin of the galaxies, etc. By the author of the popular *Great Ideas of Modern Mathematics.* A gifted popularizer of science, he makes the most difficult abstractions crystal-clear even to the most non-mathematical reader. Index. xii + 276pp. 5⅜ x 8½. Paperbound \$2.00

GREAT IDEAS OF MODERN MATHEMATICS: THEIR NATURE AND USE, *Jagjit Singh*

Reader with only high school math will understand main mathematical ideas of modern physics, astronomy, genetics, psychology, evolution, etc., better than many who use them as tools, but comprehend little of their basic structure. Author uses his wide knowledge of non-mathematical fields in brilliant exposition of differential equations, matrices, group theory, logic, statistics, problems of mathematical foundations, imaginary numbers, vectors, etc. Original publications, 2 appendices. 2 indexes. 65 illustr. 322pp. 5⅜ x 8. Paperbound \$2.00

THE MATHEMATICS OF GREAT AMATEURS, *Julian L. Coolidge*

Great discoveries made by poets, theologians, philosophers, artists and other non-mathematicians: Omar Khayyam, Leonardo da Vinci, Albrecht Dürer, John Napier, Pascal, Diderot, Bolzano, etc. Surprising accounts of what can result from a non-professional preoccupation with the oldest of sciences. 56 figures. viii + 211pp. 5⅜ x 8½. Paperbound \$1.50

COLLEGE ALGEBRA, *H. B. Fine*
Standard college text that gives a systematic and deductive structure to algebra; comprehensive, connected, with emphasis on theory. Discusses the commutative, associative, and distributive laws of number in unusual detail, and goes on with undetermined coefficients, quadratic equations, progressions, logarithms, permutations, probability, power series, and much more. Still most valuable elementary-intermediate text on the science and structure of algebra. Index. 1560 problems, all with answers. x + 631pp. 5⅜ x 8. Paperbound $2.75

HIGHER MATHEMATICS FOR STUDENTS OF CHEMISTRY AND PHYSICS, *J. W. Mellor*
Not abstract, but practical, building its problems out of familiar laboratory material, this covers differential calculus, coordinate, analytical geometry, functions, integral calculus, infinite series, numerical equations, differential equations, Fourier's theorem, probability, theory of errors, calculus of variations, determinants. "If the reader is not familiar with this book, it will repay him to examine it," *Chem. & Engineering News.* 800 problems. 189 figures. Bibliography. xxi + 641pp. 5⅜ x 8. Paperbound $2.50

TRIGONOMETRY REFRESHER FOR TECHNICAL MEN, *A. A. Klaf*
A modern question and answer text on plane and spherical trigonometry. Part I covers plane trigonometry: angles, quadrants, trigonometrical functions, graphical representation, interpolation, equations, logarithms, solution of triangles, slide rules, etc. Part II discusses applications to navigation, surveying, elasticity, architecture, and engineering. Small angles, periodic functions, vectors, polar coordinates, De Moivre's theorem, fully covered. Part III is devoted to spherical trigonometry and the solution of spherical triangles, with applications to terrestrial and astronomical problems. Special time-savers for numerical calculation. 913 questions answered for you! 1738 problems; answers to odd numbers. 494 figures. 14 pages of functions, formulae. Index. x + 629pp. 5⅜ x 8. Paperbound $2.00

CALCULUS REFRESHER FOR TECHNICAL MEN, *A. A. Klaf*
Not an ordinary textbook but a unique refresher for engineers, technicians, and students. An examination of the most important aspects of differential and integral calculus by means of 756 key questions. Part I covers simple differential calculus: constants, variables, functions, increments, derivatives, logarithms, curvature, etc. Part II treats fundamental concepts of integration: inspection, substitution, transformation, reduction, areas and volumes, mean value, successive and partial integration, double and triple integration. Stresses practical aspects! A 50 page section gives applications to civil and nautical engineering, electricity, stress and strain, elasticity, industrial engineering, and similar fields. 756 questions answered. 556 problems; solutions to odd numbers. 36 pages of constants, formulae. Index. v + 431pp. 5⅜ x 8. Paperbound $2.00

INTRODUCTION TO THE THEORY OF GROUPS OF FINITE ORDER, *R. Carmichael*
Examines fundamental theorems and their application. Beginning with sets, systems, permutations, etc., it progresses in easy stages through important types of groups: Abelian, prime power, permutation, etc. Except 1 chapter where matrices are desirable, no higher math needed. 783 exercises, problems. Index. xvi + 447pp. 5⅜ x 8. Paperbound $3.00

NUMERICAL SOLUTIONS OF DIFFERENTIAL EQUATIONS,
H. Levy & E. A. Baggott

Comprehensive collection of methods for solving ordinary differential equations of first and higher order. All must pass 2 requirements: easy to grasp and practical, more rapid than school methods. Partial contents: graphical integration of differential equations, graphical methods for detailed solution. Numerical solution. Simultaneous equations and equations of 2nd and higher orders. "Should be in the hands of all in research in applied mathematics, teaching," *Nature*. 21 figures. viii + 238pp. 5⅜ x 8. Paperbound $1.85

ELEMENTARY STATISTICS, WITH APPLICATIONS IN MEDICINE AND THE BIOLOGICAL SCIENCES, *F. E. Croxton*

A sound introduction to statistics for anyone in the physical sciences, assuming no prior acquaintance and requiring only a modest knowledge of math. All basic formulas carefully explained and illustrated; all necessary reference tables included. From basic terms and concepts, the study proceeds to frequency distribution, linear, non-linear, and multiple correlation, skewness, kurtosis, etc. A large section deals with reliability and significance of statistical methods. Containing concrete examples from medicine and biology, this book will prove unusually helpful to workers in those fields who increasingly must evaluate, check, and interpret statistics. Formerly titled "Elementary Statistics with Applications in Medicine." 101 charts. 57 tables. 14 appendices. Index. vi + 376pp. 5⅜ x 8. Paperbound $2.00

INTRODUCTION TO SYMBOLIC LOGIC,
S. Langer

No special knowledge of math required — probably the clearest book ever written on symbolic logic, suitable for the layman, general scientist, and philosopher. You start with simple symbols and advance to a knowledge of the Boole-Schroeder and Russell-Whitehead systems. Forms, logical structure, classes, the calculus of propositions, logic of the syllogism, etc. are all covered. "One of the clearest and simplest introductions," *Mathematics Gazette*. Second enlarged, revised edition. 368pp. 5⅜ x 8. Paperbound $2.00

A SHORT ACCOUNT OF THE HISTORY OF MATHEMATICS,
W. W. R. Ball

Most readable non-technical history of mathematics treats lives, discoveries of every important figure from Egyptian, Phoenician, mathematicians to late 19th century. Discusses schools of Ionia, Pythagoras, Athens, Cyzicus, Alexandria, Byzantium, systems of numeration; primitive arithmetic; Middle Ages, Renaissance, including Arabs, Bacon, Regiomontanus, Tartaglia, Cardan, Stevinus, Galileo, Kepler; modern mathematics of Descartes, Pascal, Wallis, Huygens, Newton, Leibnitz, d'Alembert, Euler, Lambert, Laplace, Legendre, Gauss, Hermite, Weierstrass, scores more. Index. 25 figures. 546pp. 5⅜ x 8.
Paperbound $2.25

INTRODUCTION TO NONLINEAR DIFFERENTIAL AND INTEGRAL EQUATIONS,
Harold T. Davis

Aspects of the problem of nonlinear equations, transformations that lead to equations solvable by classical means, results in special cases, and useful generalizations. Thorough, but easily followed by mathematically sophisticated reader who knows little about non-linear equations. 137 problems for student to solve. xv + 566pp. 5⅜ x 8½. Paperbound $2.00

AN INTRODUCTION TO THE GEOMETRY OF N DIMENSIONS, *D. H. Y. Sommerville*
An introduction presupposing no prior knowledge of the field, the only book in English devoted exclusively to higher dimensional geometry. Discusses fundamental ideas of incidence, parallelism, perpendicularity, angles between linear space; enumerative geometry; analytical geometry from projective and metric points of view; polytopes; elementary ideas in analysis situs; content of hyper-spacial figures. Bibliography. Index. 60 diagrams. 196pp. 5⅜ x 8.
Paperbound $1.50

ELEMENTARY CONCEPTS OF TOPOLOGY, *P. Alexandroff*
First English translation of the famous brief introduction to topology for the beginner or for the mathematician not undertaking extensive study. This unusually useful intuitive approach deals primarily with the concepts of complex, cycle, and homology, and is wholly consistent with current investigations. Ranges from basic concepts of set-theoretic topology to the concept of Betti groups. "Glowing example of harmony between intuition and thought," David Hilbert. Translated by A. E. Farley. Introduction by D. Hilbert. Index. 25 figures. 73pp. 5⅜ x 8. Paperbound $1.00

ELEMENTS OF NON-EUCLIDEAN GEOMETRY, *D. M. Y. Sommerville*
Unique in proceeding step-by-step, in the manner of traditional geometry. Enables the student with only a good knowledge of high school algebra and geometry to grasp elementary hyperbolic, elliptic, analytic non-Euclidean geometries; space curvature and its philosophical implications; theory of radical axes; homothetic centres and systems of circles; parataxy and parallelism; absolute measure; Gauss' proof of the defect area theorem; geodesic representation; much more, all with exceptional clarity. 126 problems at chapter endings provide progressive practice and familiarity. 133 figures. Index. xvi + 274pp. 5⅜ x 8. Paperbound $2.00

INTRODUCTION TO THE THEORY OF NUMBERS, *L. E. Dickson*
Thorough, comprehensive approach with adequate coverage of classical literature, an introductory volume beginners can follow. Chapters on divisibility, congruences, quadratic residues & reciprocity. Diophantine equations, etc. Full treatment of binary quadratic forms without usual restriction to integral coefficients. Covers infinitude of primes, least residues. Fermat's theorem. Euler's phi function, Legendre's symbol, Gauss's lemma, automorphs, reduced forms, recent theorems of Thue & Siegel, many more. Much material not readily available elsewhere. 239 problems. Index. I figure. viii + 183pp. 5⅜ x 8.
Paperbound $1.75

MATHEMATICAL TABLES AND FORMULAS, *compiled by Robert D. Carmichael and Edwin R. Smith*
Valuable collection for students, etc. Contains all tables necessary in college algebra and trigonometry, such as five-place common logarithms, logarithmic sines and tangents of small angles, logarithmic trigonometric functions, natural trigonometric functions, four-place antilogarithms, tables for changing from sexagesimal to circular and from circular to sexagesimal measure of angles, etc. Also many tables and formulas not ordinarily accessible, including powers, roots, and reciprocals, exponential and hyperbolic functions, ten-place logarithms of prime numbers, and formulas and theorems from analytical and elementary geometry and from calculus. Explanatory introduction. viii + 269pp. 5⅜ x 8½. Paperbound $1.25

MATHEMATICAL PHYSICS, *D. H. Menzel*
Thorough one-volume treatment of the mathematical techniques vital for classical mechanics, electromagnetic theory, quantum theory, and relativity. Written by the Harvard Professor of Astrophysics for junior, senior, and graduate courses, it gives clear explanations of all those aspects of function theory, vectors, matrices, dyadics, tensors, partial differential equations, etc., necessary for the understanding of the various physical theories. Electron theory, relativity, and other topics seldom presented appear here in considerable detail. Scores of definition, conversion factors, dimensional constants, etc. "More detailed than normal for an advanced text . . . excellent set of sections on Dyadics, Matrices, and Tensors," *Journal of the Franklin Institute*. Index. 193 problems, with answers. x + 412pp. 5⅜ x 8. Paperbound $2.50

THE THEORY OF SOUND, *Lord Rayleigh*
Most vibrating systems likely to be encountered in practice can be tackled successfully by the methods set forth by the great Nobel laureate, Lord Rayleigh. Complete coverage of experimental, mathematical aspects of sound theory. Partial contents: Harmonic motions, vibrating systems in general, lateral vibrations of bars, curved plates or shells, applications of Laplace's functions to acoustical problems, fluid friction, plane vortex-sheet, vibrations of solid bodies, etc. This is the first inexpensive edition of this great reference and study work. Bibliography, Historical introduction by R. B. Lindsay. Total of 1040pp. 97 figures. 5⅜ x 8. Vol. 1 Paperbound $2.50, Vol. 2 Paperbound $2.50, The set $5.00

HYDRODYNAMICS, *Horace Lamb*
Internationally famous complete coverage of standard reference work on dynamics of liquids & gases. Fundamental theorems, equations, methods, solutions, background, for classical hydrodynamics. Chapters include Equations of Motion, Integration of Equations in Special Gases, Irrotational Motion, Motion of Liquid in 2 Dimensions, Motion of Solids through Liquid-Dynamical Theory, Vortex Motion, Tidal Waves, Surface Waves, Waves of Expansion, Viscosity, Rotating Masses of Liquids. Excellently planned, arranged; clear, lucid presentation. 6th enlarged, revised edition. Index. Over 900 footnotes, mostly bibliographical. 119 figures. xv + 738pp. 6⅛ x 9¼. Paperbound $4.00

DYNAMICAL THEORY OF GASES, *James Jeans*
Divided into mathematical and physical chapters for the convenience of those not expert in mathematics, this volume discusses the mathematical theory of gas in a steady state, thermodynamics, Boltzmann and Maxwell, kinetic theory, quantum theory, exponentials, etc. 4th enlarged edition, with new material on quantum theory, quantum dynamics, etc. Indexes. 28 figures. 444pp. 6⅛ x 9¼. Paperbound $2.75

THERMODYNAMICS, *Enrico Fermi*
Unabridged reproduction of 1937 edition. Elementary in treatment; remarkable for clarity, organization. Requires no knowledge of advanced math beyond calculus, only familiarity with fundamentals of thermometry, calorimetry. Partial Contents: Thermodynamic systems; First & Second laws of thermodynamics; Entropy; Thermodynamic potentials: phase rule, reversible electric cell; Gaseous reactions: van't Hoff reaction box, principle of LeChatelier; Thermodynamics of dilute solutions: osmotic & vapor pressures, boiling & freezing points; Entropy constant. Index. 25 problems. 24 illustrations. x + 160pp. 5⅜ x 8. Paperbound $1.75

Sec 1.9 1 & 2
pg 83 #1